# Condition Monitoring and Failure Prevention of Electric Machines

# Condition Monitoring and Failure Prevention of Electric Machines

Editors

**Yuling He**
**David Gerada**
**Conggan Ma**
**Haisen Zhao**

MDPI

Basel • Beijing • Wuhan • Barcelona • Belgrade • Novi Sad • Cluj • Manchester

*Editors*

Yuling He
Department of
Mechanical Engineering
North China Electric
Power University
Baoding
China

David Gerada
PEMC Group
University of Nottingham
Nottingham
United Kingdom

Conggan Ma
School of Automotive Engineering
Harbin Institute of Technology
Weihai
China

Haisen Zhao
School of Electrical and
Electronics Engineering
North China Electric Power
University
Beijing
China

*Editorial Office*
MDPI
St. Alban-Anlage 66
4052 Basel, Switzerland

This is a reprint of articles from the Special Issue published online in the open access journal *Energies* (ISSN 1996-1073) (available at: www.mdpi.com/journal/energies/special_issues/condition_failure_electric_machines).

For citation purposes, cite each article independently as indicated on the article page online and as indicated below:

Lastname, A.A.; Lastname, B.B. Article Title. *Journal Name* **Year**, *Volume Number*, Page Range.

**ISBN 978-3-0365-9397-5 (Hbk)**
**ISBN 978-3-0365-9396-8 (PDF)**
**doi.org/10.3390/books978-3-0365-9396-8**

# Contents

# About the Editors

**Yuling He**

Yu-Ling was born in Longyan City, Fujian Province, P.R. China, in 1984. He received two B.S. degrees in Mechanical Engineering and Electrical Engineering, respectively, his M.S. degree in Mechatronics Engineering, and his Ph.D. in Power Machinery and Engineering, all from the North China Electric Power University, Baoding, China, in 2007, 2009, and 2012, respectively.

He is now a Professor within the Department of Mechanical Engineering, North China Electric Power University, Baoding, China. His research interests include condition monitoring and failure prevention of electric machines, mathematical modeling for electromechanical systems, and intelligent equipment development.

The proffessor is the Vice Dean of Hebei Provincial Engineering Research Center of Advanced Manufacturing and Intelligent Maintenance on Electric Machinery and the Vice President and Secretary General of Heibei Society for Vibration Engineering. He was named the Top Youth Talent of Hebei Province, the Second Class Talent of Hebei Provincial 3-3-3 Talent Project, and the Excellent Expert of Baoding City, Hebei Province.

**David Gerada**

David Gerada received his Ph.D. in High-Speed Electrical Machines from the University of Nottingham, Nottingham, UK, in 2012. From 2007 to 2016, he was working within the R&D Department at Cummins Inc.

At Cummins, he pioneered the design and development of high-speed electrical machines, transforming a challenging technology into a reliable one suitable for the transportation market, and establishing industry-wide metrics used for such machinery. In 2016, he joined the University of Nottingham, where he is currently a Professor, responsible for developing state-of-the-art electrical machines for future transportation that pushes the existing technology boundaries, while propelling new technologies to high technology readiness levels (TRLs).

Prof. Gerada is a Chartered Engineer in the U.K. and a Fellow of the Institution of Engineering and Technology. His research interests include high-speed machines, novel materials and their applications to electromechanical energy conversion, and traction machines.

**Conggan Ma**

Conggan Ma was born in Sichuan, China, in 1987. He received his B.S.E. and Ph.D. degrees in Automotive Engineering from Tongji University, Shanghai, China, in 2010 and 2014, respectively.

Since 2019, he has been a Professor with the Automotive Engineering College, Harbin Institute of Technology-Weihai, Weihai, China, where he was an Associate Professor from 2014 to 2019. His research interests include vehicle system dynamics and control, vehicle vibration and noise control, and vibration and noise of electrical machines.

Prof. Ma is a senior member of the Chinese Society for Vibration and Engineering, a member of the IEEE Vehicular Technology Society, a member of the Chinese Society of Theoretical and Applied Mechanics, and a member of the Society of Automotive Engineering of China.

**Haisen Zhao**

Haisen Zhao received his B.E. degree in Agricultural Electrification and Automation from the Agriculture University of Hebei, Baoding, China, in 2004, and his M.E. and Ph.D. degrees in Electric Machines and Apparatus from North China Electric Power University (NCEPU), Beijing, China, in 2007 and 2011, respectively.

Dr. Zhao is currently a Professor with the School of Electrical and Electronics Engineering, North China Electric Power University, Beijing, China. From December 2018 to December 2019, he was a Visiting Scholar with the Energy Systems Research Laboratory, Department of Electrical and Computer Engineering, Florida International University, Miami, FL, USA. He has authored and coauthored more than 100 articles in peer-reviewed journals and major international conferences, and has 20 patents awarded in the area of electric machine design, control, and energy-saving technologies. His research interests include electrical motors design, diagnosis, energy analysis, and energy-saving technologies of electric machines and drive systems, as well as wireless power transfer.

# Preface

Electric machines are key components for both power generation and industrial propulsion. Electric machines include not only rotating motors/generators, but also all kinds of electric energy conversion/transformation machines (power transformers, linear motors, etc.). The development tendency of current electric machines is moving towards a larger capacity, higher power density, lower mass, and better strength. This tendency places stricter demands for the stable and reliable operation of the components in electric machines.

Since the unexpected breakdown of electric machines leads to considerable economic loss and even disaster, the condition monitoring and failure prevention of electric machines is pretty significant. With the rapid development of computer science and intelligent technologies, more and more novel monitoring and diagnosis approaches/methods are being developed. To provide a qualified gathering for readers/researchers on this topic, this reprint contains 11 recent studies. The authors are from China, Russia, USA, and Mexico. We believe these 11 recent studies will offer a good reference for scholars and technicians working in the field of electric machine monitoring and failure prevention.

Last but not least, we appreciate the contributors of the 11 articles for their excellent work. Also, we wish to make the following acknowledgements:

National Natural Science Foundation of China (52177042);

The Chinese Fundamental Research Funds for the Central Universities (2023MS128);

The Top Youth Talent Support Program of Hebei Province ([2018]-27);

The Hebei Provincial High-Level Talent Funding Project (B20231006);

The Suzhou Social Developing Innovation Project of Science and Technology (SS202134).

**Yuling He, David Gerada, Conggan Ma, and Haisen Zhao**

*Editors*

energies

*Article*

# Analysis of Direct Torque Control Response to Stator and Rotor Faults in Permanent Magnet Synchronous Machines

**Ibrahim M. Allafi and Shanelle N. Foster ***

Department of Electrical and Computer Engineering, Michigan State University, East Lansing, MI 48824, USA; allafiib@msu.edu
* Correspondence: hogansha@egr.msu.edu

**Abstract:** Direct-torque-control-driven permanent magnet synchronous machines eliminate the need for a position sensor while providing improved torque dynamics. However, the structure, regulation principle and nature of compensation of hysteresis-based controllers used in direct torque control impacts performance under faulty operating conditions. This paper analyzes the reaction of direct torque control to the presence of various faults that occur in permanent magnet synchronous machines. The analysis presented reveals that the direct torque control injects a negative sequence voltage and manipulates the torque angle to meet the control objectives when a fault occurs. The co-simulation of finite element analysis and a multi-physic circuit simulator is used to validate the response of the hysteresis-based controller to the machine health. The results indicate that the hysteresis comparators have the ability to mask the impact of the faults in the direct-torque-control-driven permanent magnet synchronous machines.

**Keywords:** demagnetization; direct torque control; eccentricity; field-oriented control; high resistance connection; symmetrical components; turn-to-turn short circuit; permanent magnet synchronous machine

## 1. Introduction

Permanent magnet synchronous machines (PMSMs) are widely used in various industries such as transportation, manufacturing and renewable energy [1,2]. The simple structure of direct torque control (DTC), coupled with its encoderless operation and fast dynamics, are of great interest for PMSM [3]. Nevertheless, the occurrence of faults, such as turn-to-turn short circuit (TTSC), high resistance contact (HRC), static eccentricity and partial demagnetization, remains a concern [4]. As PMSMs are increasingly utilized in safety critical applications, it is important that the controller can maintain a safe, stable operation independent of the machine health state [5]. As discussed in reference [6], variations in stator resistance and measurement errors can degrade performance and result in torque oscillations at the electrical frequency. Under heavy load or high speed, it has been shown in reference [7] that DTC cannot satisfy both the flux and torque control requirements simultaneously, resulting in the torque losing control. Considering that the presence of faults impacts the characteristic parameters of the PMSM, faults can prevent the smooth drive operation of DTC and potentially lead to catastrophic losses if not detected and mitigated in their early phases. Hence, fault diagnosis in DTC-driven PMSM is paramount to ensuring reliable drive operation.

Most of the fault diagnosis algorithms for inverter-driven PMSMs were developed for field-oriented control (FOC) drives [8–13]. However, it is important to consider that the different structure, regulation principle and nature of compensation for controllers can affect the fault diagnosis. As shown in reference [4], the controller type impacts the use of motor voltage signature analysis (MVSA) for fault diagnosis. The MVSA diagnostic strategy has lower accuracy for the DTC-driven PMSM than the FOC-driven PMSM. The lack of a consistent trend in the frequency spectrum of the voltage command for the DTC-

**Citation:** Allafi, I.M.; Foster, S.N. Analysis of Direct Torque Control Response to Stator and Rotor Faults in Permanent Magnet Synchronous Machines. *Energies* **2023**, *16*, 6940. https://doi.org/10.3390/en16196940

Academic Editor: Anna Richelli

Received: 30 June 2023
Revised: 27 September 2023
Accepted: 2 October 2023
Published: 3 October 2023

driven PMSM under faulty operating conditions leads to higher incidence of incorrect fault classification.

Understanding the impact of faults on drive operation and its corresponding reaction is a fundamental aspect of developing effective fault detection and separation methods. This becomes especially critical in the context of inverter-driven PMSM, where the compensatory capability of the drive system impacts the fault detectability [14]. In the case of DTC-driven PMSM, the electromagnetic torque error and stator flux linkage error are compensated directly through hysteresis comparators. Therefore, it is essential to comprehensively examine the nonlinear behavior that arises from the utilization of flux and torque hysteresis comparators, both in healthy and faulty conditions. Some efforts have been made to develop fault diagnosis techniques suitable for DTC-driven PMSMs [15–22]; however, the existing literature still lacks a comprehensive understanding of the potential ramifications of these faults on the operation of DTC-driven PMSMs.

This work analyzes the operation of DTC-driven PMSMs under four common faults: TTSC, HRC, static eccentricity and partial demagnetization. The analysis provides insight into the response of DTC to faulty motor conditions and its effect on fault detection, fault separation and control stability. The major contributions of this work are as follows:

(1) Its thorough analysis of the DTC reaction to the faulty operation of a PMSM; and
(2) Its summary of the challenges in stability for DTC-driven PMSMs.

These contributions facilitate the further development of DTC schemes that are robust to the presence of faults in PMSMs.

## 2. Faulty Operation of DTC-Driven PMSM

The schematic diagram of DTC drive, developed for PMSMs in reference [23], is shown in Figure 1. Measured currents ($i_s = [i_\alpha, i_\beta]$) and estimated voltages ($v_s = [v_\alpha, v_\beta]$) in the stationary frame are required to estimate the stator flux linkages ($\lambda_s = [\lambda_\alpha, \lambda_\beta]$), as described in (1).

$$\lambda_s = \int (v_s - r_s i_s)dt + \lambda_{s_0} \tag{1}$$

where, ($r_s$) is the phase resistance and ($\lambda_{s_0}$) is the initial condition for flux linkage estimation. Consequently, the electromagnetic torque ($t_e$) is estimated using the cross product of the current and stator flux linkages, as described in (2), where ($P_n$) is the number of pole pairs.

$$t_e = \frac{3P_n}{2}(i_\beta \lambda_\alpha - i_\alpha \lambda_\beta) \tag{2}$$

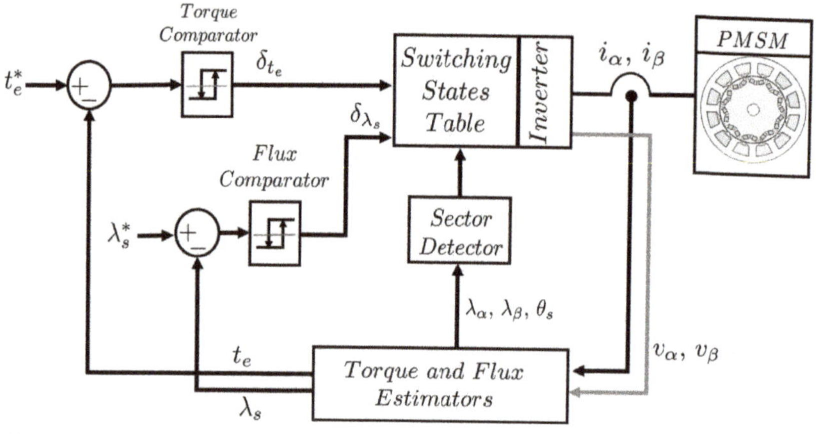


**Figure 1.** Electromagnetic torque and stator flux linkage control loops in DTC-driven PMSM.

Machine currents or voltages will include additional harmonics as a result of the comparator's response to the fault occurrence. Therefore, a thorough analysis of the reaction of the flux and torque hysteresis comparators upon a fault occurrence is essential. However, it is known that the nonlinear nature of the hysteresis comparators in DTC poses challenges to addressing this problem analytically [24–26]. Here, the behavior of the current under faulty conditions is described analytically to investigate the DTC reaction.

The initial consequence of the fault occurrence is the generation of a negative sequence component in addition to the positive sequence of the machine stator currents. The resulting machine current space vector ($\vec{i_s}$) is given in (3).

$$\vec{i_s} = \left|I_P^f\right| e^{j(\omega_s t+\phi_{iP})} + \left|I_N^f\right| e^{-j(\omega_s t+\phi_{iN})} \quad (3)$$

The variables $I_P^f$ and $I_N^f$ represent the magnitudes of the positive and negative sequence components of the machine currents, while $\phi_{iP}$ and $\phi_{iN}$ represent their phase angles, respectively. As the stator current is required for the flux linkage estimation in (1), the resulting stator flux linkage space vector is described in (4).

$$\vec{\lambda_s^f} = \left|\lambda_P^f\right| e^{j(\omega_s t+\alpha_P)} + \left|\lambda_N^f\right| e^{-j(\omega_s t+\alpha_N)} + \lambda_0 \quad (4)$$

where, $\lambda_P^f$ and $\lambda_N^f$ are the positive and negative sequence component of the stator flux linkages while $\alpha_P$ and $\alpha_N$ are their phase angles, respectively. Since the electromagnetic torque is a function of the stator currents and flux linkages of the machine, as described in (2), the resulting electromagnetic torque would be the sum of three parts, given in (5).

$$t_e = t_{mechanical} + t_{static} + t_{dynamic} \quad (5)$$

The machine output torque, $t_{mechanical}$, is produced by the interaction of the positive sequence components of the machine current and flux linkage, described in (6).

$$t_{mechanical} = \frac{3P_n}{2}\left(\left|I_P^f\right|\left|\lambda_P^f\right| \sin(\phi_{iP} - \alpha_P)\right) \quad (6)$$

The DC component of the torque $t_{static}$, described in (7), is a function of the negative sequences introduced by the fault occurrence.

$$t_{static} = \frac{3P_n}{2}\left(\left|I_N^f\right|\left|\lambda_N^f\right| \sin(\alpha_N - \phi_{iN})\right) \quad (7)$$

Finally, $t_{dynamic}$ is a torque component oscillating at double fundamental frequency ($2f_s$), as shown in (8); it is introduced in the machine output torque by the fault.

$$t_{dynamic} = \frac{3P_n}{2}\left(\left|I_P^f\right|\left|\lambda_N^f\right| \sin(2\omega t + \phi_{iP} + \alpha_N) + \left|I_N^f\right|\left|\lambda_P^f\right| \sin(2\omega t + \phi_{iN} + \alpha_P)\right) \quad (8)$$

It is worth mentioning that the errors in the stator flux linkage and machine torque due to a fault occurring, described in (4) and (5), are derived prior to the DTC reaction and compensation for them.

## 3. Response of Flux and Torque Comparators

The DTC technique employs hysteresis comparators to regulate the magnitude of stator flux linkage and electromagnetic torque. The bandwidth in the DTC scenario is determined by the torque and flux hysteresis bands ($\Delta T$ and $\Delta\lambda$), respectively. Figure 2 presents the torque and flux variation within the hysteresis bands.

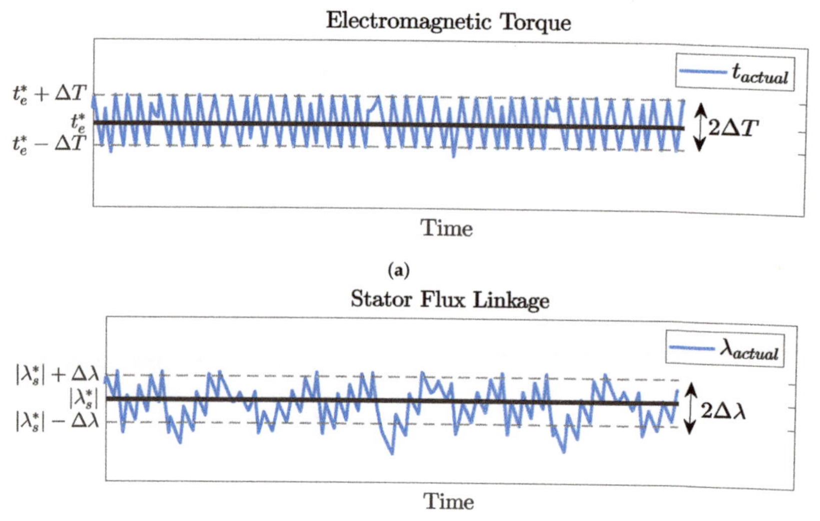


**Figure 2.** Torque and stator flux linkage variations due to hysteresis comparators. (**a**) Actual torque variation within torque hysteresis bands. (**b**) Actual flux variation within flux hysteresis bands.

The bands can either have fixed values, that remain constant regardless of operating conditions [27], or they can be percentages of a preset value based on operating conditions [28]. Alternatively, they can be variables that enable nearly constant switching frequency [29]. The utilization of wider hysteresis bands leads to a reduction in bandwidth and switching frequency due to the infrequent violation of the hysteresis bands during the control cycle. Conversely, a high controller bandwidth necessitates narrow bands, leading to increased switching frequency due to the likelihood of torque and flux errors exceeding the hysteresis bands during a significant portion of the control cycle. In general, DTC permits deviations in the estimation of flux and torque, provided that they fall within the predetermined tolerance margin established by the hysteresis band. Nonetheless, the estimated torque and flux linkage should follow the reference ones regardless of the machine condition, whether it is healthy or faulty.

### 3.1. Flux Comparator Response

Since a negative sequence component appears in the flux linkages due to its presence on the current, the flux comparator should cancel it by imposing another negative sequence component ($\vec{\lambda_c}$) in the flux linkage, as described in (9).

$$\vec{\lambda_c} = \left|\lambda_N^f\right| e^{-j(\omega_s t + \alpha_N)} \tag{9}$$

As a result, even in the presence of the fault, the actual flux linkage of the machine would follow the reference flux linkage:

$$\vec{\lambda_s^h} = \vec{\lambda_s^f} - \vec{\lambda_c} = \left|\lambda_P^h\right| e^{j(\omega_s t + \phi_{\lambda_s})} + \lambda_0 \tag{10}$$

Considering (1), the DTC drive injects a negative sequence component on the reference voltages to inject the cancellation term ($\vec{\lambda_c}$). As a result of (9), the static torque error in (7) and the first term of (8) are cancelled. However, the second term of the torque dynamic error in (8) remains.

The flux hysteresis band limits the maximum acceptable change in the stator flux linkage, which, in turn, limits the negative flux linkage sequence component that can be generated by applying a negative voltage sequence, as expressed in (11).

$$\frac{\left|\lambda_N^f\right|}{\left|\lambda_P^f\right|} \approx \frac{2\Delta\lambda}{|\lambda_s|} \tag{11}$$

The upper limit of the negative voltage sequence that the DTC flux comparator can apply in response to the flux linkage error is directly related to the flux hysteresis band, as described by (12).

$$\frac{|V_N|}{|V_P|} \approx \frac{2\Delta\lambda}{|\lambda_s|} \tag{12}$$

### 3.2. *Torque Comparator Response*

The regulation of output torque in DTC drive is achieved through the use of a torque hysteresis comparator. As a result, it is expected that the output torque will conform to the torque reference regardless of the condition of the machine. However, the flux comparator response does not eliminate all the consequences of the fault, as the second term of the torque dynamic error in (8) remains. This error is tolerated by DTC as long as this error falls within the hysteresis band of the torque comparator. However, if the error exceeds the maximum allowable torque error, the torque comparator must find an alternative method to compensate for it.

The variables that DTC can adjust to maintain the desired torque can be known by considering the torque equation given in (13).

$$t_e = \frac{3P_n}{8L_d L_q}\lambda_s\left[2\lambda_{PM}L_q\sin(\delta) + \lambda_s(L_d - L_q)\sin(2\delta)\right] \tag{13}$$

where, $(L_d, L_q)$ are the inductances in the synchronous rotating frame; $(\lambda_{PM})$ is the magnet flux linkage; and $(\delta)$ is the torque angle, which represents the difference between the stator flux linkage angle $(\theta_s)$ and the rotor flux linkage angle $(\theta_r)$. It is clear that the two variables that DTC can adjust are $\lambda_s$ and $\delta$. DTC has the capability to adjust the stator flux linkage vector during the transient conditions to either accelerate it to increase the torque angle and generate additional torque, or decelerate it to reduce the torque angle and produce less torque. Under steady-state conditions, the stator flux linkage vector rotates at the same speed as the rotor flux linkage vector, such that the torque is constant. However, $\lambda_s$ is almost fixed as it is controlled by the flux comparator. Therefore, the only option for DTC to respond to the torque error is by manipulating the torque angle.

According to [30], the PMSM torque ripple exhibits harmonic orders that are multiples of the sixth harmonic, $f_{t_{ripple}} = 6 \cdot k \cdot f_s$, where $f_s$ is the machine electric frequency and $k = 1, 2, 3, \ldots$. The frequency spectrum, obtained through Fast Fourier Transform (FFT), of the torque produced by an IPMSM driven by Field-Oriented Control (FOC) and DTC are provided in Figure 3. It is evident that the FOC driven PMSM exhibits significant magnitudes of the 6th and 12th harmonics in its output torque, whereas such high magnitudes are not observable in the case of DTC. However, if the magnitude of the second term in (8) increases the torque beyond the hysteresis bands $(2\Delta T)$, DTC will attempt to compensate for it in order to maintain the torque error within the band. Since the DTC drive provides the ability to adjust the torque angle $(\delta)$, any induced harmonics are expected to manifest in the torque angle instead of the torque. This can be observed in the frequency spectrum of the torque angle depicted in Figure 4.

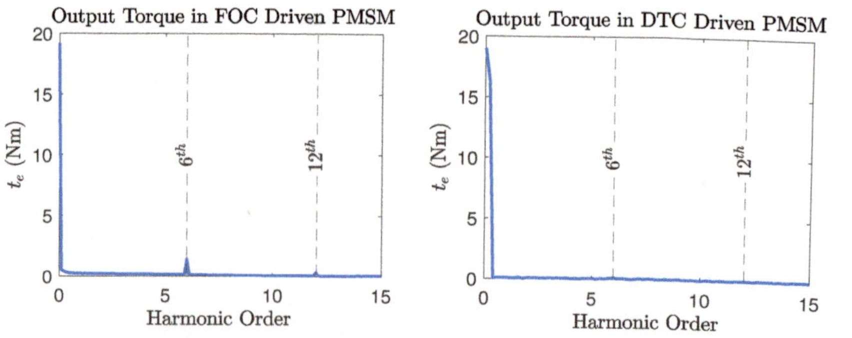


**Figure 3.** Frequency spectrum of the IPMSM torque magnitude under both FOC (**left**) and DTC (**right**) drives. The 6th and 12th harmonics are not observed in the DTC drive.

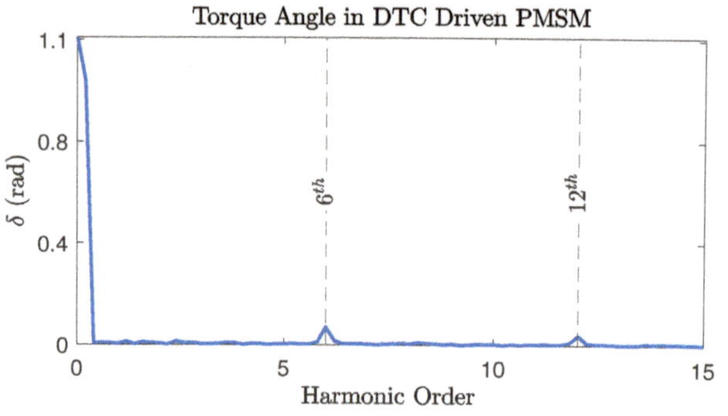


**Figure 4.** Frequency spectrum of the torque angle in DTC-driven PMSM.

The torque comprises two components: the magnetic torque ($t_m$) and the reluctance torque ($t_{re}$), as shown in (14).

$$\begin{aligned} t_e &= t_m + t_{re} \\ t_m &= A \cdot \sin(\delta) = Im\left[Ae^{j\delta}\right] \\ t_{re} &= B \cdot \sin(2\delta) = Im\left[Be^{j2\delta}\right] \\ A &= \frac{3P_n\lambda_s\lambda_{PM}}{4L_d} \\ B &= \frac{3P_n\lambda_s^2(L_d - L_q)}{8L_dL_q} \end{aligned} \tag{14}$$

The 6th and 12th harmonics are introduced into the torque angle for both the magnetic and reluctance torque equations given in (14), as shown in (15) and (16).

$$\begin{aligned} t_{m_1} &= A \cdot \sin(\delta + m) = Im\left[Ae^{j(\delta+m)}\right] \\ m &= x_6\cos(6\delta + \gamma_6) + x_{12}\cos(12\delta + \gamma_{12}) \end{aligned} \tag{15}$$

$$\begin{aligned} t_{re_1} &= B \cdot \sin(2\delta + d) = Im\left[Be^{j(2\delta+d)}\right] \\ d &= 2x_6\cos(6\delta + \gamma_6) + 2x_{12}\cos(12\delta + \gamma_{12}) \end{aligned} \tag{16}$$

The terms $x_6$, $x_{12}$, $\gamma_6$, and $\gamma_{12}$ correspond to the amplitudes and phases of the 6th and 12th harmonics present in the frequency spectrum of the torque angle. The trigonometric identity described in (17) is used to expand the electromagnetic and reluctance torque equations as shown in (18) and (19).

$$\sin(x+y) = \sin(x)\cos(y) + \sin(y)\cos(x) \tag{17}$$

$$t_{m_1} = t_m \cos(m) + A \cdot \cos(\delta)\sin(m) \tag{18}$$

$$t_{re_1} = t_{re} \cos(d) + B \cdot \cos(2\delta)\sin(d) \tag{19}$$

By considering the presence of the 6th and 12th harmonics in the torque angle frequency spectrum, it can be observed that the magnitudes of $t_m$ and $t_{re}$ are subject to multiplication by $\cos(m)$ and $\cos(d)$, respectively. Besides that, additional terms are present in $t_m$ and $t_{re}$. These offset terms are $A \cdot \cos(\delta)\sin(m)$ and $B \cdot \cos(2\delta)\sin(d)$. In the presence of the fault, the torque of the machine would attempt to follow the reference torque through variations in the torque angle.

## 4. Impact of Variations in Stator Flux Linkage Angle

The robustness and tolerance of DTC can also be attributed to its discrete control action, which involves selecting a finite set of possible actions rather than continuous modulation. The finite set of actions, namely the voltage vectors, is chosen based on heuristics and system dynamics, whereby an exact position of the stator flux linkage vector is not necessary. Figure 5 presents the sector partitions and voltage vectors available for selection in DTC. It consists of six voltage vectors ($V_1 - V_6$) within six sectors ($S_1 - S_6$) that are displaced by 60°. The appropriate voltage vector is chosen as shown in Table 1.

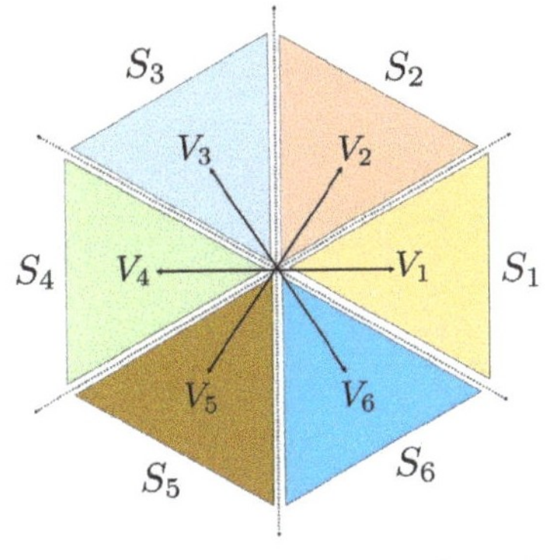


**Figure 5.** Sector partitions and the available set of voltage vectors in the DTC drive.

**Table 1.** Voltage vector selection table used in DTC-driven PMSM.

| | | Sector | | | | | |
|---|---|---|---|---|---|---|---|
| $\delta_{\lambda_s}$ | $\delta_{t_e}$ | $S_1$ | $S_2$ | $S_3$ | $S_4$ | $S_5$ | $S_6$ |
| +1 | +1 | $V_2$ | $V_3$ | $V_4$ | $V_5$ | $V_6$ | $V_1$ |
| +1 | −1 | $V_6$ | $V_1$ | $V_2$ | $V_3$ | $V_4$ | $V_5$ |
| −1 | +1 | $V_3$ | $V_4$ | $V_5$ | $V_6$ | $V_1$ | $V_2$ |
| −1 | −1 | $V_5$ | $V_6$ | $V_1$ | $V_2$ | $V_3$ | $V_4$ |

As described in reference [18], the accurate estimation of the flux of a faulty PMSM requires additional terms that are not included in (1). Provided that the actual and estimated stator flux linkage vectors are in the same sector, identical voltage vectors will be applied. However, if they are situated at the boundaries between sectors, different voltage vectors are applied, which will lead to a deviation in both desired torque and flux. For illustration purposes, the error in estimating the flux caused by the fault has been assumed to result in the estimated stator flux linkage vector ($|\lambda|_{est}$) being in the second sector, while it is assumed that the actual vector ($|\lambda|_{act}$) is in the third sector, as depicted in Figure 6. It is clear that the estimated value of $|\lambda_s|$ is higher than the reference value, which requires the selection of a voltage vector that can reduce the estimated value of $|\lambda_s|$. By assuming that the electromagnetic torque should be increased at this instant, the voltage vector ($V_4$) should be applied to the estimated stator flux linkage based on Table 1 since it lies in the second sector. Figure 7 presents the trace of ($|\lambda|_{est}$) where ($V_4$) results in moving the vector from the second sector to the third one. Subsequently, a varying pattern is introduced now between $V_4$ and $V_5$ to the estimated stator flux linkage vector, which is located in the third sector, in order to align it with the reference vector.

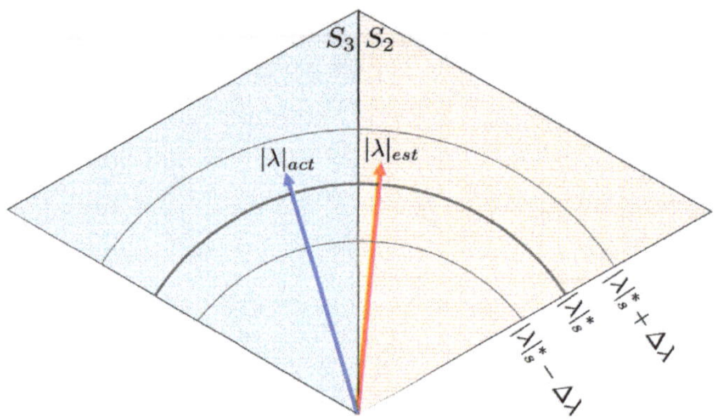


**Figure 6.** Illustration of the error between the actual and estimated stator flux linkages after the occurrence of a fault.

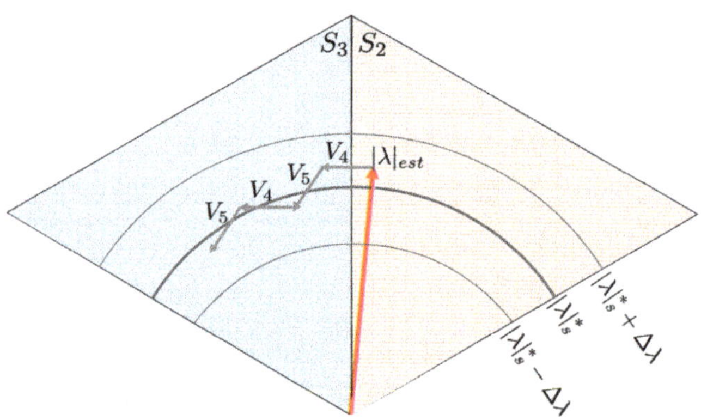


**Figure 7.** Path of the estimated stator flux linkage based on the voltage vector selection table shown in Table 1.

Nonetheless, when dealing with the actual stator flux linkage vector, the voltage vector ($V_5$) must be utilized because the flux vector is situated in the third sector. This decision is supported by the information provided in Table 1, where a decrease in flux linkage is required while simultaneously increasing torque. Figure 8 depicts the trace of the actual stator flux linkage vector in the case that the voltage vector ($V_5$) is applied first.

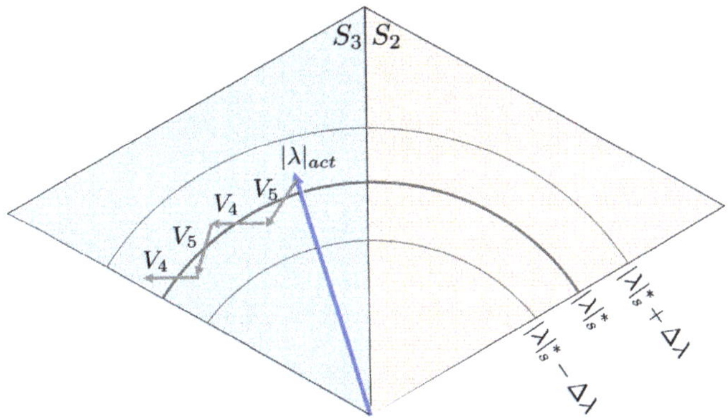


**Figure 8.** Optimal path of the actual stator flux linkage based on the voltage vector selection table shown in Table 1.

As the feedback loop of DTC relies on the estimated stator flux linkage signal instead of the actual one, the actual stator flux linkage vector will follow the same voltage vector pattern illustrated in Figure 7. However, this can cause inaccuracies in the control process, leading to differences between the actual stator flux linkage magnitude and the desired reference value, as shown in Figure 9, even if the estimated value is shown to be properly regulated. The actual stator flux linkage will be adjusted to a new reference position that takes into account the offset caused by the erroneous control action. Nonetheless, if the error is not substantial, it may be within the hysteresis bands. Hence, the unmodeled fault dynamics in the flux estimation lead to unobservable effects of the fault that may be tolerated by DTC. However, if the changes in the PMSM characteristic parameters due to the fault violate (20), the controller may become unstable, as discussed in reference [31].

$$\lambda_s < \left| \frac{L_q}{L_q - L_d} \right| \lambda_{pm} \tag{20}$$

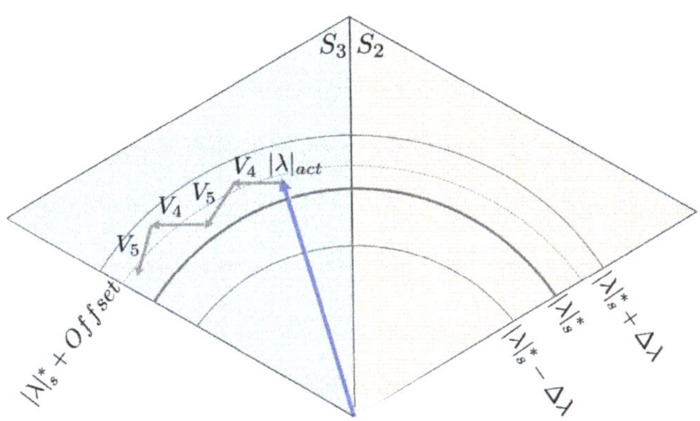


**Figure 9.** Path of the actual stator flux linkage based on the voltage vector selection table shown in Table 1.

## 5. Numerical Results

The validation of the DTC response to the presence of faults is conducted numerically. A co-simulation study of the DTC drive and finite element model of a PMSM within ANSYS software is conducted.

### 5.1. Numerical Setup

The DTC-driven PMSM is simulated for both healthy and faulty cases. The specification of the finite elements model (FEM) of the simulated PMSM in MAXWELL simulator are listed in Table 2 while the settings of the implemented DTC drive in SIMPLORER simulator are listed in Table 3.

**Table 2.** Specifications for the IPMSM used in the co-simulation.

| Parameter | Value |
|---|---|
| Rated power | 3.8 kW |
| Rated RMS phase current | 18 A |
| Rated RMS line voltage | 480 V |
| Phases/Poles/Slots | 3/10/12 |
| Turns per slot | 150 |
| Phase resistance | 2 Ω |
| Magnet flux linkage | 0.287 Wb |
| Residual flux density | 1.2 T |
| Air gap length | 1 mm |

**Table 3.** Settings of the DTC-Driven PMSM

| Controller Settings | Symbol | Value |
|---|---|---|
| Flux hysteresis band | $\Delta\lambda$ | $0.05 \cdot \lvert\lambda_s^*\rvert$ |
| Torque hysteresis band | $\Delta T$ | $0.09 \cdot t_e^*$ |
| Sampling frequency | $f_{sample}$ | 25 µs |
| Initial conditions | $(\lambda_{\alpha_0}, \lambda_{\beta_0})$ | (0.287, 0) Wb |
| Operating points | $(t_e^*, \lvert\lambda_s^*\rvert)$ | (20 Nm, 0.52 Wb) |

To simulate the faulty conditions of DTC-driven PMSM, modifications are made to both the geometry of the studied machine and its drive circuit. To model the TTSC fault occurring within the coils of the phase A winding, the coil area is divided into two separate parts: one comprising the healthy turns and the other consisting of the shorted turns, as described in Figure 10. The shaded section within the slot, allocated for phase A winding, represents the turns, located near the air gap, shorted through a fault resistance. The TTSC severity is indicated by the number of faulty turns ($N_f$) and the short circuit resistance ($R_f$). The non-shaded section within the slot indicates the remaining healthy turns of the phase A winding. To mimic the HRC fault, a resistance ($\Delta r_s$) is connected in series with the phase resistance ($r_s$). Here, the fault is implemented in phase B, and the severity of the fault is indicated by varying the magnitude of $\Delta r_s$. To model the partial demagnetization fault in the PMSM, the remnant flux density of selected magnets is reduced. Each pole in the studied machine includes two magnets of V-shape. The material of the demagnetized magnets is replaced with a new material having zero flux density, while the other properties remain unchanged. To implement the static rotor eccentricity, the rotor and its rotation axis are moved in the direction of the fault while keeping the stator coordinates fixed. The fault severity is determined by shifting the components along the positive Y-axis, using different values ($\epsilon$) relative to the length of the air gap ($g_a$). The severity levels of each fault are given in Table 4.

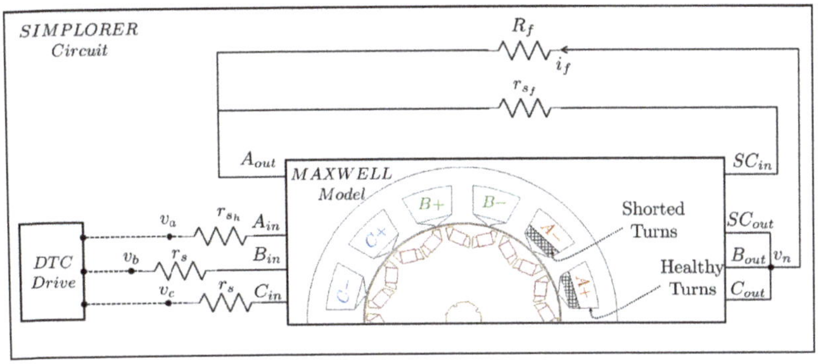


**Figure 10.** Machine model in MAXWELL coupled with DTC circuit in SIMPLORER including the TTSC fault circuit on phase A winding.

**Table 4.** Severity levels of the studied faults.

| Fault Type | Severity | Value |
|---|---|---|
| TTSC | $(N_f, R_f)$ | (15, 0.5 Ω)<br>(30, 0.25 Ω) |
| HRC | $(\Delta r_s / r_s)$ | 100%<br>150% |
| Partial Demagnetization | Number of Affected Magnets | 1 Magnet<br>3 Magnets |
| Static Eccentricity | $(\epsilon / g_a)$ | 40%<br>60% |

*5.2. Flux Comparator*

To evaluate the response of the flux comparator to fault presence in DTC-driven PMSM, the symmetrical components of the commanded voltages are studied. The positive, negative, and zero sequence components of the commanded voltages $(V_P, V_N, V_0)$ are computed using (21).

$$\begin{bmatrix} V_P \\ V_N \\ V_0 \end{bmatrix} = \frac{1}{3} \begin{bmatrix} 1 & a & a^2 \\ 1 & a^2 & a \\ 1 & 1 & 1 \end{bmatrix} \begin{bmatrix} V_a \\ V_b \\ V_c \end{bmatrix} \tag{21}$$

where $a = 1\angle 120^\circ$. Under normal operating conditions, the negative sequence component is absent because the commanded voltages are balanced, as shown in Figure 11.

In the event of a fault, the commanded voltages become unbalanced, which causes the negative sequence voltage component to appear for TTSC, HRC and demagnetization faults, as shown in Figures 12–14. This component helps to mitigate the effect of the fault on the stator flux linkage magnitude. However, the negative sequence voltage is nearly zero under eccentricity fault, as shown in Figure 15. The absence of the negative sequence voltage under eccentricity fault is similar to the healthy case, Figure 11. It is expected that the operation of the DTC-driven PMSM will remain stable despite the presence of an eccentricity fault. Nonetheless, this may lead to challenges in detecting the eccentricity fault in DTC-driven PMSM.

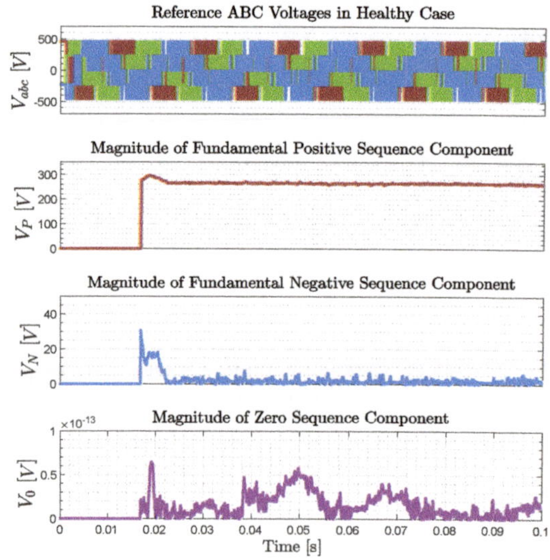


**Figure 11.** Sequence components of the commanded voltages in DTC-driven PMSM under healthy operating conditions.

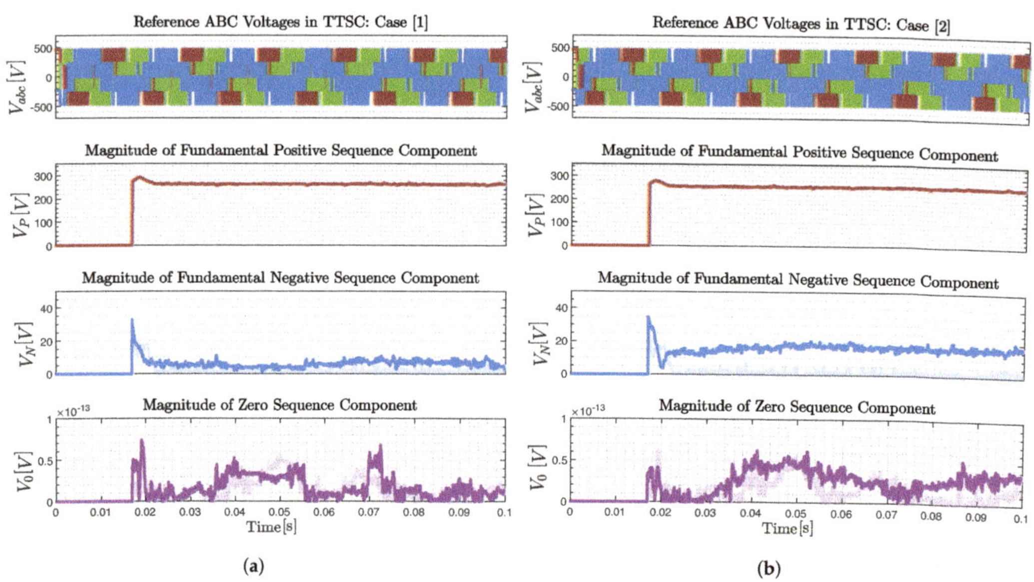


**Figure 12.** Sequence components of the commanded voltages in DTC-driven PMSM under TTSC fault with different severity levels. The healthy sequence components, from Figure 11, are shown in the faded color. (**a**) Case 1: (15, 0.5 Ω). (**b**) Case 2: (30, 0.25 Ω).

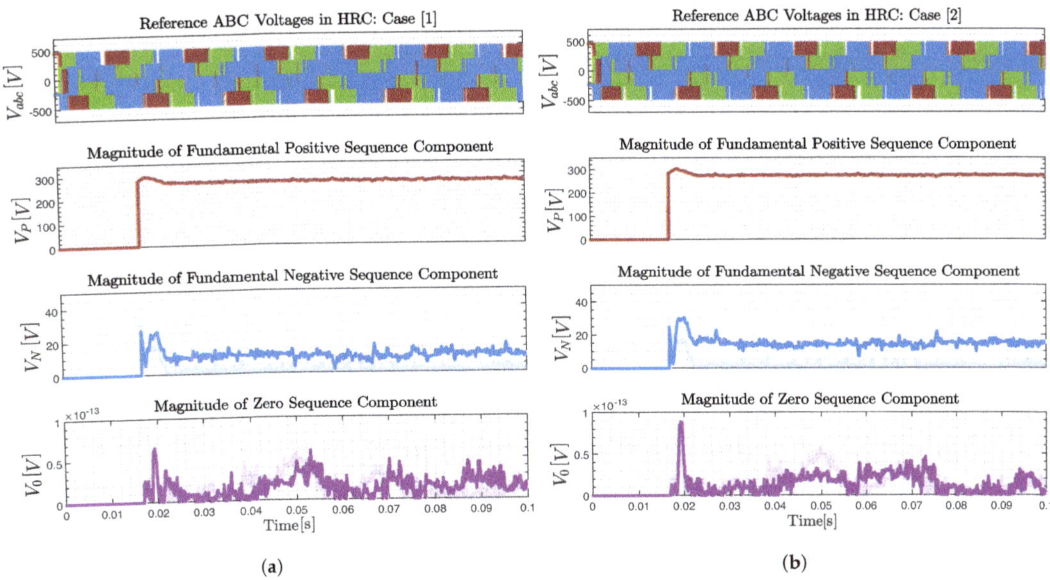


**Figure 13.** Sequence components of the commanded voltages in DTC-driven PMSM under HRC fault with different severity levels. The healthy sequence components, from Figure 11, are shown in the faded color. (**a**) Case 1: 100%. (**b**) Case 2: 150%.

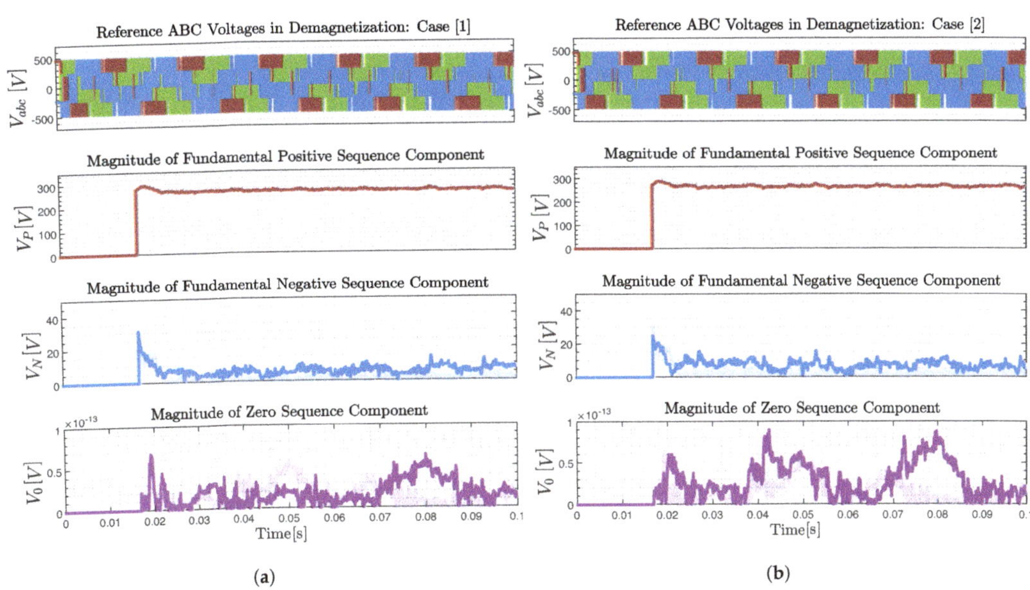


**Figure 14.** Sequence components of the commanded voltages in DTC-driven PMSM under demagnetization fault with different severity levels. The healthy sequence components, from Figure 11, are shown in the faded color. (**a**) Case 1: 1 Magnet. (**b**) Case 2: 3 Magnets.

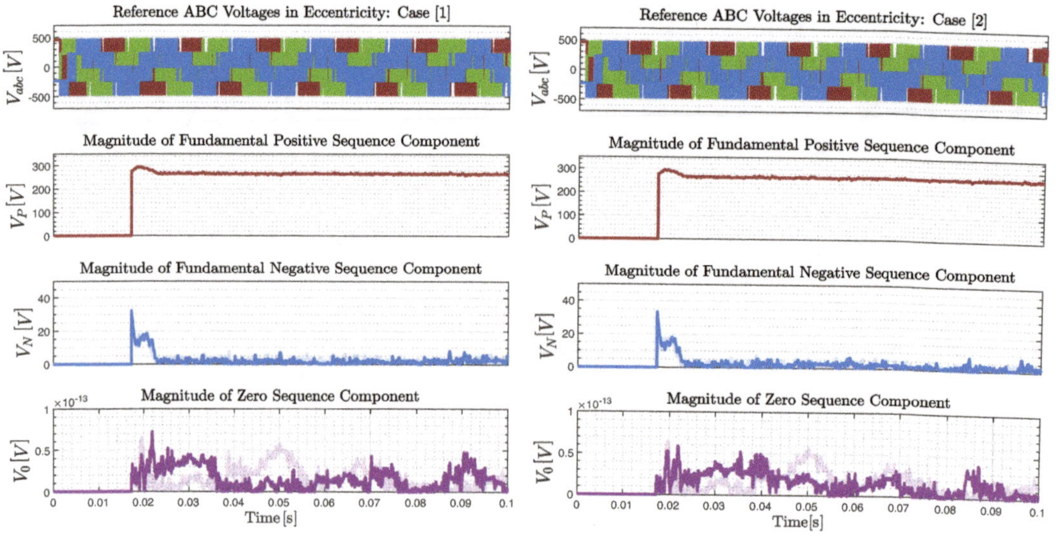


**Figure 15.** Sequence components of the commanded voltages in DTC-driven PMSM under eccentricity fault with different severity levels. The healthy sequence components, from Figure 11, are shown in the faded color. (**a**) Case 1: 40%. (**b**) Case 2: 60%.

### 5.3. Torque Comparator

To investigate the response of the torque comparator to fault presence in DTC-driven PMSM, the multiplication terms and offset terms on the torque signal given in (18) and (19) are studied. The multiplication terms are $\cos(m)$ and $\cos(d)$ and they are shown in Figure 16. It can be observed that their effect on $t_m$ and $t_{re}$ is negligible, as their magnitude is near unity.

The offset terms $A \cdot \cos(\delta)\sin(m)$ and $B \cdot \cos(2\delta)\sin(d)$ are shown in Figure 17. It can be noticed that these terms mostly offset each other and, hence, their overall effect on the torque is negligible, as depicted in Figure 17.

In summary, the DTC drive has the capability to incorporate the 6th and 12th harmonics in the torque angle instead of the torque signal itself. Although these additional harmonics are present, they have a negligible impact on the torque signal. As a result, the DTC drive can maintain the torque error within the torque hysteresis bands. This nonlinear behavior of the torque hysteresis comparator justifies the compensation of the second term of dynamic torque error in (8) due to the fault presence through the manipulation of the torque angle.

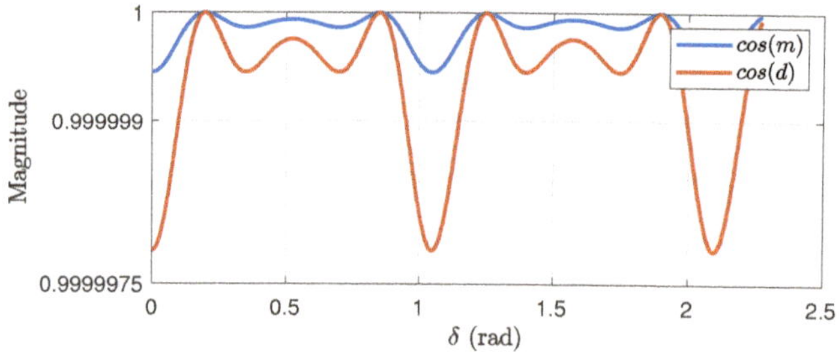


**Figure 16.** The magnitude of the magnet and reluctance torque multipliers, cos (m) and cos (d), in (18) and (19).

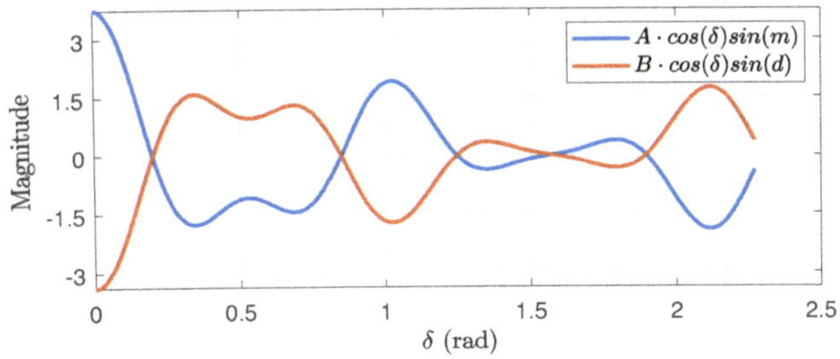


**Figure 17.** The magnitude of the offset error in (18) and (19).

## 6. Conclusions

The impact of various faults on the performance of DTC-driven PMSMs has been studied. Four types of faults were studied: TTSC, HRC, partial demagnetization, and static eccentricity. An analytical description of the stator current behavior under faulty conditions was used to evaluate the response of the flux and torque hysteresis comparators. Faults introduce negative sequence components in the stator current that impact flux linkage estimation. The presence of the fault also introduces (1) a DC offset and (2) a double frequency component in the torque. The flux comparator adjusts the applied voltage vector to maintain the desired flux linkage by applying a negative sequence voltage to compensate for the fault effect. However, the negative sequence voltage applied to correct the error due to the fault is limited by the flux hysteresis bandwidth. If the bandwidth for the flux comparator is adequate, the DC offset and a portion of the double frequency component of the torque will be suppressed. The torque comparator manipulates the torque angle to keep the machine torque within the specified hysteresis bands.

Initially, it may seem advantageous for DTC to compensate for the flux linkage and torque errors introduced by the presence of the fault. However, excessive variations in the torque angle and unmodeled dynamics may lead to instability. Ideally, the torque angle should remain relatively constant to maintain synchronization between the stator and rotor flux linkage vectors. Additionally, violations of (20) due to the presence of the fault may cause the controller to become unstable.

As shown in Figures 12–14, low severity faults result in minimal changes to the negative sequence component of the voltage. As the fault severity increases, the negative sequence component of the voltage increases. As long as the flux comparator is able to keep the flux within the specified hysteresis bands, the torque angle variations are minimal. Moreover, it is observed that the impact of the static eccentricity fault on overall drive operation is minimal compared to other faults. This indicates that the flux and torque hysteresis comparators employed in DTC can also mask the impact of faults. Therefore, the compensatory nature of DTC necessitates the development of fault detection and separation techniques that are suitable for the DTC structure to deploy mitigation schemes and maintain stability in the drive system. The results of this work are as follows:

(1) Provided insight into the DTC reaction to faulty operation of a PMSM; and
(2) Identified challenges in stability for DTC-driven PMSMs.

Future research should, therefore, focus on other factors that may affect fault detectability, such as the controller bandwidth, machine saturation level and sampling frequency.

**Author Contributions:** Conceptualization, I.M.A. and S.N.F.; methodology, I.M.A.; software, I.M.A.; validation, I.M.A.; formal analysis, I.M.A.; investigation, I.M.A.; resources, S.N.F.; data curation, I.M.A.; writing—original draft preparation, I.M.A.; writing—review and editing, S.N.F.; visualization, I.M.A.; supervision, S.N.F. All authors have read and agreed to the published version of the manuscript.

**Funding:** This work was partially funded by Michigan State University (MSU), USA, and the Jordan University of Science and Technology (JUST), Jordan. The APC was funded by Michigan State University (MSU).

**Data Availability Statement:** Data will be made available on request.

**Acknowledgments:** Ibrahim M. Allafi sincerely appreciates the Jordan University of Science and Technology (JUST) for sponsoring his graduate studies.

**Conflicts of Interest:** The authors declare no conflict of interest.

## Abbreviations

The following abbreviations are used in this manuscript:

| | |
|---|---|
| DTC | Direct Torque Control |
| FEM | Finite Element Model |
| FFT | Fast Fourier Transform |
| FOC | Field-Oriented Control |
| HRC | High Resistance Contact |
| PMSM | Permanent Magnet Synchronous Machine |
| TTSC | Turn-to-Turn Short Circuit |
| ZSVC | Zero-Sequence Voltage Component |

## References

1. Thangavel, S.; Mohanraj, D.; Girijaprasanna, T.; Raju, S.; Dhanamjayulu, C.; Muyeen, S.M. A Comprehensive Review on Electric Vehicle: Battery Management System, Charging Station, Traction Motors. *IEEE Access* **2023**, *11*, 20994–21019. [CrossRef]
2. González-Cagigal, M.A.; Rosendo-Macías, J.A.; Gómez-Expósito, A. Parameter Estimation of Wind Turbines With PMSM Using Cubature Kalman Filters. *IEEE Trans. Power Syst.* **2020**, *35*, 1796–1804. [CrossRef]
3. Wang, W.; Liu, C.; Zhao, H.; Song, Z. Improved Deadbeat-Direct Torque and Flux Control for PMSM With Less Computation and Enhanced Robustness. *IEEE Trans. Ind. Electron.* **2023**, *70*, 2254–2263. [CrossRef]
4. Allafi, I.M.; Foster, S.N. Condition Monitoring Accuracy in Inverter-Driven Permanent Magnet Synchronous Machines Based on Motor Voltage Signature Analysis. *Energies* **2023**, *16*, 1477. [CrossRef]
5. Bhuiyan, E.A.; Akhand, M.M.A.; Das, S.K.; Ali, M.F.; Tasneem, Z.; Islam, M.R.; Saha, D.K.; Badal, F.R.; Ahamed, M.H.; Moyeen, S.I. A Survey on Fault Diagnosis and Fault Tolerant Methodologies for Permanent Magnet Synchronous Machines. *Int. J. Autom. Comput.* **2020**, *17*, 763–787. [CrossRef]
6. Kadjoudj, M.; Golea, N.; Benbouzid, M. Problems of Stator Flux Estimation in DTC of PMSM Drives. *J. Electr. Eng. Technol.* **2007**, *2*, 468–477. [CrossRef]
7. Feng, C.; Chaoying, X.; Xiaoxin, H. Study on the losing control problem of direct torque control in permanent magnet synchronous motor drive. *Trans. Inst. Meas. Control* **2019**, *41*, 504–515.
8. Haddad, R.Z.; Strangas, E.G. On the Accuracy of Fault Detection and Separation in Permanent Magnet Synchronous Machines Using MCSA/MVSA and LDA. *IEEE Trans. Energy Convers.* **2016**, *31*, 924–934. [CrossRef]
9. Haddad, R.Z.; Lopez, C.A.; Foster, S.N.; Strangas, E.G. A Voltage-Based Approach for Fault Detection and Separation in Permanent Magnet Synchronous Machines. *IEEE Trans. Ind. Appl.* **2017**, *53*, 5305–5314. [CrossRef]
10. Zanardelli, W.G.; Strangas, E.G.; Aviyente, S. Identification of Intermittent Electrical and Mechanical Faults in Permanent-Magnet AC Drives Based on Time–Frequency Analysis. *IEEE Trans. Ind. Appl.* **2007**, *43*, 971–980. . [CrossRef]
11. Chen, H.; He, J.; Guan, X.; Demerdash, N.A.O.; EL-Refaie, A.M.; Lee, C.H.T. High-Resistance Connection Diagnosis in Five-Phase PMSMs Based on the Method of Magnetic Field Pendulous Oscillation and Symmetrical Components. *IEEE Trans. Ind. Electron.* **2022**, *69*, 2288–2299. [CrossRef]
12. Ebrahimi, B.M.; Faiz, J. Feature Extraction for Short-Circuit Fault Detection in Permanent-Magnet Synchronous Motors Using Stator-Current Monitoring. *IEEE Trans. Power Electron.* **2010**, *25*, 2673–2682. [CrossRef]
13. Ebrahimi, B.M.; Faiz, J.; Roshtkhari, M.J. Static-, Dynamic-, and Mixed-Eccentricity Fault Diagnoses in Permanent-Magnet Synchronous Motors. *IEEE Trans. Ind. Electron.* **2009**, *56*, 4727–4739. [CrossRef]
14. Wang, B.; Wang, J.; Griffo, A.; Sen, B. Stator Turn Fault Detection by Second Harmonic in Instantaneous Power for a Triple-Redundant Fault-Tolerant PM Drive. *IEEE Trans. Ind. Electron.* **2018**, *65*, 7279–7289. [CrossRef]
15. Allafi, I.M.; Foster, S.N. Fault Detection and Identification for Inverter-Driven Permanent Magnet Synchronous Machines. In Proceedings of the 2021 IEEE 13th International Symposium on Diagnostics for Electrical Machines, Power Electronics and Drives (SDEMPED), Dallas, TX, USA, 22–25 August 2021; Volume 1, pp. 358–364. . [CrossRef]
16. Hang, J.; Ding, S.; Ren, X.; Hu, Q.; Huang, Y.; Hua, W.; Wang, Q. Integration of Interturn Fault Diagnosis and Torque Ripple Minimization Control for Direct-Torque-Controlled SPMSM Drive System. *IEEE Trans. Power Electron.* **2021**, *36*, 11124–11134. [CrossRef]

17. Hang, J.; Wu, H.; Ding, S.; Hua, W.; Wang, Q. A DC-Flux-Injection Method for Fault Diagnosis of High-Resistance Connection in Direct-Torque-Controlled PMSM Drive System. *IEEE Trans. Power Electron.* **2020**, *35*, 3029–3042. [CrossRef]
18. Allafi, I.M.; Foster, S.N. On the Accuracy of Frequency Based Fault Diagnosis for DTC-driven PMSM. In Proceedings of the 2022 International Conference on Electrical Machines (ICEM), Valencia, Spain, 5–8 September 2022; pp. 1628–1634. [CrossRef]
19. Allafi, I.M.; Foster, S.N. Condition Monitoring of Direct Torque Controlled Permanent Magnet Synchronous Machines. In Proceedings of the 2022 IEEE Energy Conversion Congress and Exposition (ECCE), Detroit, MI, USA, 9–13 October 2022; pp. 1–7. [CrossRef]
20. Hadef, M.; Djerdir, A.; Mekideche, M.R.; N'Diaye, A.O. Diagnosis of stator winding short circuit faults in a direct torque controlled interior permanent magnet synchronous motor. In Proceedings of the 2011 IEEE Vehicle Power and Propulsion Conference, Chicago, IL, USA, 6–9 September 2011; pp. 1–8. [CrossRef]
21. Teng, Q.; Zhu, J.; Wang, T.; Lei, G. Fault tolerant direct torque control of three-phase permanent magnet synchronous motors. *WSEAS Trans. Syst.* **2012**, *11*, 465–476.
22. Son, D.I.; Han, J.S.; Park, J.S.; Lim, H.S.; Lee, G.H. Performance Improvement of DTC-SVM of PMSM with Compensation for the Dead Time Effect and Power Switch Loss Based on Extended Kalman Filter. *Electronics* **2023**, *12*, 966. [CrossRef]
23. Rahman, M.; Zhong, L.; Hu, W.; Lim, K.; Rahman, M. An investigation of direct and indirect torque controllers for PM synchronous motor drives. In Proceedings of the Second International Conference on Power Electronics and Drive Systems, Singapore, 26–29 May 1997; Volume 2, pp. 519–523. [CrossRef]
24. Ojaghi, M.; Faiz, J. An experimental/simulation investigation to mixed eccentricity fault diagnosis of induction motors under DTC. In Proceedings of the 2014 IEEE International Conference on Industrial Technology (ICIT), Busan, Republic of Korea, 26 February–1 March 2014; pp. 143–148. [CrossRef]
25. Cruz, S.; Cardoso, A. Diagnosis of stator inter-turn short circuits in DTC induction motor drives. *IEEE Trans. Ind. Appl.* **2004**, *40*, 1349–1360. [CrossRef]
26. Berzoy, A.; Mohammed, O.A.; Restrepo, J. Analysis of the Impact of Stator Interturn Short-Circuit Faults on Induction Machines Driven by Direct Torque Control. *IEEE Trans. Energy Convers.* **2018**, *33*, 1463–1474. [CrossRef]
27. Niu, F.; Wang, B.; Babel, A.S.; Li, K.; Strangas, E.G. Comparative Evaluation of Direct Torque Control Strategies for Permanent Magnet Synchronous Machines. *IEEE Trans. Power Electron.* **2016**, *31*, 1408–1424. [CrossRef]
28. Alsofyani, I.M.; Idris, N.R.N.; Lee, K.B. Dynamic Hysteresis Torque Band for Improving the Performance of Lookup-Table-Based DTC of Induction Machines. *IEEE Trans. Power Electron.* **2018**, *33*, 7959–7970. [CrossRef]
29. Mathapati, S.; Bocker, J. Analytical and Offline Approach to Select Optimal Hysteresis Bands of DTC for PMSM. *IEEE Trans. Ind. Electron.* **2013**, *60*, 885–895. [CrossRef]
30. Chen, X.; Hu, J.; Chen, K.; Peng, Z. Modeling of electromagnetic torque considering saturation and magnetic field harmonics in permanent magnet synchronous motor for HEV. *Simul. Model. Pract. Theory* **2016**, *66*, 212–225. .: 10.1016/j.simpat.2016.02.012. [CrossRef]
31. Zhong, L.; Rahman, M.; Hu, W.; Lim, K. Analysis of direct torque control in permanent magnet synchronous motor drives. *IEEE Trans. Power Electron.* **1997**, *12*, 528–536. [CrossRef]

*energies*

MDPI

*Article*

# Electromagnetic Torque Fluctuating Properties under Dynamic RISC Fault in Turbogenerators

**Yuling He [1], Minghao Qiu [1], Xinghua Yuan [1], Haipeng Wang [1,*], Mengya Jiang [1], Chris Gerada [2] and Shuting Wan [1]**

1 Hebei Key Lab of Electric Machninery Maintenance and Failure Prevention, Department of Mechanical Engineering, North China Electric Power University, Baoding 071003, China; heyuling1@ncepu.edu.cn (Y.H.); 220192224062@ncepu.edu.cn (M.Q.); 52451713@ncepu.edu.cn (X.Y.); 220212224028@ncepu.edu.cn (M.J.); 52450809@ncepu.edu.cn (S.W.)

2 Power Electronics, Machines and Control (PEMC) Research Group, University of Nottingham, Nottingham NG7 2RD, UK; chris.gerada@nottingham.ac.uk

* Correspondence: 52452403@ncepu.edu.cn

**Abstract:** This paper analyzes the electromagnetic torque (EMT) fluctuation characteristics in synchronous generators under rotor interturn short-circuit (DRISC) fault. The novelty of this paper is that the DRISC fault is proposed based on the intermittent interturn short circuit existing in the actual operation and compared with the static rotor interturn short-circuit (SRISC) fault. In the work, by studying the influence of DRISC with different positions and different short-circuit degrees, the fluctuation characteristic of the EMT is analyzed and verified. The results show that when the DRISC5% fails, the location is in slot 3, the amplitude of first harmonic decreases by 7.2%, second harmonic amplitude increases by 33.4%, third harmonic decreases by 4.3%, and fourth harmonic increases by 26.8%. As the degree increased and positioned away from the large tooth of the DRISC, the overall EMT amplitude and reverse pulse increased, first and third harmonics decreased, and second and fourth harmonics increased.

**Keywords:** synchronous generator; dynamic rotor interturn short circuit (DRISC); electromagnetic torque (EMT); magneto-motive force (MMF); harmonic component

**Citation:** He, Y.; Qiu, M.; Yuan, X.; Wang, H.; Jiang, M.; Gerada, C.; Wan, S. Electromagnetic Torque Fluctuating Properties under Dynamic RISC Fault in Turbogenerators. *Energies* **2022**, *15*, 3821. https://doi.org/10.3390/en15103821

Academic Editor: Djaffar Ould-Abdeslam

Received: 11 April 2022
Accepted: 19 May 2022
Published: 23 May 2022



## 1. Introduction

Rotor interturn short circuit (RISC) is a common fault, and can usually be divided into dynamic rotor interturn short circuit (DRISC) and static rotor interturn short circuit (SRISC). This kind of fault is usually due to the small radius, and the external insulation part will be broken and cause a short circuit. On the other hand, in the operation of the generator, the high temperature makes the rotor winding expand, causing deformation and displacement. Meanwhile, the generator vibration will lead to insulation damage. Further, the rotor short-circuit failure occurs [1–3].

At present, the domestic and foreign RISC diagnosis method is mainly based on electrical and mechanical parameters. Electrical parameters include stator voltage, stator current [4–6], and parallel branch circulation characteristics [7–9]. The mechanical parameters include unbalanced magnetic tension, stator core vibration, winding vibration characteristics, electromagnetic torque, etc. [10,11].

In terms of the electrical characteristics, Huang et al.'s study concluded that when RISC appears, it will reduce the excitation current, resulting in reactive power [12]. D'Angelo proposed a simulation model of a generator rotor winding detection system, which is a mathematical model established by using the divine network MLP to study the operating power of generator set when RISC appears [13]. Junqing Li et al. analyzed the mathematical expressions of the magnetic momentum and current under RISC conditions and presented the relationship between different harmonic changes of the stator branch current and the RISC short-circuit position [14]. In [15,16], the stator current and circulation characteristics of the composite failure of the generator are analyzed, and the stator voltage RMS decreases when the RISC occurs. Hao et al. used the multi-loop theorem to obtain an expression for the current of each loop. Both the experiment and simulation met the theoretical analysis, and therefore a method of online fault monitoring using a multi-loop current was obtained [6]. Mazzoletti et al. calculated the stator current difference, obtained the fault correspondence of the residual current vector, and realized the fault diagnosis of RISC for the permanent magnet generator [17].

According to the mechanical characteristics, Albright, D. R. measured the change rate of the air-gap flux density of the rotor by a fixed search coil, which can obtain the short-circuit position more sensitively [18]. In [15], the expression of the unbalanced magnetic pull is obtained by the analysis of the magnetic flux density. Through simulation and experimental verification, the unbalanced magnetic tension makes the generator rotor vibrate at the base frequency under the RISC condition. When RISC occurs, the influence of electromagnetic torque on motor vibration is more obvious. By analyzing the electromagnetic torque under the composite fault of rotor short circuit and eccentric, it was concluded that the generator rotor vibrates at three times the frequency, and the vibration intensifies with the deepening of the fault [16]. Hao used the virtual displacement method to analyze the steady-state harmonic characteristics after EMT failure, and concluded that RISC produces AC pulsation components, which are closely related to the stator winding structure [3]. Hang, J detected the positions of the interturn fault based on the amplitude and initial phase angle of the zero-order voltage component (ZSVC), and proposed a method to directly diagnose the interturn fault by directly collecting the electromagnetic torque [19].

The above literature establishes the basis for the diagnosis of RISC, but rarely studies DRISC. As an improvement, this paper presents a comprehensive study on the EMT fluctuating properties via the new DRISC model with different degrees and positions in synchronous generators. On the basis of comparative studies, variable differences between SRISC and DRISC are further obtained.

## 2. Analysis Model of DRISC

When the rotor short circuit occurs, the effective turns of the rotor winding decrease, which will affect the decrease of the air-gap magnetic potential. The excitation current in the rotor winding no longer circulates through the fault site and creates a new loop. There will be a reverse current in this loop to create a reverse magnetic potential.

The DRISC from rotor insulation damage over time is investigated by analyzing fault schematics and simplified models, see Figure 1.

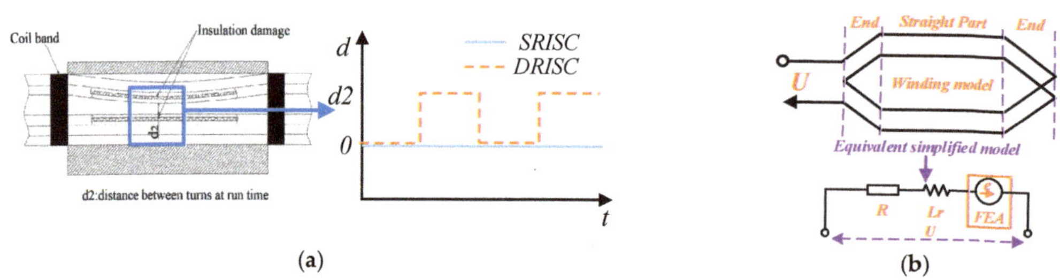


**Figure 1.** (**a**) Fault schematic diagram, (**b**) simplified model.

At the initial stage of short-circuit failure in the generator, intermittent contact occurs at the coil insulation damage. Therefore, in a steady-state excitation circuit, a new circuit occurs in the form of a pulse signal. Moreover, in the late stage of short-circuit fault, the insulation is seriously damaged, and the exposed line is completely touching. Meanwhile, a reverse loop occurs in the excitation circuit, and the reverse signal appears in the form of a step, such as:

$$\text{step}\begin{cases} 0\cdots\cdots t<0 \\ R_0\cdots\cdots t\geq 0 \end{cases}$$
$$\text{pulse}\begin{cases} 0\cdots\cdots t<0, t>\varepsilon \\ \frac{H}{\varepsilon}\cdots\cdots 0\leq t\leq\varepsilon \end{cases} \quad (1)$$

The response of the pulse and step signals to a stable system is shown in Figure 2.

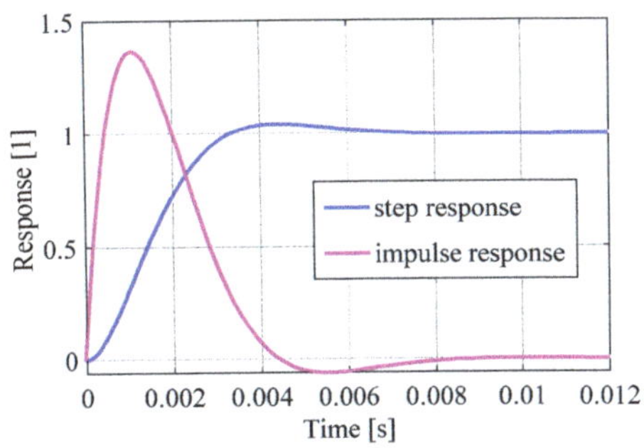


**Figure 2.** Signal response map.

It is known that the rotor dynamic short circuit is a transient short-circuit process, which can be equivalent to the reverse pulse current generated by the short circuit, resulting in the addition of a reverse pulse magnetic potential based on the original magnetic potential. The static short circuit can be understood as a reverse magnetic potential in generating a step, as shown in Figure 3.

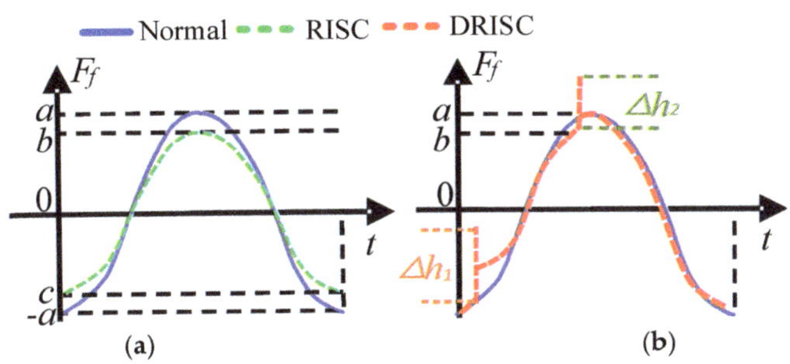


**Figure 3.** (**a**) Rotor MMF in SRISC case, (**b**) rotor MMF in DRISC case.

In DRISC, $T$ is the fault cycle, $k$ is the duty cycle of the short circuit, and $h$ is the peak generated by the short circuit. When the short circuit starts and ends, the $h$ pulse is infinite.

According to the Gaussian magnetic flux theorem, the reverse magnetic potential generated by the reverse current is:

$$F_d(\theta_r) = \begin{cases} -\frac{I_f n_m (2\pi-\alpha_r)}{2\pi}\cdots\cdots\cdots \beta' \leq \theta_r \leq \beta' + \alpha_r \\ \frac{I_f n_m \alpha_r}{2\pi}\cdots\cdots\cdots\cdots\cdots\cdots \theta_r < \beta', \theta_r > \beta' + \alpha_r \end{cases} \quad (2)$$

In (2), $F_d$ is the reverse MMF, $I_f$ is the excitation current, $n_m$ is the number of short-circuit turns, $\theta_r$ is the circumferential angle of the rotor surface, $\beta'$ is the starting angle of the groove, where the interturn short circuit occurs, and $\alpha_r$ is the angle of the two grooves, and the angle where the short circuit turns are located.

For a better understanding, the schematic DRISC and the rotor magnetic potential vector diagram are displayed in Figure 4.

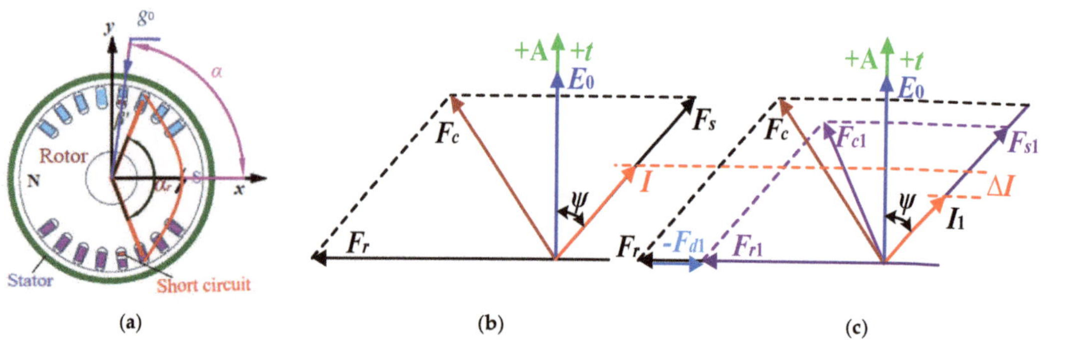


**Figure 4.** (**a**) Schematic diagram of short-circuit principle, (**b**) normal MMF vector, (**c**) MMF vector under short-circuit conditions.

In Figure 4, $F_r$ is the normal base principal magnetic potential, $F_s$ is the normal armature reaction magnetic potential, and $F_c$ is the synthetic magnetic potential for the composite magnetic potential of the gap base wave. $E_0$ is the empty load potential, $I$ is the positive phase current, and $\psi$ is the internal power angle of the generator. In Figure 4c, the sub-angle marked 1 indicates the parameters after the interturn rotor.

### 2.1. Impact of DRISC on MFF

The reverse magnetic potential may be expanded by a Fourier series as:

$$\begin{cases} F_d(\theta_r) = A_0 + \sum\limits_{n=1}^{\infty} [A_n \cos(n\theta_r) + B_n \sin(n\theta_r)] \\ A_0 = \frac{1}{2\pi}\int_0^{2\pi} F_d(\theta_r)d\theta_r = 0 \\ A_n = \frac{1}{\pi}\int_0^{2\pi} F_d(\theta_r)\cos(n\theta_r)d\theta_r = -\frac{I_f n_m [\sin(n(\alpha_r+\beta')) - \sin(n\beta')]}{n\pi} \\ B_n = \frac{1}{\pi}\int_0^{2\pi} F_d(\theta_r)\sin(n\theta_r)d\theta_r = \frac{I_f n_m [\cos(n(\alpha_r+\beta')) + \cos(n\beta')]}{n\pi} \end{cases} \tag{3}$$

Moreover, $F_d$ can be represented as:

$$\begin{cases} F_d(\theta_r) = \sum\limits_{n=1}^{\infty} F_{dn}\cos(n\theta_r - \phi) = \sum\limits_{n=1}^{\infty} F_{dn}\cos(n\omega t - \phi_n) \\ F_{dn} = \sqrt{A_n^2 + B_n^2} = \frac{\sqrt{2} I_f n_m}{n\pi}\sqrt{1-\cos(n\alpha_r)} = \frac{2I_f n_m}{n\pi}\sin\frac{n\alpha_r}{2} \\ \phi_n = \arctan\frac{B_n}{A_n} = \frac{\cos[n(\alpha_r+\beta)] - \cos(n\beta)}{\sin(n\beta) - \sin[n(\alpha_r+\beta)]} \end{cases} \tag{4}$$

In order to simplify the calculation model, the higher harmonics can be ignored, and only the cases when $n = 1$ (indicating odd harmonic) and $n = 2$ (representing even harmonic) are considered. After the DRISC, the air-gap magnetic potential of the generator is as shown in Figure 4c. $F_{d1}$ and $F_{d2}$ are the base amplitude and second harmonic amplitude of the reverse magnetic potential after the DRISC. Therefore, the magnetic potential after the DRISC can be expressed as:

$$f(\alpha_m, t) = \begin{cases} F_s\cos(\omega t - \alpha_m - \psi - \frac{\pi}{2}) + F_r\cos(\omega t - \alpha_m) \\ \quad = F_c\cos(\omega t - \alpha_m - \gamma_1) \cdots\cdots \text{Normal} \\ F_{s1}\cos(\omega t - \alpha_m - \psi - \frac{\pi}{2}) + (F_r - F_{d1})\cos(\omega t - \alpha_m) \\ -F_{d2}\cos 2(\omega t - \alpha_m - \phi_2) \\ \quad = F_{c1}\cos(\omega t - \alpha_m - \gamma_2) - F_{d2}\cos 2(\omega t - \alpha_m - \phi_2) \cdots \text{SRISC} \\ \begin{cases} F_c\cos(\omega t - p\alpha_m) \cdots\cdots (i-1)T < t < (i-1+k)T \\ \Delta h_1 \cdots\cdots t = (i-1+k)T \\ F_{c1}\cos(\omega t - p\alpha_m) - F_{d2}\cos 2(\omega t - p\alpha_m) \\ \cdots\cdots (i-1+k)T < t < iT \\ \Delta h_2 \cdots\cdots t = iT \end{cases} \cdots \text{DRISC} \end{cases} \tag{5}$$

It can be seen that when DRISC occurs in the generator, the even harmonic component increases. In the DRISC cycle, during the short circuit and normal state switch, the RMS of the magnetic potential should be between normal and SRISC (SRISC can be considered a special case of DRISC with a duty cycle of 1). Due to the DRISC period of the normal state, the fluctuation of the magnetic potential amplitude will change accordingly.

### 2.2. Impact of DRISC on MFD

When DRISC occurs, the magnetic tension per unit area remains unchanged, and the magnetic flux density is:

$$B(\alpha_m,t)=\begin{cases} F_c\cos(\omega t-p\alpha_m)\Lambda_0\cdots\cdots\text{Normal}\\ [F_{c1}\cos(\omega t-p\alpha_m)-F_{d2}\cos 2(\omega t-p\alpha_m)]\Lambda_0\cdots\cdots\text{SRISC}\\ \begin{cases} F_c\cos(n\omega t-p\alpha_m)\Lambda_0\cdots\cdots(i-1)T<t<(i-1+k)T\\ \Delta h_1\Lambda_0\cdots\cdots t=(i-1+k)T\\ [F_{c1}\cos(\omega t-p\alpha_m)-F_{d2}\cos 2(\omega t-p\alpha_m)]\Lambda_0\cdots(i-1+k)T<t<iT\\ \Delta h_2\Lambda_0\cdots\cdots t=iT \end{cases}\\ \cdots\cdots\text{DRISC} \end{cases} \quad (6)$$

Combined with the previous analysis, the $B$ amplitude of the DRISC of the rotor can be seen from (6). In addition to the original odd harmonic component, the harmonic component also increases. With the increase of the short circuit, the amplitude of the original odd subharmonic decreases, and the amplitude of the newly generated even subharmonic increases. There is a pulse at the short-circuit and normal state transition moments, and the overall magnetic density amplitude of DRISC is between normal and SRISC.

As mentioned above, both location and degree will affect the change of the air-gap magnetic density, so the influence on the air-gap magnetic density can be obtained by Formula (7):

$$\begin{cases} F_{d1}=\frac{2I_f}{\pi}n_m\sin\frac{\alpha_r}{2}=\frac{2I_fN}{\pi}\frac{n_m}{N}\sin\frac{\alpha_r}{2}=0.637F_rP_s\sin\frac{\alpha_r}{2}\\ F_{d2}=\frac{I_f}{\pi}n_m\sin\alpha_r=\frac{I_fN}{\pi}\frac{n_m}{N}\sin\alpha_r=0.3185F_rP_s\sin\alpha_r\\ P_s=\frac{n_m}{N} \end{cases} \quad (7)$$

where $P_s$ is the percentage of short circuit (the ratio of short circuit turns, $n_m$, to the total turns, $N$), which can be obtained from Formula (7). As the degree of RISC increases, both $F_{d1}$ and $F_{d2}$ will increase. When the short circuit is away from the large tooth, $F_{d1}$ increases, and $F_{d2}$ increases first and then decreases.

### 2.3. Impact of DRISC on EMT

According to the virtual displacement principle, the EMT studied here can be expressed as:

$$\begin{cases} T=p\frac{\partial W}{\partial \psi}\\ W=\int_v\frac{[B(\alpha_m,t)]^2}{2\mu_0}dv \end{cases} \quad (8)$$

The EMT depends on the square of the MFD. Qualitatively, the trend of EMT should be consistent with MFD, and the trend of the squared operation will be more obvious. Consequently: (1) The appearance of DRISC will reduce the RMS value of EMT, and with the increase of the degree of DRISC, the RMS value of EMT is smaller. (2) The DRISC position also affects the electromagnetic torque, while the farther the DRISC is from the large tooth, the smaller the EMT will be. (3) The appearance of DRISC will bring additional odd harmonics in the electromagnetic torque, because even harmonics are newly produced in MFD, and normally only even harmonics (usually only odd harmonics in MFD, because the DC excitation current can only produce odd harmonics, and the square of MFD only contains even harmonics). The DC component and even harmonics decrease, while the odd harmonics increase. (4) The time domain curve of EMT will be affected by the influence of the dynamic short-circuit pulse. (5) Under the same position and degree, the SRISC fault has more of an effect on the RMS of EMT than DRISC.

### 3. FEA and Experimental Validation

In this paper, the CS-5 hidden pole synchronous generator is studied. The finite element calculation and the experimental verification were conducted on the dynamic and static rotor. Parameters are listed in Table 1.

**Table 1.** Parameters of the CS-5 prototype generator.

| Parameters | Values | Parameters | Values |
|---|---|---|---|
| rated power | 5 kW | stator core length | 130 mm |
| pole-pairs | 1 | stator coil turns per slot | 22 |
| Power factor ($\cos\varphi$) | 0.8 | rotor slots | 16 |
| radial air-gap length | 1.2 mm | rotor core outer diameter | 142.6 mm |
| stator slots | 36 | rotor core inner diameter | 40 mm |
| stator outer diameter | 250.5 mm | rotor coil turns per slot | 60 |
| stator inner diameter | 145 mm | Internal power factor ($\cos\psi$) | 0.62 |

On the generator, there is a plate with different short-circuit taps. By connecting different taps, varied short-circuit degrees and positions can be simulated, respectively. The tap settings are indicated in Figure 5a. During the experiment, four group tests were performed: (1) common condition without RISC, (2) 5% RISC in slot 1 (L1–L2), (3) 10% RISC in slot 1 (L1–L3), and (4) 10% RISC in slot 2 (L2–L4). Tests (1)–(3) formed a comparison for different short-circuit positions, while (3) and (4) were employed for varied short-circuit degrees.

This prototype generator was specifically designed and manufactured by us, and is able to simulate RISC. This paper establishes the FEA model in ANSYS Maxwell according to the CS-5 hidden pole synchronous generator design parameters, and verifies the stability and convergence of the model through previous calculations.

For vividly simulated failures in the experiment, we treated the photoelectric coupler in the DC solid-state relay as a short-circuit coil. When the photoelectric coupler moves, the short-circuit fault starts. Otherwise, the short-circuit fault ends. For a schematic diagram of the DC solid-state relay, see Figure 5d. During the experiment, the start of the DC solid-state relay was controlled by the PWM. Moreover, the corresponding short-circuit taps were connected by the DC solid-state relay to realize DRISC. In this study, square-wave pulses with a 20 ms period and a forward voltage duty cycle of 15% were generated using PWM. When the square-wave voltage exceeds the DC solid-state relay trigger threshold, the above method can realize the high-frequency DRISC simulation of the generator.

To match of finite element analysis and the experimental settings, the physical model in finite elements was coupled to an external circuit. The external coupling circuit of the generator excitation winding and the armature winding is shown in Figure 5e. S_A1 is a short-circuit trigger switch, setting the trigger voltage interval [$V_{on}$, $V_{off}$] to control the short-circuit switch with the pulse voltage source. The dynamic and static rotor interturn short-circuit model can be realized. The pulse time setting of the DRISC is shown in Figure 5c.

To reasonably simulate the short-circuit behavior, the section of the short-circuit winding bar was divided into two components in the 3D model. One is to represent the short-circuit part, and the other is to represent the normal part (see Figure 5b). The short-circuit degree was set by changing the short-circuit turn-numbers of the slot as well as the values of $R_f$ and $R_w$ (see Figure 5e). During simulation, the band rotation was 3000 rpm, the start time was 0 s, the termination time was 0.2 s, and the step size was 0.0002 s.

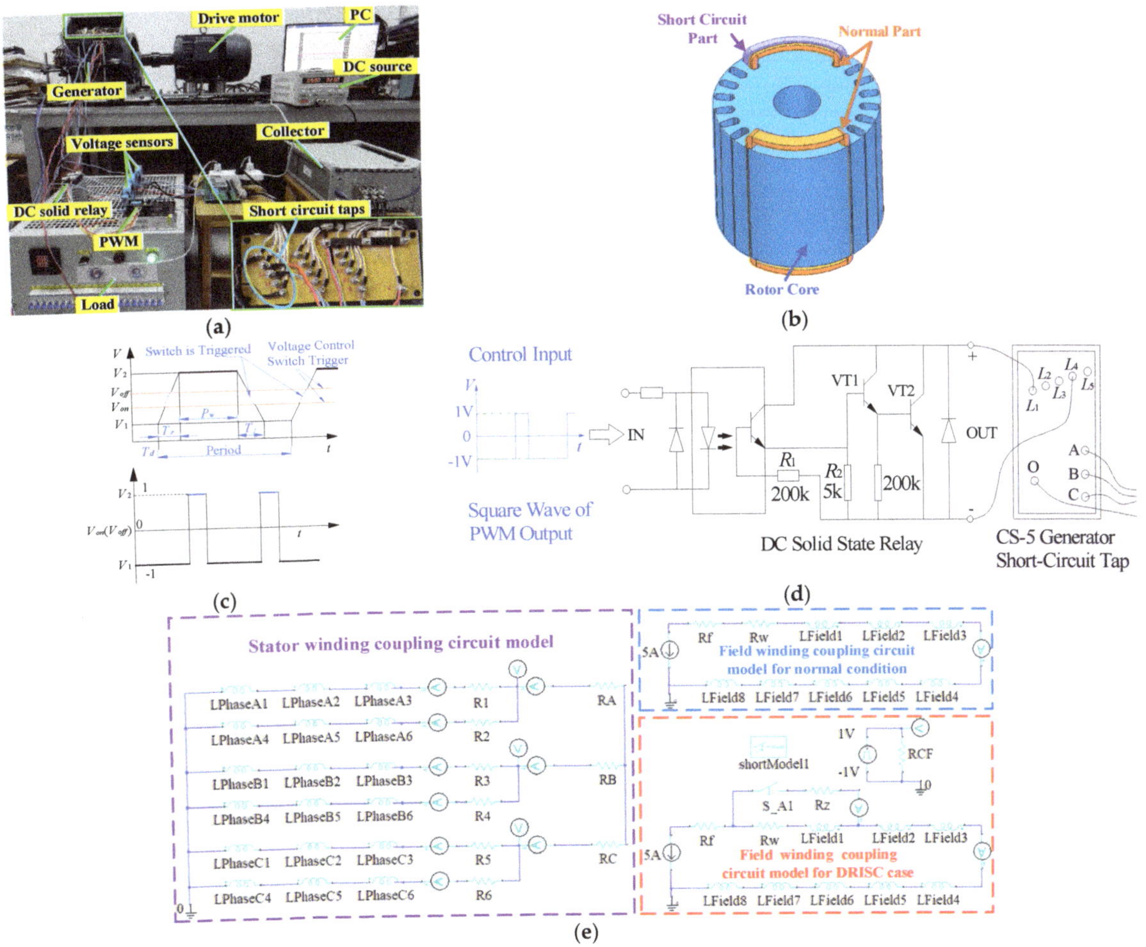


**Figure 5.** CS-5 generator: (**a**) picture, (**b**) FEA model, (**c**) short-circuit switch control, (**d**) DRISC trigger circuit, (**e**) external coupling circuit model.

During the finite element analysis, the external circuit contains parameters such as normal and short-circuit winding and changes the resistance values of the corresponding resistors $R_w$ and $R_f$. Different degrees of fault simulation can be achieved. In Figure 5c, $T_d$ is the delay time, $T_r$ is the rise time, $T_f$ is the fall time, $P_w$ is the pulse width, the DRISC cycle corresponds to the pulse cycle, and $V_{on}$ is the control switch of the voltage trigger voltage. When the voltage value of the pulse voltage source is greater than $V_{on}$, the switch S_A1 acts and the excitation circuit is a short circuit. $V_{off}$ represents the blocking voltage of the voltage control switch. When the voltage value of the pulse voltage source is less than $V_{off}$, the switch S_A1 is turned off and the excitation circuit is normal. Both the duty cycle of the short-circuit portion and the DRISC frequency can be changed by adjusting the $P_w$ and the period. The short-circuit portion produced an external circuit of the DRISC with a period of 20 ms (15% duty cycle) under the same settings used in the experiment.

In the experiment, DRISC simulation with different degrees and at different positions can be realized by short-connecting short-circuit taps to different degrees. In the finite element simulation, it corresponds to the experiment by changing the finite element model and the external coupled circuit parameters.

The fault types described in this paper are shown in Table 2. This paper analyzed the harmonic growth ratio on the basis of the normal harmonic amplitude value.

**Table 2.** Abbreviation of different cases.

| Full Name | Abbreviation | Full Name | Abbreviation |
|---|---|---|---|
| Normal | N | DRISC 5%0 | D5-0 |
| SRISC 5%3 | S5-3 | DRISC 5%5 | D5-5 |
| SRISC 10%3 | S10-3 | Short Circuit 5%0 | 5%0 |
| DRISC 5%3 | D5-3 | Short Circuit 5%3 | 5%3 |
| DRISC 10%3 | D10-3 | Short Circuit 5%5 | 5%5 |
| SRISC 5%0 | S5-0 | Short Circuit 10%3 | 10%3 |
| SRISC 5%5 | S5-5 | | |

1. Effect of Short-Circuit Degree

The MFD results at different RISC degrees by finite element analysis are shown in Figure 6. As shown in Figure 6a, the appearance and increase of DRISC decreased the MFD amplitude compared to normal, and the odd harmonic decreased and the even harmonic increased. Compared with SRISC, the appearance of DRISC caused the amplitude fluctuation of the reverse pulse produced by the short circuit. As the degree of the short circuit intensified, the pulse impact became greater. The equivalent degree of DRISC failure had less effect on harmonics than SRISC, but with the same trend (see Figure 6b). This result is consistent with a previous theoretical analysis (see Equation (15)).The EMT results of the FEM model and experiments established in this paper are shown in Figure 7. It is shown that odd harmonic and amplitude pulses occurred when DRISC occurred. As shown by the FEA curve in Figure 7d,e, when the short circuit 5% occurred, the amplitude of first harmonic decreased by 7.2%, second harmonic amplitude increased by 33.4%, third harmonic decreased by 4.3%, and fourth harmonic increased by 26.8%. With the increase of DRISC degree, the total electromagnetic torque amplitude, constant component amplitude, and second harmonic amplitude decreased, while the fourth harmonic and pulse amplitude increased. The reason for the amplitude fluctuation spikes in EMT is that DRISC generated reverse pulses in the original stable excitation circuit. The reverse pulse increased as the short circuit increased.

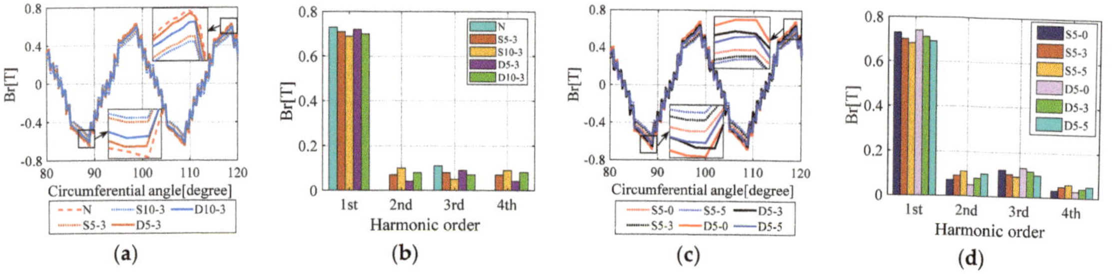


**Figure 6.** MFD variations: (**a**,**b**) varied dynamic short-circuit degrees, (**c**,**d**) varied dynamic short-circuit positions.

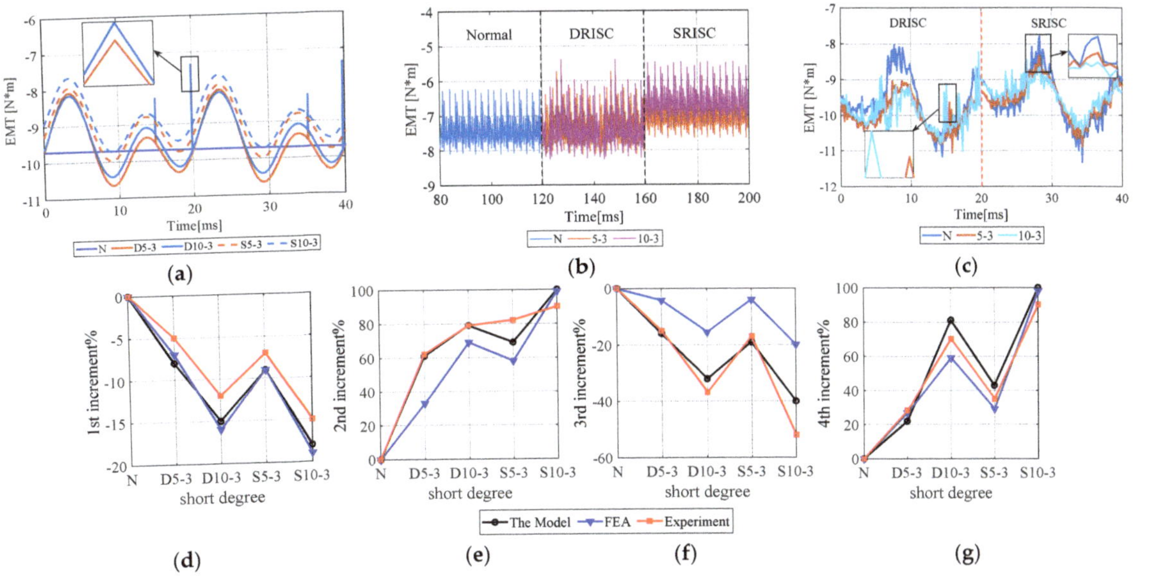


**Figure 7.** EMT in varied DRISC degree cases: (**a**–**c**) waves by model, FEA, and experiment, and (**d**–**g**) harmonic variations.

Compared to SRISC, the amplitude of the EMT second harmonics decreased by 17.6% and fourth harmonic decreased by 10.8% when the same degree of DRISC failure occurred. The theoretical analysis, FEA, and experiment had the same trends.

2. Effect of Short-Circuit Position

The results of the finite element model analysis are shown in Figure 6. When DRISC occurred and the position was far from the large tooth, the reverse magnetic potential generated by the short circuit had a greater influence on the rotor excitation potential. The overall amplitude of the MFD was compressed, with the odd harmonic component increasing and the even harmonic component decreasing. When the DRISC occurred, the MFD curve had a smaller amplitude pulse compared with the SRISC.

The EMT obtained from the proposed model and from the finite element analysis and experiments are shown in Figure 8. It can be seen that the odd harmonic component appeared when the DRISC fault occurred. When the DRISC5% fault position changed from slot 3 to slot 5, the amplitude of first harmonic decreased by 42.5%, second harmonic amplitude increased by 11.4%, third harmonic decreased by 19.4%, and fourth harmonic increased by 32.3%. With the short-circuit position away from the large tooth, the total electromagnetic torque amplitude, constant component amplitude, and second harmonic component amplitude will decrease, while the fourth harmonic and reverse amplitude pulse value will increase. This is consistent with the qualitative analysis conclusion obtained by (19).

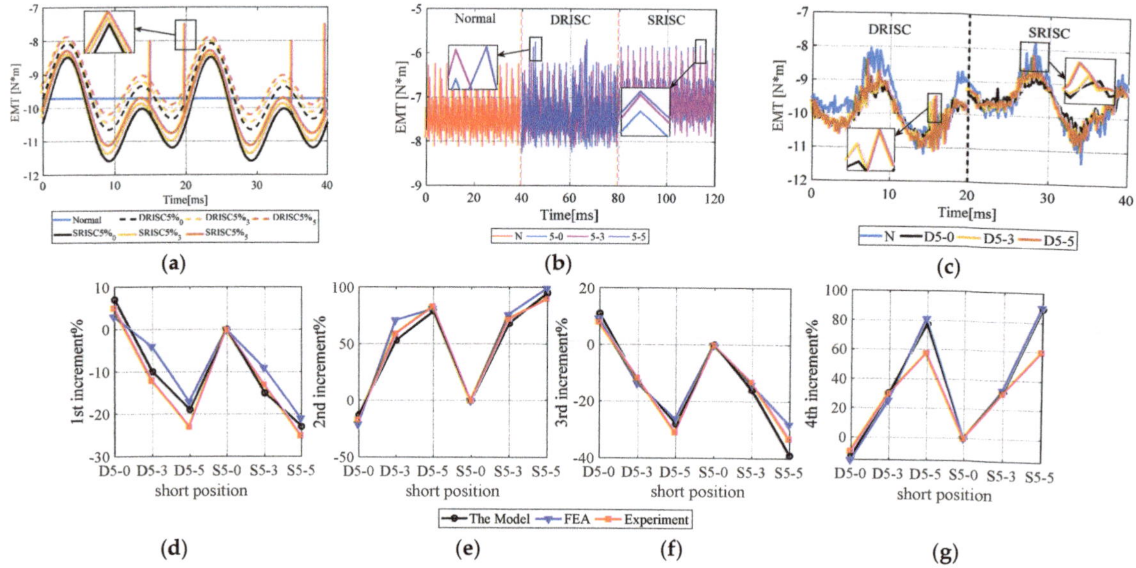

**Figure 8.** EMT in varied DRISC position cases: **(a–c)** waves by model, FEA, and experiment, and **(d–g)** harmonic variations.

## 4. Conclusions

This paper presented a new analytical model for RISC fault in synchronous generators. Further, the EMT fluctuations were comprehensively studied based on the proposed DRISC model with different degrees and positions. Using the CS-5 fault simulation generator for the analysis, the FEA calculation and the experimental tests confirmed the validity and the eligibility of the proposed model. The results showed that:

(1) When the DRISC5% failed, the location was in slot 3, the amplitude of first harmonic decreased by 7.2%, second harmonic amplitude increased by 33.4%, third harmonic decreased by 4.3%, and fourth harmonic increased by 26.8%. As the degree of the DRISC increased, the overall EMT amplitude and reverse pulse increased, first and third harmonics decreased, and second and fourth harmonics increased.

(2) When the DRISC5% fault position changed from slot 3 to slot 5, the amplitude of first harmonic decreased by 42.5%, second harmonic amplitude increased by 11.4%, third harmonic decreased by 19.4%, and fourth harmonic increased by 32.3%. At the DRISC position away from the large tooth, the overall EMT amplitude and reverse pulse increased, first and third harmonics decreased, and second and fourth harmonics increased.

(3) Compared with SRISC, DRISC had less influence on the EMT amplitude and each harmonic, but the electromagnetic torque curve showed obvious amplitude pulse.

(4) When a DRISC failure occurred, significant amplitude spikes appeared in the EMT curve. As the DRISC failure intensified and the fault was away from the large tooth, the reverse pulse generated by the short circuit became larger. Therefore, the vibration amplitude of the EMT became larger.

**Author Contributions:** Writing—original draft, M.Q.; Writing—review & editing, Y.H., X.Y., H.W., M.J., C.G. and S.W. All authors have read and agreed to the published version of the manuscript.

**Funding:** This research was funded by [National Natural Science Foundation of China] grant number [51777074] and [National Natural Science Foundation of China] grant number [52177042].

**Conflicts of Interest:** The authors declare no conflict of interest.

## References


1. Dirani, H.C.; Merkhouf, A.; Kedjar, B.; Giroux, A.M.; Al-Haddad, K. Rotor interturn short circuit impact on large hydrogenerator magnetic quantities. *IEEE Trans. Ind. Appl.* **2018**, *54*, 3702–3711. [CrossRef]
2. Valavi, M.; Jørstad, K.G.; Nysveen, A. Electromagnetic analysis and electrical signature-based detection of rotor inter-turn faults in salient-pole synchronous machine. *IEEE Trans. Magn.* **2018**, *54*, 1–9. [CrossRef]
3. Hao, L.; Wu, J.; Zhou, Y. Theoretical analysis and calculation model of the electromagnetic torque of nonsalient-pole synchronous machines with interturn short circuit in field windings. *IEEE Trans. Energy Convers.* **2014**, *30*, 110–121. [CrossRef]
4. Hang, J.; Ding, S.; Zhang, J.; Cheng, M.; Chen, W.; Wang, Q. Detection of Interturn Short-Circuit Fault for PMSM With Simple Fault Indicator. *IEEE Trans. Energy Convers.* **2016**, *31*, 1697–1699. [CrossRef]
5. Park, Y.; Jeong, M.; Lee, S.B.; Antonino-Daviu, J.A.; Teska, M. Influence of Blade Pass Frequency Vibrations on MCSA-Based Rotor Fault Detection of Induction Motors. *IEEE Trans. Ind. Appl.* **2017**, *53*, 2049–2058. [CrossRef]
6. Hao, L.; Sun, Y.; Qiu, A.; Wang, X. Steady-State Calculation and Online Monitoring of Interturn Short Circuit of Field Windings in Synchronous Machines. *IEEE Trans. Energy Convers.* **2011**, *27*, 128–138. [CrossRef]
7. Li, G.-J.; Hloui, S.; Ojeda, J.; Hoang, E.; Lecrivain, M.; Gabsi, M.; Zhu, Z.Q. Excitation Winding Short-Circuits in Hybrid Excitation Permanent Magnet Motor. *IEEE Trans. Energy Convers.* **2014**, *29*, 567–575. [CrossRef]
8. Dorrell, D.G.; Makhoba, K. Detection of Inter-Turn Stator Faults in Induction Motors Using Short Term Averaging of Forwards and Backwards Rotating Stator Current Phasors for Fast Prognostics. *IEEE Trans. Magn.* **2017**, *53*, 1–7. [CrossRef]
9. Yucai, W.; Yonggang, L.; Heming, L.; Wenjing, Z. An Analysis of the Impact of Rotor Winding Interturn Short Circuits on Turbine Generator Operating Variables. *Electr. Power Compon. Syst.* **2015**, *43*, 674–684. [CrossRef]
10. Chai, F.; Li, Y.; Pei, Y.; Yu, Y. Analysis of Radial Vibration Caused by Magnetic Force and Torque Pulsation in Interior Permanent Magnet Synchronous Motors Considering Air-Gap Deformations. *IEEE Trans. Ind. Electron.* **2018**, *66*, 6703–6714. [CrossRef]
11. He, Y.-L.; Deng, W.-Q.; Tang, G.-J. Analysis and Simulation on UMP and EMT Characters of Turbogenerator under Axial Air-Gap Eccentricity. *Int. J. Rotating Mach.* **2015**, *2015*, 1–10. [CrossRef]
12. Huang, H.; Zhang, K.; Zhang, Y. Detection of turbine generator field winding serious inter-tum short circuit based on the rotor vibration feature. In Proceedings of the International Universities Power Engineering, Glasgow, UK, 1–4 September 2009; pp. 1–5.
13. D'Angelo, M.F.; Costa, P.P. Detection of shorted turns in the field winding of turbogenerators using the neural network MLP. In Proceedings of the IEEE International Conference on Systems, Tucson, AZ, USA, 7–10 October 2001; pp. 1930–1935.
14. Li, J.; Shi, W.; Li, Q. Research on interturn short circuit fault location of rotor winding in synchronous electric machines. In Proceedings of the International Conference on Electrical Machines and Systems, Sydney, NSW, Australia, 11–14 August 2017; pp. 1–4. [CrossRef]
15. He, Y.L.; Wang, F.L.; Ke, M.Q.; Tang, G.J. Rotor vibration difference among the single and the combined faults composed by static air-gap eccentricity and rotor interturn short circuit. In *Proceedings of the 9th IFToMM International Conference on Rotor Dynamics*; Springer: Cham, Switzerland, 2015.
16. Shuting, W.; Yonggang, L.; Heming, L.; Guiji, T.A. Compositive Diagnosis Method on Turbine-Generator Rotor Winding Inter-turn Short Circuit Fault. In Proceedings of the 2006 IEEE International Symposium on Industrial, Electronics, Montreal, QC, Canada, 9–13 July 2006; pp. 1662–1666.
17. Mazzoletti, A.; Bossio, G.R.; De Angelo, C.H.; Espinoza-Trejo, D.R. A Model-Based Strategy for Interturn Short-Circuit Fault Diagnosis in PMSM. *IEEE Trans. Ind. Electron.* **2017**, *64*, 7218–7228. [CrossRef]
18. Albright, D.R. Interturn short-circuit detector for turbine generator rotor windings. *IEEE Trans. Power Appar. Syst.* **1971**, *90*, 478–483. [CrossRef]
19. Hang, J.; Ding, S.; Ren, X.; Hu, Q.; Huang, Y.; Hua, W.; Wang, Q. Integration of interturn fault diagnosis and torque ripple minimization control for direct-torque-controlled SPMSM drive system. *IEEE Trans. Power Electron.* **2021**, *36*, 11124–11134. [CrossRef]

energies

MDPI

Article

# Condition Monitoring Accuracy in Inverter-Driven Permanent Magnet Synchronous Machines Based on Motor Voltage Signature Analysis

Ibrahim M. Allafi * and Shanelle N. Foster

Department of Electrical and Computer Engineering, Michigan State University, East Lansing, MI 48824, USA
* Correspondence: allafiib@msu.edu

**Abstract:** Condition monitoring and preventative maintenance are essential for reliable and efficient operation of permanent magnet synchronous machines driven by inverters. There are two types of industrial inverter drives available: field oriented control and direct torque control. Their compensation nature and control structure are distinct and, therefore, the condition monitoring approach designed for the former control may not be applicable to the latter one. In this paper, we investigate the Motor Voltage Signature Analysis approach for both inverter drives under healthy and faulty conditions. Four typical fault conditions are addressed: turn-to-turn short circuit, high resistance contact, static eccentricity, and local demagnetization. High fidelity cosimulation is developed by coupling the finite element machine model with both control drives. The spectral elements of the commanded stator voltage are utilized as indicators for supervised classification to identify, categorize, and estimate the severity of faults. Linear discriminate analysis, k-nearest neighbor, and support vector machines are the classification techniques used. Results indicate that the condition monitoring based on the Motor Voltage Signature Analysis performs adequately in field oriented control. Nevertheless, the utilized monitoring scheme does not exhibit satisfactory performance in direct torque control owing to the nonlinear characteristics and tolerance nature of this drive.

**Keywords:** condition monitoring; demagnetization; direct torque control; eccentricity; fault diagnosis; field oriented control; high resistance connection; turn-to-turn short circuit; permanent magnet synchronous machine; supervised classification

**Citation:** Allafi, I.M.; Foster, S.N. Condition Monitoring Accuracy in Inverter-Driven Permanent Magnet Synchronous Machines Based on Motor Voltage Signature Analysis. *Energies* **2023**, *16*, 1477. https://doi.org/10.3390/en16031477

Academic Editor: Chunhua Liu

Received: 27 December 2022
Revised: 23 January 2023
Accepted: 28 January 2023
Published: 2 Febuary 2023



## 1. Introduction

The use of Permanent Magnet Synchronous Machines (PMSMs) has increased dramatically during the last three decades. This is due to the development of new magnets with high residual flux density, high energy product, and considerable demagnetization resilience, such as the Neodymium-Iron-Boron (NdFeB) magnet [1]. Therefore, PMSMs outperform induction machines in terms of efficiency, rotor mass, and torque density [2,3]. Such advantages make PMSMs more prevalent in different critical industries such as automotive, aerospace, and renewable energy applications [4]. In these critical applications, where safety is a must, highly reliable and efficient electricity-driven PMSMs systems are required. However, a failure may occur unexpectedly due to material aging, assembly defects, poor installation, or inappropriate operation, all of which may result in costly shutdowns and tragic loss of human lives [5–7]. Therefore, inverter-driven PMSMs systems necessitate health monitoring and fault diagnosis algorithms to guarantee the highest level of safety and reliability.

Fault diagnosis is considerably demanding and challenging in inverter-driven PMSMs as compared to the line-fed ones. This is evidenced by the fact that fault detectability in inverter-driven PMSM systems is diminished by the compensatory capability of the drive system [8]. The two most common drives for PMSMs are Field Oriented Control (FOC) and Direct Torque Control (DTC). FOC was introduced in the late 1960s for alternating current machines and then first adopted by the Toshiba industry a decade later. The main concept of FOC is to control the stator current vector in the rotor flux linkage frame at fixed switching frequency [9,10]. DTC, on the other hand, aims to directly control the stator flux linkage and the electromagnetic torque through hysteresis comparators. Thus, the switching frequency is variable and a position sensor is not necessary to achieve the complex current orientation into the rotor flux linkage frame [11]. It was proposed in the early 1970s and then commercialized by the ABB industry two decades later [12,13]. Both control schemes have a simple structure and provide satisfactory performance with better steady-state behavior in the case of FOC and faster torque dynamics in DTC. However, one control scheme can be more appropriate than the other based on the drive requirements of the application.

Intensive research on fault diagnosis has been conducted in the literature for FOC-driven PMSMs under different faults [14–16]. For the most precise fault detection in various operating conditions, the Machine Voltage or Current Signature Approach (MVSA/MCSA) has been commonly used at high loads. This method allows remote machine monitoring through the Motor Control Center (MCC) with no extra hardware for implementation. Turn-to-Turn Short Circuit (TTSC) and eccentricity faults cause changes in the sideband harmonics of the stator current; therefore, they could be used as features for fault presence [17,18]. As the Signal-to-Noise Ratio (SNR) of the utilized features drops in the presence of increasing noise levels, so does the reliability of the detection method. Therefore, the main spectral elements in the stator current or voltage have been considered because of their higher SNR for identifying and classifying TTSC, eccentricity, and demagnetization faults using Linear Discriminate Analysis (LDA) classifier [19]. Harmonics in the stator current or voltage waveforms, however, are sensitive to the bandwidth of the controller. Thus, it is recommended that the harmonics in the voltage waveform be used for fault diagnosis since they are prominent when a high bandwidth is utilized to obtain better control performance [20]. Despite the fact that the MVSA method has had widespread use in the research community for FOC drives, it has not been examined for DTC-driven PMSMs.

The change in commanded voltages in the rotor flux linkage frame, according to [14], may be utilized to identify and discriminate eccentricity, demagnetization, and TTSC faults. Monitoring magnitude variations of these voltage signals present in the FOC drives allows for nonintrusive fault detection. Due to the DTC drive being executed in the stationary reference frame, where these voltages are not accessible and the position sensor is unnecessary, this method could not be extended for DTC.

Few contemporary studies examine condition monitoring for DTC-driven PMSMs [20–24]. Zero-sequence voltage component (ZSVC) magnitude and its initial phase have been proposed by authors in [21] for TTSC fault detection. The negative influence of TTSC on DTC performance is then compensated by using torque injection and an enhanced flux observer. To calculate the zero-sequence network, however, neutral point access is required by paralleling a balanced three-phase resistive load with the machine. In [22], the High Resistance Contact (HRC) fault has been investigated by including a fixed flux deviation in the stator flux linkage estimate necessary for DTC drive operation. However, further analysis is needed to ensure reliable DTC operation considering the importance of accurate flux estimation to DTC. To accommodate PMSM drives with open phase fault, recent research adapted the DTC system for use as a fault tolerant control [25–27].

Fault identification is required to enhance drive performance considering multiple faults to avoid false alarms. Fault diagnosis methods of PMSMs that exist in the literature based on the FOC scheme are not suitable for DTC-driven PMSMs as they have a different nature of compensation, a different structure, and a different regulation principle. There is a clear gap in the research on the reliability of DTC-driven PMSMs. The major contributions of this work are: (1) Reliable multiple fault detection and separation is introduced by adopting multifeature analysis to avoid misclassification and false alarms, as the vast majority of the existing detection methods in the literature consider single fault. (2) Evaluating the supervised learning techniques in fault diagnosis for both FOC and DTC-driven PMSMs. (3) Reducing the research gap in the reliability of DTC-driven PMSMs by addressing the MVSA approach and examining its performance in detecting and discriminating faults for DTC-driven PMSMs. (4) The challenges in fault diagnosis for DTC-driven PMSMs are highlighted in order to pave the path for further development in monitoring approaches that suit the DTC scheme.

## 2. Theoretical Background

### 2.1. PMSM Faults and MVSA Approach

Stator faults are most commonplace in low and medium power machines [28]. They could be TTSC due to the degradation in winding insulation or HRC due to the loose connection between the machine and inverter terminals. Machine overloading in harsh environments and electric stress from switching devices lead to electrical stator faults eventually. TTSC fault propagates faster than the other faults and results in further insulation degradation and motor outage in the end [29]. HRC results in excessive heat at the motor–inverter joints and can lead to open phase fault if left unaddressed [22]. Therefore, stator fault detection and discrimination help in reducing the risk of motor outage, reducing the maintenance cost, and selecting the proper mitigation scheme.

Rotor faults occur more commonly in high power machines; they are 49% of failure distribution [30]. Eccentricity faults are an example of the mechanical rotor failures that inherently exist at the manufacturing phase. It increases the Noise, Vibration, and Harshness (NVH) and can lead to rotor rubbing the stator laminations [18]. Additionally, demagnetization is another rotor fault that could take place within driven PMSMs if the electromagnetic constraints are exceeded by improper operation in the field weakening region or due to the severe TTSC fault occurrence [31]. As a result, the rotor magnets get demagnetized causing asymmetric flux distribution in the air gap. Reduction in the average torque and increase in the torque ripple are the main consequences of the demagnetization fault. Therefore, rotor fault detection is crucial to avoid any damage to the machine structure and improve the drive performance.

Stator and rotor faults manifest themselves in the spectrum of the machine voltage signal and leave either distinct or similar trends. Proper utilization of these trends, which is known as MVSA, does not help only in identifying whether the PMSM is healthy or faulty but also helps in fault classification and severity estimation. Several signal processing techniques are available to obtain the frequency spectrum of the machine voltage signal [32]. The spectrum is obtained at a steady state using Fast Fourier Transform (FFT), or under nonstationary conditions using time–frequency analysis methods like Short Time Fourier transform (STFT). Variations in magnitude and/or phase of the obtained spectral elements could be utilized as features for supervised classifiers to detect faults.

### 2.2. PMSM Drives

#### 2.2.1. Field Oriented Control

Magnetizing and torque current components in the direct and quadrature axes ($i_d$ and $i_q$) of the rotor flux linkage frame are regulated separately using a proportional–integral ($PI$) controllers in FOC so that the output torque is regulated indirectly. Hence, FOC is also called indirect torque control. Figure 1 depicts the implementation of FOC-driven PMSM, where an encoder is needed to estimate the rotor position ($\theta_r$) required for current vector decoupling. The widely applied modulation technique to control the inverter switching scheme in FOC is the Space Vector Pulse Width Modulation (SVPWM) due to its high utilization of the DC link and low harmonic distortion [33]. The principle behind SVPWM is to synthesize a reference space voltage vector using timely applied active and zero voltage vectors at a fixed switching frequency.

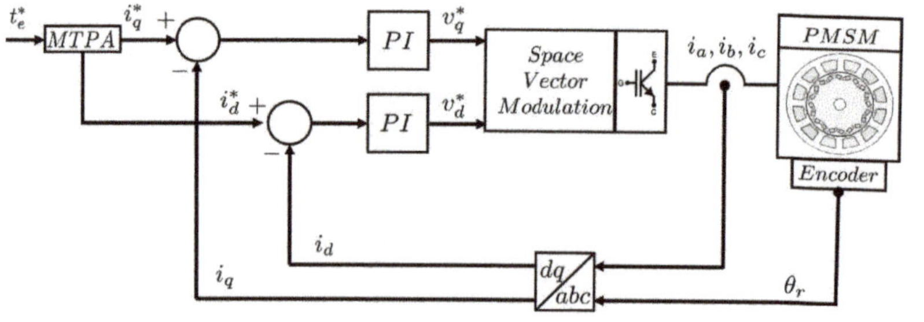


**Figure 1.** Stator current regulation loop in FOC drive.

The expressions of $dq$ currents and the corresponding machine torque as a function of both currents are shown in (1) and (2), respectively.

$$
\begin{aligned}
i_q &= |i_s|cos(\beta) \\
i_d &= -|i_s|sin(\beta)
\end{aligned} \tag{1}
$$

$$
t_e = \frac{3P_n}{4} i_q \left[ \lambda_{PM} + (L_d - L_q) i_d \right] \tag{2}
$$

where $|i_s|$ and $\beta$ are the stator current magnitude and angle measured from the $q$-axis, respectively. $t_e$ is the machine electromagnetic torque. $P_n$ is the number of rotor poles in PMSM. $L_d$ and $L_q$ are the inductance of the direct and quadrature axes. $\lambda_{PM}$ is the magnet flux linkage.

By controlling $|i_s|$ and $\beta$, different $(i_d,i_q)$ combinations can be generated and different control modes can be achieved as follows: Maximum Torque Per Ampere (MTPA), field weakening, maximum torque per voltage, and unity power factor [34]. The control mode selection is based on the PMSM type and operating speed. Below the base speed, the MTPA control is utilized to generate the maximum torque for a given current in order to minimize the copper losses [35,36]. The excitation angle in MTPA region $(\beta_{T_{max}})$ is expressed in (3). PMSM control in MTPA region is used in this work. Here, the excitation angle $(\beta)$, to get the optimum $dq$ currents, for a given torque is found.

$$
\beta_{T_{max}} = sin^{-1}\left[ \frac{\sqrt{8(L_q - L_d)^2 |i_s|^2 + \lambda_{PM}^2} - \lambda_{PM}}{4|i_s|(L_q - L_d)} \right] \tag{3}
$$

PMSM control in MTPA region is used in this work. Here, the excitation angle $(\beta)$, to get the optimum $dq$ currents, for a given torque is found.

### 2.2.2. Direct Torque Control

The nonlinear form of DTC drive, which employs hysteresis comparators, was first presented for PMSMs in [37]. Figure 2 depicts the basic block diagram utilized for DTC implementation where $\delta_{t_e}$ and $\delta_{\lambda_s}$ are the outputs of hysteresis comparators.

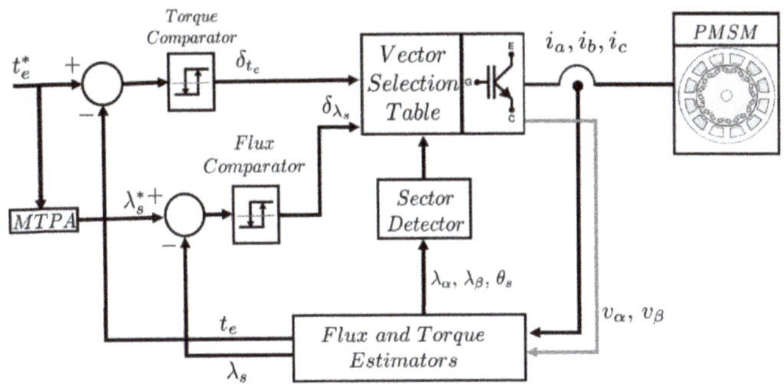


**Figure 2.** Torque and flux regulation loop in DTC drive.

The amplitude of the torque output ($t_e$) and the stator flux linkage ($\lambda_s$) are both adjusted by using two hysteresis comparators. Thus, the feedback loop needs flux and torque estimates. These estimates are typically computed in the stationary ($\alpha\beta$) frame, as explained in (4) and (5), to obviate the requirement for a position sensor.

$$\lambda_{\alpha\beta} = \lambda_{\alpha_0\beta_0} + \int (v_{\alpha\beta} - r_s i_{\alpha\beta})dt \tag{4}$$

$$t_e = \frac{3P_n(i_\beta\lambda_\alpha - i_\alpha\lambda_\beta)}{4} \tag{5}$$

where $i_\alpha$ and $i_\beta$ are the currents in the stationary frame. $r_s$ is the stator phase resistance. $\lambda_{\alpha_0}$ and $\lambda_{\beta_0}$ represent the initial flux linkage estimates based on the frame alignment. The voltages $v_\alpha$ and $v_\beta$ in stationary frame are calculated as in (6) where $(sw_A, sw_B, sw_C)$ are the switching signals to the inverter legs and $(v_{link})$ is the bus link voltage.

$$\begin{aligned} v_\alpha &= \frac{v_{link}[2sw_A - sw_B - sw_C]}{3} \\ v_\beta &= \frac{v_{link}[sw_B - sw_C]}{\sqrt{3}} \end{aligned} \tag{6}$$

### 2.3. Supervised Classification

The method of developing a prediction model using specified data input is known as supervised machine learning [38]. A class may be anticipated for a sample that has not yet been classified using this data-driven method. Three classification methods are utilized in this paper: LDA, k-nearest neighbor ($k$-NN), and support vector machines (SVM).

The training data in LDA are assumed to have a normal distribution with a fixed class-independent covariance matrix. The decision border partitioning classes in a two-indicator situation is linear using LDA. Nevertheless, this discriminant technique is highly biased if the assumption of normal data distribution is violated [39]. Here, the dataset is split to ($K$) classes, with every class containing a number of samples belonging to the same class. Each class is associated with weighting factors that are used to determine the discriminant function for that class. The discriminant function ($Y_k$) for $k$th class is calculated as following:

$$Y_k(X) = \alpha_{1k}x_1 + \alpha_{2k}x_2 + \cdots + \alpha_{Nk}x_N \tag{7}$$

where $X = [x_1, x_2, \ldots, x_N]$ is the $N$ dimensional observation vector and $[\alpha_{1k}, \alpha_{2k}, \ldots, \alpha_{Nk}]$ is the matrix of weighting elements of the $k^{th}$ group. A sample will be placed in a given class if and only if its discriminant function is greater than those of other classes. For example, an unidentified sample $X_a$ will be assigned to a class $b$ if

$$Y_b(X_a) \geq Y_k(X_a), \quad \forall b \neq k \tag{8}$$

Both $k$-NN and SVM may be used to eliminate the reliance on the assumption of normal data distribution, but their classification capability is proportional to the amount of training data provided. An assessment of $k$-NN and SVM classifiers is carried out in [40] for the purpose of identifying bearing and uneven shaft rotation failures in directly line connected induction motors. The $k$-NN method classifies an unidentified sample according to the majority vote of its closest neighbors. Its performance is determined by the distance measure being used to locate the neighbors and the number of these neighbors. The Euclidean distance measure and the closest four neighbors demonstrate the highest level of classification accuracy for this work when employing $k$-NN. The Euclidean distance is given by:

$$D_{XY} = \sqrt{\sum_{i=1}^{n}(x_{mi} - y_{pi})^2} \tag{9}$$

where $X = [x_{m1}, \ldots, x_{mn}]$ is the tested samples vector, $Y = [y_{p1}, \ldots, y_{pn}]$ is the vector of training data, $p = 1, \ldots, l$ is the sample counter for unidentified samples and $m = 1, \ldots, j$ is the sample counter for labeled samples.

Another type of supervised machine learning is SVM, which is notable for its resistance to any bias or noise in the training sample space [41]. The aim of the SVM classification model is to generate the optimal hyperplane using support vectors that results in the best discrimination between two distinct data sets. Support vectors are the outlying samples closest to the discrimination boundary. The optimal hyperplane obtained by SVM is used to classify an unidentified sample by measuring the perpendicular distance between the sample and the hyperplane. SVM training will be a constrained optimization problem in which the weighting vector $\eta(X)$ is minimized so that every sample satisfies the following criteria:

$$\eta(X) = \eta_o + v^T X \geq 0 \tag{10}$$

where $v$ is the vector containing each class weight and $\eta_o$ is a threshold setpoint for classifier modeling. SVM incorporates kernel functions if the data classification cannot be achieved linearly in the original dimensional space of the training data. As a consequence, SVM turns the training data into higher-dimensional spaces where it may be separated linearly. It is possible that the SVM classification model might be modified to handle situations involving more than one class by pitting one class against all of the other classes or by pitting one class against another class at a time. The latter method is the one that is used to investigate the various fault classes in this research.

One way to evaluate the effectiveness of the abovementioned classifiers is by using the leave-one-out method, in which a selected sample from the training dataset is removed and thereafter treated as an unidentified sample to be predicted. This procedure is repeated until all samples in all classes have been processed. The classifier accuracy is then evaluated as follows:

$$\Lambda = \left(\frac{N_{labeled}}{N_{total}}\right) \times 100\% \tag{11}$$

where $\Lambda$ represents classifier accuracy and $N_{labeled}$ represents the number of properly labeled samples out of the total samples count in the training pool $N_{total}$.

## 3. Finite Element Analysis Simulations

Coupling the finite element model (FEM) of an electric machine with a control circuit demonstrates the capability of simulating the machine harmonics due to the slotting, saturation, and switching frequency. Such a model would allow the efficient representation of the PMSM, considering the aforementioned faults, at different operating states. Besides that, this high-fidelity simulation would allow the maintenance and reliability engineers to study the extreme fault severities on the machine without destructive or dangerous experimentation. PMSM modeling, control, and fault implementation are discussed in this section. FOC- and DTC-driven PMSM are assessed in both healthy and faulty situations by integrating the control circuit of FOC or DTC in SIMPLORER simulator with the FEM model of the studied PMSM in MAXWELL simulator within the ANSYS software. Four faulty situations are considered: TTSC, HRC, static eccentricity, and local demagnetization.

### 3.1. *Healthy Machine Modeling and Control*

To develop a fault diagnosis approach, it is necessary to model the PMSM in the healthy case and use the obtained measurement as a reference in case of fault occurs. Any deviation from the processed healthy data could be considered as fault symptoms. In order to implement the aforementioned faults in the FEM of the studied PMSM, geometric and electrical modifications should be made. Table 1 lists the main electric and geometrical parameters of the investigated PMSM.

**Table 1.** Nameplate data of the studied PMSM.

| Machine Specification | Symbol | Value | Machine Specification | Symbol | Value |
|---|---|---|---|---|---|
| Pole pairs | $P_n$ | 5 | $d$-axis inductance | $L_d$ | 31.3 mH |
| Slots | $Q$ | 12 | $q$-axis inductance | $L_q$ | 62.4 mH |
| Turns per slot | $N_s$ | 150 | Mutual inductance | $M_s$ | ≈0 H |
| Air gap length | $g$ | 1 mm | Rated phase current | $I_{rms}$ | 18 A |
| Residual flux density | $B_r$ | 1.2 T | Rated line voltage | $V_{rms}$ | 480 V |
| Magnet flux linkage | $\lambda_{PM}$ | 287 mWb | Rated torque | $T_e$ | 65 N·m |
| Phase resistance | $r_s$ | 1.5 Ω | Rated speed | $n_m$ | 800 rpm |

The FEM of the simulated PMSM in ANSYS MAXWELL software is depicted in Figure 3. It could be observed that the machine winding of each phase are wound around one slot. This winding configuration is referred as concentrated winding machine where the winding are physically isolated and, therefore, the mutual inductance is negligible.

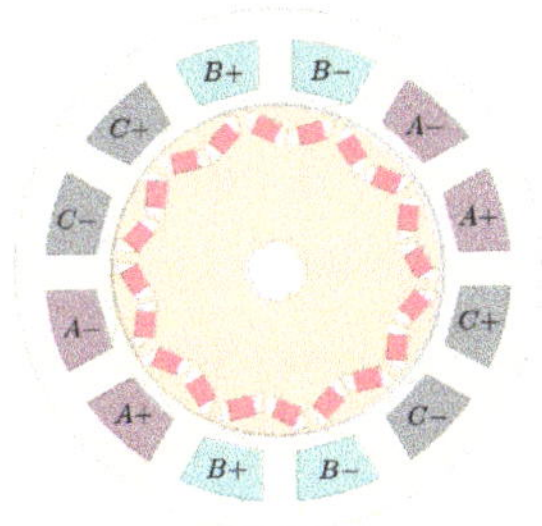


(a)

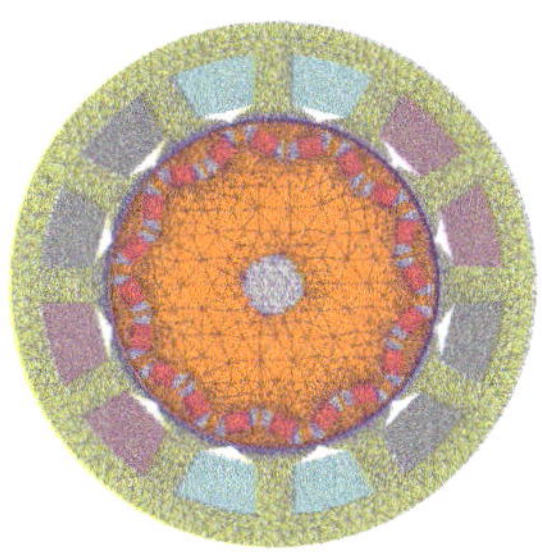

(b)

**Figure 3.** The electromagnetic simulation of studied PMSM machine: (**a**) FEM of the studied machine and (**b**) Mesh of the machine model.

The MTPA profile was first obtained by characterizing the PMSM using the approach described in [35]. Figure 4a depicts the MTPA profile for various operating currents. For FOC-driven PMSM, it can be noticed that the excitation angle should be $\beta_{T_{max}} = 30°$ to get the MTPA operation below the rated speed. For DTC, the stator flux linkage ($\lambda_s$) was changed in step of 0.25 Wb from the permanent magnet flux linkage to the rated one at different operating torque levels. A look-up table stores the stator flux linkage that yields the minimal current for different torque levels. Figure 4b shows the machine MTPA profile for the operating torque range.

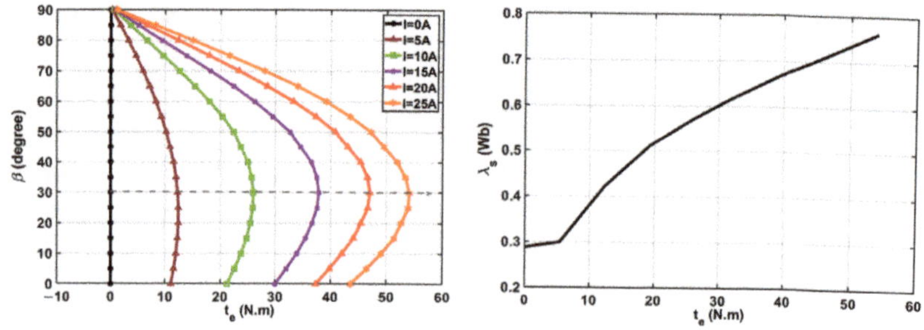


(**a**) (**b**)

**Figure 4.** The MTPA profile of the driven PMSM: (**a**) In FOC drives and (**b**) In DTC drives.

### 3.2. *Turn-to-Turn Short Circuit Fault Implementation*

The implementation of TTSC fault requires modification to the geometrical model in MAXWELL by dividing the coil region into two: (1) a region reflecting the healthy turns and (2) one proportional to the faulty turns. Figure 5 depicts the PMSM model as well as the phase A circuit with a TTSC fault, assuming that phase A is the one that is faulty. After splitting the coil region of the faulty phase in MAXWELL, additional terminals appear in the PMSM model in SIMPLORER to provide access to the shorted turns, as shown in Figure 5b. Consequently, the electric circuit in SIMPLORER has been modified by inserting a short circuit path with fault resistance ($R_f$) over the shorted turns of the faulted phase.

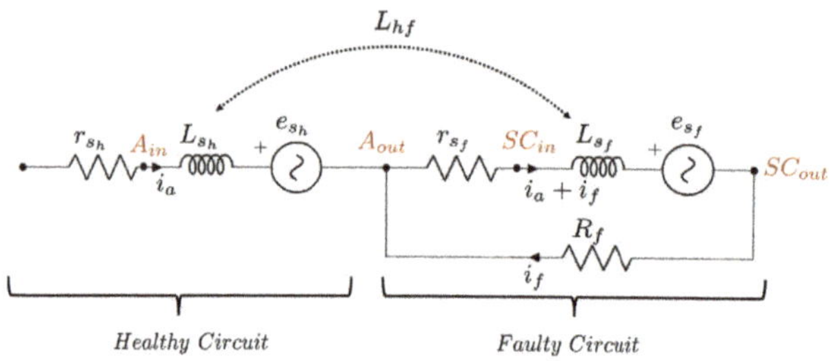


(**a**) Electrical model of phase A with TTSC fault.

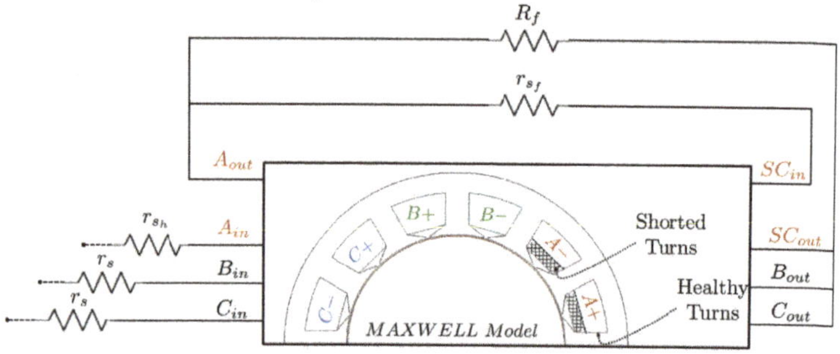


(**b**) External coupling circuit of the geometrical model of phase A with TTSC.

**Figure 5.** TTSC fault simulation in the PMSM model.

where $r_{s_h}$, $L_{s_h}$, and $e_{s_h}$ are the resistance, inductance, and back-EMF of the healthy turns. $r_{s_f}$, $L_{s_f}$, and $e_{s_f}$ are the resistance, inductance, and back-EMF of the shorted turns. $L_{hf}$ is the mutual inductance between healthy and faulty turns of the faulted phase. $i_f$ is the fault circulating current in the short circuit path between the healthy and faulted turns. Early detection of an incipient TTSC fault enables better treatment and prevents total machine failure. The worst-case scenario for a TTSC fault is a bold short circuit with almost zero fault resistance while an incipient short circuit fault is represented with higher fault resistances [42]. As indicated in Table 2, different severities of incipient TTSC fault are examined by adjusting the faulted turns $N_f$ and the resistance shorting them $R_f$. The fault resistance $R_f$ is selected high, in this work, to mimic incipient stages of TTSC fault.

**Table 2.** Severity levels of TTSC fault.

| Case | $N_f$ | $R_f$ |
|---|---|---|
| SC1 | 15 | 0.5 Ω |
| SC2 | 15 | 0.25 Ω |
| SC3 | 30 | 0.5 Ω |
| SC4 | 30 | 0.25 Ω |

### 3.3. *High Resistance Contact Fault Implementation*

The implementation of the HRC fault involves simply a change to the SIMPLORER circuit in which the affected phase circuit is linked in series with an adjustable fault resistance ($\Delta r_s$) proportional to the fault severity. The extreme case of HRC fault is when a relatively large fault resistance is used with the faulted phase circuit. This situation is considered as an open phase fault. Figure 6 depicts the HRC fault implementation in phase B circuit of the faulted driven PMSM.

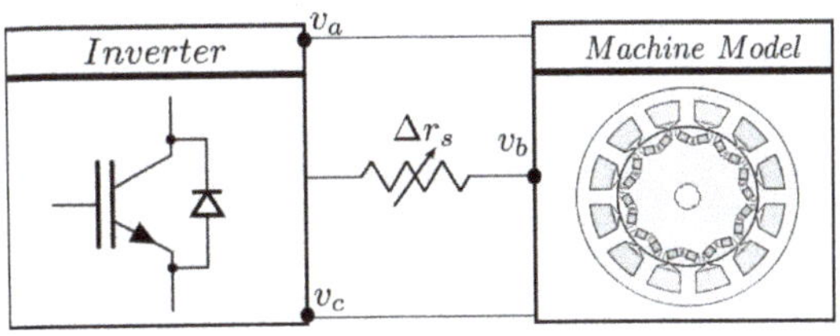


**Figure 6.** Electrical model of inverter-driven PMSM with HRC fault in phase B.

Three different fault severity levels of HRC are analyzed as given in Table 3, all based on the assumption that the fault occurred in the phase B circuit.

**Table 3.** Severity levels of HRC fault.

| Case | $\frac{\Delta r_s}{r_s}\%$ |
|---|---|
| HRC1 | 50% |
| HRC2 | 100% |
| HRC3 | 150% |

### 3.4. *Static Eccentricity Fault Implementation*

The machine is considered healthy if the stator and rotor geometrical centers are concentric with the rotational axis. If the rotor center and rotational axis are shifted from the stator center by a constant value, the machine is considered faulty with static eccentricity. The coordinate center of the stator remains unchanged while the rotor and its axis of rotation move in the direction of the fault as shown in Figure 7.

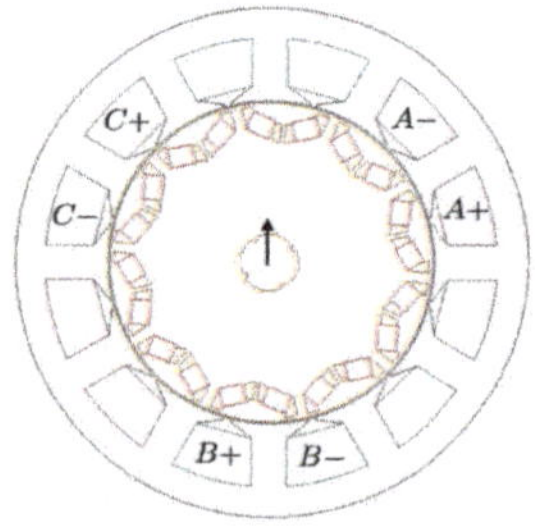


**Figure 7.** Shift direction of the static eccentricity fault.

The fault is simulated along the positive Y-axis by modifying the shift ($\epsilon$) value, taking the three severities shown in Table 4 into account. Since the eccentricity fault is inherently available during the manufacturing process, the machine is considered healthy if the severity level of static eccentricity is below than 10%.

**Table 4.** Severity levels of static eccentricity fault.

| **Case** | $\frac{\epsilon}{g}$% |
|---|---|
| ECC1 | 10% |
| ECC2 | 40% |
| ECC3 | 60% |
| ECC4 | 80% |

*3.5. Local Demagnetization Fault Implementation*

The demagnetization fault is applied in FEM by lowering the density of the remnant flux in the affected magnets. This could be uniform demagnetization in all magnets or locally in some magnets. Three magnets have their top corners (near to the air gap) replaced with an identical material with flux density reduced to 1 T. This is the most prevalent case for demagnetization failures in PMSM. Figure 8 shows the affected magnets by demagnetization fault.

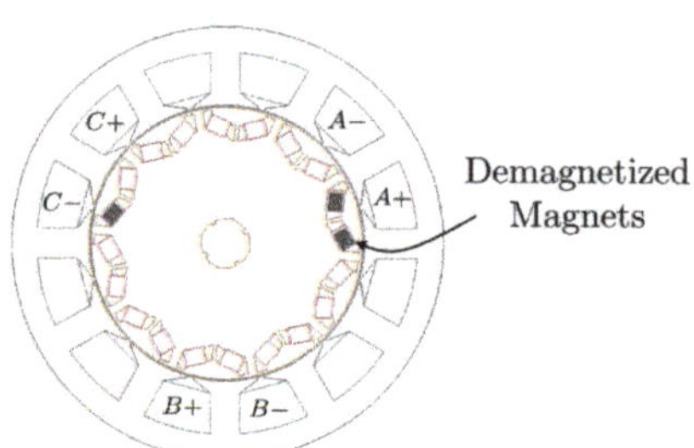


**Figure 8.** Demagnetized magnets.

Changes in the number of demagnetized magnets in the modeled machine provide different fault severities, as shown in Table 5. The local demagnetization fault is considered in this study as some magnets get demagnetized in the machine model.

**Table 5.** Severity levels of demagnetization fault.

| Case | Severity |
|---|---|
| Demag1 | One Magnet |
| Demag2 | Two Adjacent Magnets |
| Demag3 | Three Nonadjacent Magnets |

## 4. Numerical Results

### *4.1. Analysis in FOC-Driven PMSM*

The spectral domain of the commanded phase A voltage is obtained using FFT for different machine health conditions. The sampling period is set to 400 μs. The fundamental, 5th, and 7th harmonics are used here to show the fault effect on the voltage spectrum at different severity levels. For FOC, the results are shown for the considered PMSM when it is loaded with ($I = 10$ A). The machine is running at 600 rpm; therefore, the electrical frequency of the machine is (50 Hz). Figure 9 shows the commanded voltage of phase A in FOC drive for the healthy case.

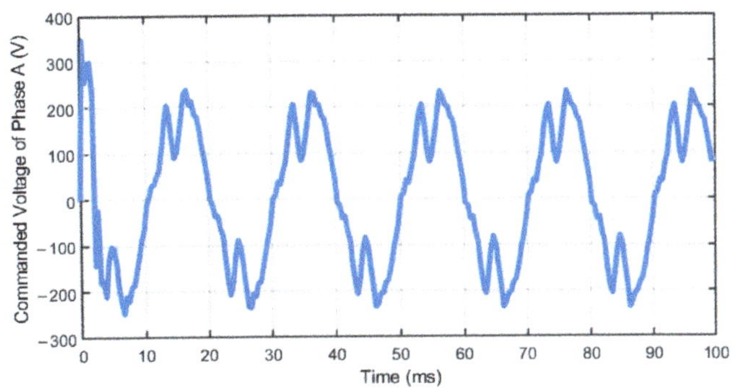


**Figure 9.** The commanded voltage of phase A in FOC drive.

The stator voltage spectrum is shown in Figure 10 for healthy and faulty machine under different severity levels of TTSC fault.

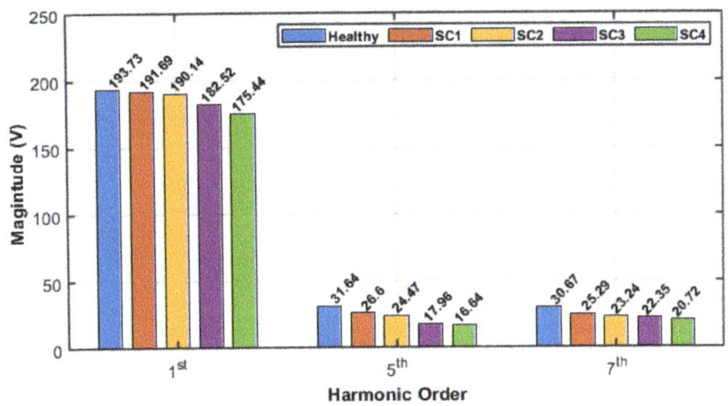


**Figure 10.** Variations in the harmonic content of stator voltage spectrum for healthy and faulty machine under TTSC fault in FOC drive.

It could be observed that the magnitude of the fundamental, 5th, and 7th harmonics in the voltage spectrum is reduced when TTSC fault is present and becomes more severe. The magnitude changes in the spectrum are significant to distinguish between the healthy case and TTSC fault. Besides that, it has been noticed that there is an increase in the

2nd harmonic when TTSC is present. However, this analysis will show the fault effect on the fundamental, 5th, and 7th harmonics in the voltage spectrum. Nevertheless, the first eight harmonics will be used as features for diagnosis, as will be described in the following section.

The stator voltage spectrum is shown in Figure 11 for healthy and faulty machine under different severity levels of HRC fault.

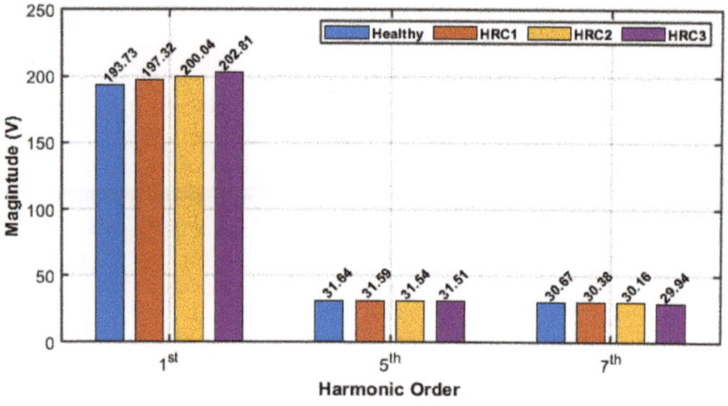


**Figure 11.** Variations in the harmonic content of stator voltage spectrum for healthy and faulty machine under HRC fault in FOC drive.

It could be observed that the magnitude of the fundamental harmonic in the voltage spectrum is increasing significantly when HRC fault is present and become more severe. On the other hand, the magnitude of the 5th and 7th harmonics is reduced, but these magnitude changes might be too small to distinguish between the healthy case and the HRC fault. However, the considered severity levels of HRC are relatively low, and this would justify the insignificant changes in the magnitude of the 5th and 7th harmonics. Nevertheless, these changes would be higher once the fault was present at a higher severity level.

The stator voltage spectrum is shown in Figure 12 for healthy and faulty machine under different severity levels of eccentricity fault.

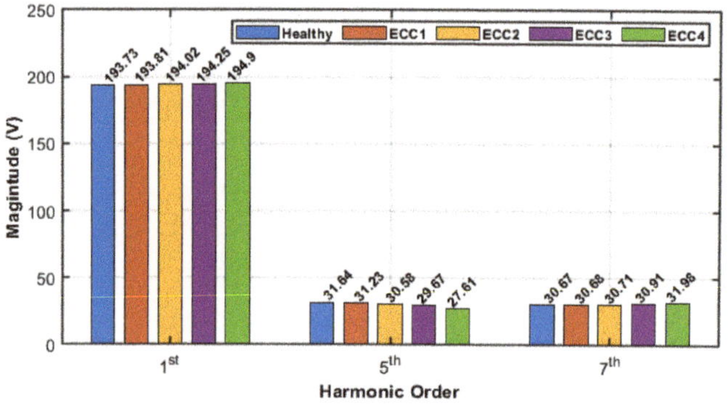


**Figure 12.** Variations in the harmonic content of stator voltage spectrum for healthy and faulty machine under eccentricity fault in FOC drive.

It could be observed that the magnitudes of the fundamental and seventh harmonics in the voltage spectrum are increasing when the eccentricity fault is present and becoming more severe, while the magnitude of the fifth harmonic is significantly reduced. These magnitude changes are significant to distinguish between the healthy case and eccentricity fault. Besides that, these magnitude changes due to eccentricity faults are different from those for TTSC and HRC faults. This would help in fault classification later on.

The stator voltage spectrum is shown in Figure 13 for healthy and faulty machine under different severity levels of demagnetization fault.

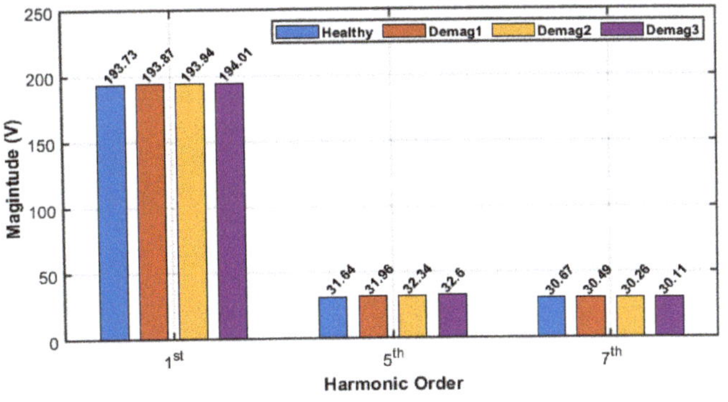


**Figure 13.** Variations in the harmonic content of stator voltage spectrum for healthy and faulty machine under demagnetization fault in FOC drive.

It could be observed that the magnitude of the fundamental and 5th harmonics in the voltage spectrum is increasing when demagnetization fault is present and becoming more severe, while the magnitude of the 7th harmonic is reduced. These magnitude changes could be used to distinguish between the healthy case and demagnetization fault. It is worth mentioning that the considered severity levels of demagnetization fault are relatively low as the outer corners of selected magnets get demagnetized only from 1.2 T to 1 T. Besides that, the trend in magnitude change of the fundamental harmonic due to demagnetization fault is similar to the trend due to HRC and eccentricity faults, while it is the opposite in case of the 5th harmonic. This would help the classifier in discriminating between these three faults, for instance.

### *4.2. Analysis in DTC-Driven PMSM*

The spectral domain of the commanded phase A voltage is obtained using FFT for different machine health conditions under DTC drive. The sampling period is set to 100 µs; DTC requires a high sampling frequency to function properly. The fundamental, 5th, and 7th harmonics are also used here to show the fault effect on the voltage spectrum at different severity levels. For DTC, the results are shown for the considered PMSM when it is loaded with ($T = 20$ N·m). The machine is running at 600 rpm; therefore, the electrical frequency of the machine is (50 Hz). Figure 14 shows the commanded voltage of phase A in DTC drive for the healthy case.

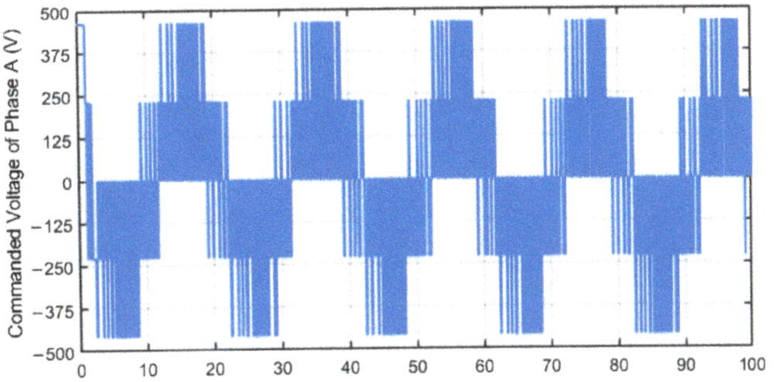


**Figure 14.** The commanded voltage of phase A in DTC drive.

The stator voltage spectrum is shown in Figure 15 for healthy and faulty machine under different severity levels of TTSC fault when the machine is driven by DTC.

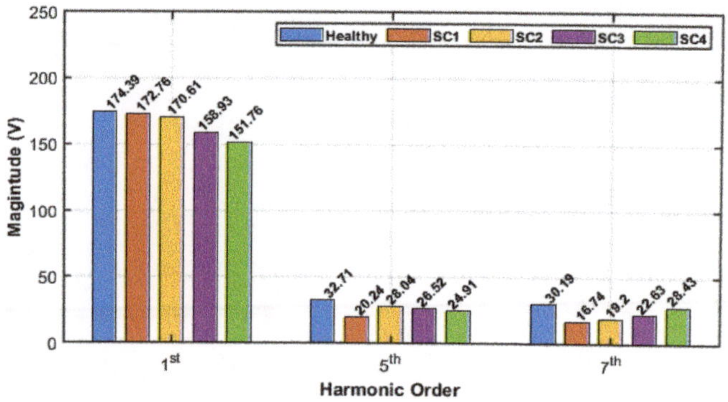


**Figure 15.** Variations in the harmonic content of stator voltage spectrum for healthy and faulty machine under TTSC fault in DTC drive.

It could be observed that the magnitude of the fundamental, 5th, and 7th harmonics in the voltage spectrum is reduced when TTSC fault is present, similar to the fault effect in the FOC drive. However, it could be noticed that the magnitude of the 7th harmonic is increased from SC1 to SC2 and from SC3 to SC4, where the number of shorted turns is the same but different fault resistance is used. This is the same case for the 5th harmonic from SC1 to SC2 but it is not from SC3 to SC4. Additionally, the magnitude of the 7th harmonic is increased from SC1 to SC3 and from SC2 to SC4, where the number of shorted turns is different but with the same fault resistance. This is the same case for the 5th harmonic from SC1 to SC3, but it is not from SC2 to SC4. If these harmonics are used as features for diagnosis approach, the classifier will be overwhelmed since the changes in the used harmonics due to different TTSC severity levels are not consistent with an increase in the fault severity.

The stator voltage spectrum is shown in Figure 16 for healthy and faulty machine under different severity levels of HRC fault when the machine is driven by DTC.

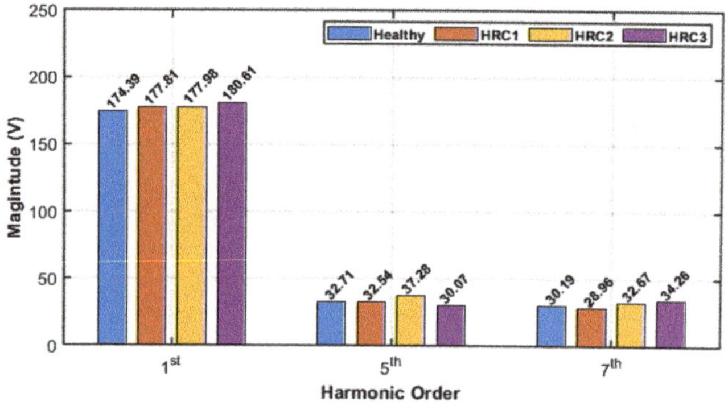


**Figure 16.** Variations in the harmonic content of stator voltage spectrum for healthy and faulty machine under HRC fault in DTC drive.

It could be observed that the magnitude of the fundamental harmonic in the voltage spectrum is increasing significantly when HRC fault is present and become more severe. On the other hand, the magnitude of the 5th has inconsistent change, where it does not change significantly at HRC1 but it increases at HRC2 and then decreases at HRC3. Similarly, the magnitude of the 7th starts decreasing at HRC1 but then starts increasing at HRC2 and

HRC3. Once again, these inconsistent changes in the voltage spectrum due to different HRC severity levels will result in high false classification results.

The stator voltage spectrum is shown in Figure 17 for healthy and faulty machine under different severity levels of eccentricity fault when the machine is driven by DTC.

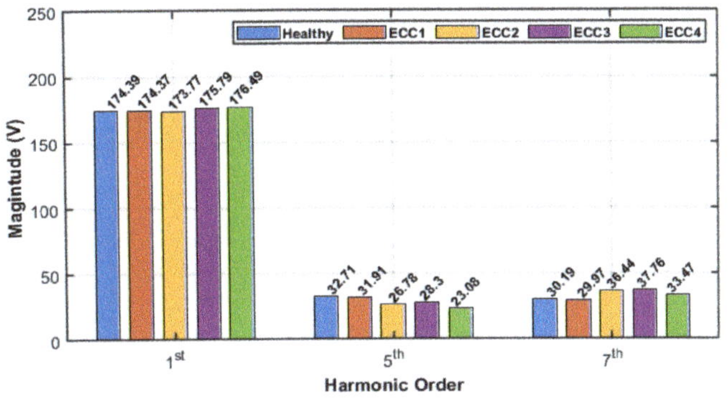


**Figure 17.** Variations in the harmonic content of stator voltage spectrum for healthy and faulty machine under eccentricity fault in DTC drive.

It could be observed that the magnitude of the fundamental harmonic in the voltage spectrum has inconsistent changes due to different severity levels of the eccentricity fault, where it is reducing at ECC1 and ECC2 while it is increasing at ECC3 and ECC4. Similarly, the magnitude of the 5th harmonic reduces at all eccentricity severity levels. However, it decreases from healthy case to ECC1 and ECC2, increases from ECC2 to ECC3, and then decreases from ECC3 to ECC4. It could be also noticed a similar inconsistent pattern happens with the 7th harmonic. These irregular changes in the spectrum will result a difficulty in performing fault detection and separation in DTC drives.

The stator voltage spectrum is shown in Figure 18 for healthy and faulty machine under different severity levels of demagnetization fault when the machine is driven by DTC.

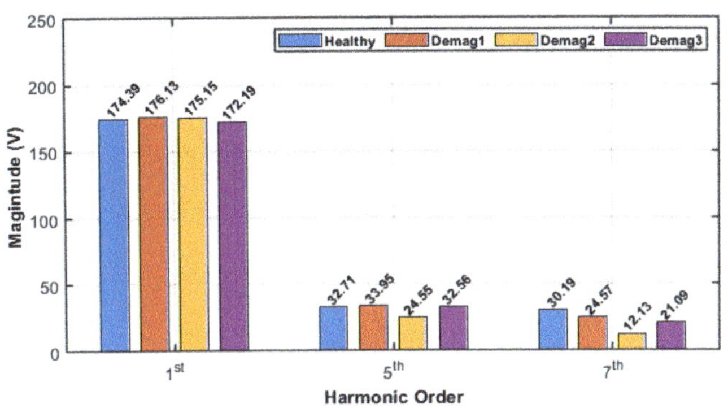


**Figure 18.** Variations in the harmonic content of stator voltage spectrum for healthy and faulty machine under demagnetization fault in DTC drive.

It could be noticed that the magnitude of the fundamental harmonic in the voltage spectrum has inconsistent changes due to different severity levels of demagnetization fault where it is increasing at Demag1 and Demag2 but then reducing at Demag3. Similarly, the magnitude of the 5th and 7th harmonics have inconsistent changes with increasing the demagnetization fault severity. In comparison with applying FOC drives, the changes in the voltage spectrum would not help the classifier in identifying fault, discriminating between them, and estimating their severity.

## 5. Diagnosis Approach

The spectrum of the commanded phase A voltage was obtained using the FFT method in MATLAB, and the amplitude of the first eight harmonics was then extracted. The inclusion of additional harmonics allows fault identification, categorization and estimate for TTSC, HRC, static eccentricity and local demagnetization without the need for extra sensors. This limits possible false indication using the proposed detection methods in the literature that rely on using only single feature for fault diagnosis.

In this section, supervised classification algorithms (LDA, *k*-NN, and SVM) are compared in regards to their capability in identifying faulty cases, discriminating them, and evaluating their severity. Figure 19 shows the flowchart of the fault detection, separation, and severity estimation algorithm.

In the stage of classification, the amplitude of considered harmonics serves as features for fault occurrence and its severity increase. The number of samples for each machine condition should be greater than the number of features that are utilized for detection in order to ensure that the training of the classifier will eventually converge to a representative classification model [43]. In this work, eight features are used for detection; therefore, ten samples are utilized. Therefore, the speed is adjusted in FEM from 250 rpm up to 700 rpm in 50 rpm increments to obtain representative samples for healthy and faulty conditions. For FOC, the magnitude of the stator current is set to be 10 A at $\beta = 30°$, while the commanded torque in DTC is set at 20 N·m at 0.57 Wb to get the MTPA operation for both controllers. The numerical setup generates from each drive system a total of 140 data samples: 10 for a healthy case, 40 for all severity levels of TTSC fault, 30 for all severity levels of HRC fault, 40 for all severity levels of static eccentricity fault, and 30 for all severity levels of local demagnetization fault.

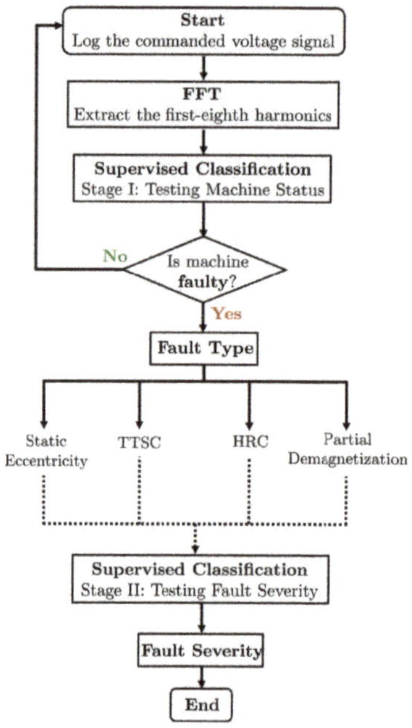


**Figure 19.** Flowchart of fault detection, separation, and severity estimation algorithm.

The first stage of the classification approach is aimed at identifying the fault type at its early level, allowing for the application of appropriate mitigation techniques to avoid any further damage, if possible. The fault severity is monitored using the second stage of the classifiers and the severity estimation could be used to determine remedial actions.

Therefore, samples of the faulty cases at the lowest severity are considered only in the training pool to perform fault detection and separation in the first stage. The overall detection accuracy may be determined by averaging the sum of the accuracy of each individual case as the sample size is ten for each case. Table 6 displays the accuracy of identifying and classifying the faults using the classifiers under consideration.

**Table 6.** Detection accuracy using MVSA approach for FOC- and DTC-driven PMSM.

| Case | FOC | | | DTC | | |
|---|---|---|---|---|---|---|
| | **LDA** | ***k*-NN** | **SVM** | **LDA** | ***k*-NN** | **SVM** |
| Healthy | 100% | 100% | 97.5% | 0% | 0% | 50% |
| TTSC | 100% | 100% | 100% | 80% | 70% | 75% |
| HRC | 70% | 100% | 80% | 40% | 40% | 67.5% |
| Eccentricity | 70% | 100% | 97.5% | 30% | 50% | 57.5% |
| Demagnetization | 80% | 70% | 87.5% | 70% | 40% | 80% |
| Overall | 84% | 94% | 92.5% | 44% | 42% | 66% |

It is clear that the considered classifiers are capable of fault detection and separation using the MVSA approach in FOC-driven PMSM while their overall performance is significantly degraded in case of DTC-driven PMSM. Moreover, in contrast to FOC drives, it can be observed that all classifiers achieve reduced detection accuracy of the healthy machine status in DTC drives. This results in a high number of erroneous indications in DTC drives and unneeded drive shutdowns. Furthermore, in both drives, the SVM classifier has the best detection and separation accuracy. Therefore, it is demonstrated that relying on the MVSA approach for fault diagnosis in DTC drives would result in poor performance and significant amount of false alarms.

Figure 20 provides the confusion matrix of LDA classifier for fault diagnosis to show the false alarms and incorrect classification in FOC and DTC drives.

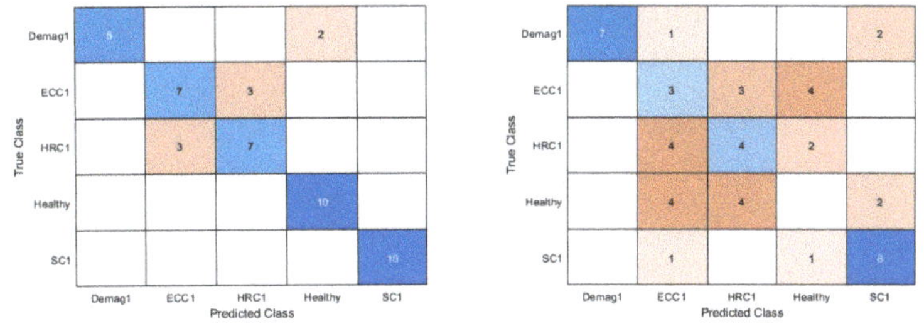


(a) (b)

**Figure 20.** Confusion matrix of LDA classifier for fault diagnosis for: (**a**) FOC drives and (**b**) DTC drives.

The accuracy of the fault severity estimation using the LDA and SVM classifiers is shown in Table 7. Once the faulty operation is indicated and fault type is detected from the first stage of the supervised classification, then the second classification stage will be dedicated for severity estimation of the detected fault. Therefore, samples of the detected fault at each severity level are considered in the training pool to perform fault severity estimation.

**Table 7.** The accuracy of fault severity estimation using MVSA approach for FOC- and DTC-driven PMSM.

| Fault Type | FOC | | DTC | |
|---|---|---|---|---|
| | LDA | SVM | LDA | SVM |
| TTSC | **85%** | **97.5%** | **42.5%** | **64.167%** |
| SC1 | 100% | 100% | 50% | 63.333% |
| SC2 | 100% | 100% | 40% | 66.667% |
| SC3 | 80% | 93.333% | 30% | 66.667% |
| SC4 | 60% | 96.667% | 50% | 60% |
| HRC | **96.667%** | **86.667%** | **30%** | **38.333%** |
| HRC1 | 100% | 100% | 30% | 35% |
| HRC2 | 90% | 80% | 20% | 30% |
| HRC3 | 100% | 100% | 40% | 50% |
| Eccentricity | **100%** | **100%** | **30%** | **62.5%** |
| ECC1 | 100% | 100% | 20% | 53.333% |
| ECC2 | 100% | 100% | 20% | 53.333% |
| ECC3 | 100% | 100% | 40% | 70% |
| ECC4 | 100% | 100% | 40% | 73.333% |
| Demagnetization | **86.667%** | **71.667%** | **46.667%** | **56.667%** |
| Demag1 | 100% | 100% | 80% | 75% |
| Demag2 | 90% | 50% | 20% | 55% |
| Demag3 | 70% | 60% | 40% | 40% |

It can be observed from Table 7 the degradation in the performance of LDA and SVM classifiers for the fault severity estimation when DTC is applied in comparison with FOC. This is related to the earlier mentioned fact of inconsistent changes in the voltage spectrum due to different severity levels of each fault in case of DTC drives. Therefore, further investigation for reliable fault diagnosis is required in DTC-driven PMSMS. It is also important to note that the classification accuracy of the studied approach is directly impacted by the density of the supervised learning. Accuracy may be enhanced by collecting more samples. Furthermore, the investigated method is appropriate for steady-state PMSM operations. Different signal processing techniques may be incorporated to study the PMSM during nonstationary operation. Nevertheless, the incorporation of extra harmonics enables fault recognition, classification, and estimation for TTSC, HRC, static eccentricity, and local demagnetization without the need for additional sensors. This reduces the likelihood of false indication utilizing detection techniques presented in the literature that depend on a single feature for fault diagnosis.

## 6. Conclusions

The accuracy of the MVSA approach and three supervised classifiers (LDA, k-NN, and SVM) for fault identification and severity estimation in inverter-driven PMSM was investigated. Features used for classifier training to identify fault type and severity were the amplitudes of harmonics in the phase voltage signal. Four faults were addressed: TTSC, HRC, static eccentricity, and local demagnetization. The results show that the studied MVSA diagnostic strategy worked well in FOC-driven PMSM while deteriorating in DTC-driven PMSM. This is due to the compensatory nature of each controller. When FOC drive is applied, there is a consistent correlation between the magnitude variations in the voltage spectrum that are caused by the incidence of faults and the rise in the severity of faults. The utilization of hysteresis comparators and variable switching frequency cause these variations to be inconsistent in the case of DTC-driven PMSM, despite the fact that these features are present. The vast majority of the existing detection and separation methods in the literature for inverter-driven PMSMs were developed either for FOC drives or without addressing the controller type. As a result, when developing fault diagnosis algorithms, the controller type and its impact on fault detectability should be addressed.

**Author Contributions:** Conceptualization, I.M.A. and S.N.F.; methodology, I.M.A.; software, I.M.A.; validation, I.M.A.; formal analysis, I.M.A.; investigation, I.M.A.; resources, S.N.F.; data curation, I.M.A.; writing—original draft preparation, I.M.A.; writing—review and editing, S.N.F.; visualization, I.M.A.; supervision, S.N.F. All authors have read and agreed to the published version of the manuscript.

**Funding:** This work was partially funded by Michigan State University (MSU), USA, and the Jordan University of Science and Technology (JUST), Jordan. The APC was funded by Michigan State University (MSU).

**Data Availability Statement:** Data will be made available on request.

**Acknowledgments:** Ibrahim M. Allafi sincerely appreciates the Jordan University of Science and Technology (JUST) for sponsoring his graduate studies.

**Conflicts of Interest:** The authors declare no conflict of interest.

## Abbreviations

The following abbreviations are used in this manuscript:

| | |
|---|---|
| DTC | Direct Torque Control |
| FEM | Finite Element Model |
| FFT | Fast Fourier Transform |
| FOC | Field Oriented Control |
| HRC | High Resistance Contact |
| k-NN | K-Nearest Neighbor |
| LDA | Linear Discriminate Analysis |
| MCC | Motor Control Center |
| MTPA | Maximum Torque Per Amperes |
| MVSA/MCSA | Machine Voltage or Current Signature Analysis |
| NdFeB | Neodymium-Iron-Boron Magnet |
| NVH | Noise, Vibration, and Harshness |
| PMSM | Permanent Magnet Synchronous Machine |
| SNR | Signal-to-Noise Ratio |
| STFT | Short Time Fourier transform |
| SVM | Support Vector Machines |
| SVPWM | Space Vector Pulse Width Modulation |
| TTSC | Turn-to-Turn Short Circuit |
| VFD | Variable Frequency Drives |
| ZSVC | Zero-Sequence Voltage Component |

## References

1. Petrov, I.; Egorov, D.; Link, J.; Stern, R.; Ruoho, S.; Pyrhönen, J. Hysteresis Losses in Different Types of Permanent Magnets Used in PMSMs. *IEEE Trans. Ind. Electron.* **2017**, *64*, 2502–2510. [CrossRef]
2. Melfi, M.J.; Evon, S.; McElveen, R. Induction versus permanent magnet motors. *IEEE Ind. Appl. Mag.* **2009**, *15*, 28–35. [CrossRef]
3. Cao, W.; Mecrow, B.C.; Atkinson, G.J.; Bennett, J.W.; Atkinson, D.J. Overview of Electric Motor Technologies Used for More Electric Aircraft (MEA). *IEEE Trans. Ind. Electron.* **2012**, *59*, 3523–3531. [CrossRef]
4. Jung, S.Y.; Mi, C.C.; Nam, K. Torque Control of IPMSM in the Field-Weakening Region With Improved DC-Link Voltage Utilization. *IEEE Trans. Ind. Electron.* **2015**, *62*, 3380–3387. [CrossRef]
5. Choi, S.; Haque, M.S.; Tarek, M.T.B.; Mulpuri, V.; Duan, Y.; Das, S.; Garg, V.; Ionel, D.M.; Masrur, M.A.; Mirafzal, B.; et al. Fault Diagnosis Techniques for Permanent Magnet AC Machine and Drives—A Review of Current State of the Art. *IEEE Trans. Transp. Electrif.* **2018**, *4*, 444–463. [CrossRef]
6. Lee, S.B.; Stone, G.C.; Antonino-Daviu, J.; Gyftakis, K.N.; Strangas, E.G.; Maussion, P.; Platero, C.A. Condition Monitoring of Industrial Electric Machines: State of the Art and Future Challenges. *IEEE Ind. Electron. Mag.* **2020**, *14*, 158–167. [CrossRef]
7. Roux, W.L.; Harley, R.G.; Habetler, T.G. Detecting faults in rotors of PM drives. *IEEE Ind. Appl. Mag.* **2008**, *14*, 23–31. [CrossRef]
8. Wang, B.; Wang, J.; Griffo, A.; Sen, B. Stator Turn Fault Detection by Second Harmonic in Instantaneous Power for a Triple-Redundant Fault-Tolerant PM Drive. *IEEE Trans. Ind. Electron.* **2018**, *65*, 7279–7289. [CrossRef]

9. Mohanraj, D.; Gopalakrishnan, J.; Chokkalingam, B.; Mihet-Popa, L. Critical Aspects of Electric Motor Drive Controllers and Mitigation of Torque Ripple—Review. *IEEE Access* **2022**, *10*, 73635–73674. [CrossRef]
10. Wang, Z.; Chen, J.; Cheng, M.; Chau, K.T. Field-Oriented Control and Direct Torque Control for Paralleled VSIs Fed PMSM Drives With Variable Switching Frequencies. *IEEE Trans. Power Electron.* **2016**, *31*, 2417–2428. [CrossRef]
11. Niu, F.; Wang, B.; Babel, A.S.; Li, K.; Strangas, E.G. Comparative Evaluation of Direct Torque Control Strategies for Permanent Magnet Synchronous Machines. *IEEE Trans. Power Electron.* **2016**, *31*, 1408–1424. [CrossRef]
12. Geyer, T.; Mastellone, S. Model Predictive Direct Torque Control of a Five-Level ANPC Converter Drive System. *IEEE Trans. Ind. Appl.* **2012**, *48*, 1565–1575. [CrossRef]
13. Geyer, T.; Papafotiou, G.; Morari, M. Model Predictive Direct Torque Control—Part I: Concept, Algorithm, and Analysis. *IEEE Trans. Ind. Electron.* **2009**, *56*, 1894–1905. [CrossRef]
14. Haddad, R.Z.; Lopez, C.A.; Foster, S.N.; Strangas, E.G. A Voltage-Based Approach for Fault Detection and Separation in Permanent Magnet Synchronous Machines. *IEEE Trans. Ind. Appl.* **2017**, *53*, 5305–5314. [CrossRef]
15. Zanardelli, W.G.; Strangas, E.G.; Aviyente, S. Identification of Intermittent Electrical and Mechanical Faults in Permanent-Magnet AC Drives Based on Time–Frequency Analysis. *IEEE Trans. Ind. Appl.* **2007**, *43*, 971–980. [CrossRef]
16. Chen, H.; He, J.; Guan, X.; Demerdash, N.A.O.; EL-Refaie, A.M.; Lee, C.H.T. High-Resistance Connection Diagnosis in Five-Phase PMSMs Based on the Method of Magnetic Field Pendulous Oscillation and Symmetrical Components. *IEEE Trans. Ind. Electron.* **2022**, *69*, 2288–2299. [CrossRef]
17. Ebrahimi, B.M.; Faiz, J. Feature Extraction for Short-Circuit Fault Detection in Permanent-Magnet Synchronous Motors Using Stator-Current Monitoring. *IEEE Trans. Power Electron.* **2010**, *25*, 2673–2682. [CrossRef]
18. Ebrahimi, B.M.; Faiz, J.; Roshtkhari, M.J. Static-, Dynamic-, and Mixed-Eccentricity Fault Diagnoses in Permanent-Magnet Synchronous Motors. *IEEE Trans. Ind. Electron.* **2009**, *56*, 4727–4739. [CrossRef]
19. Haddad, R.Z.; Strangas, E.G. On the Accuracy of Fault Detection and Separation in Permanent Magnet Synchronous Machines Using MCSA/MVSA and LDA. *IEEE Trans. Energy Convers.* **2016**, *31*, 924–934. [CrossRef]
20. Allafi, I.M.; Foster, S.N. Fault Detection and Identification for Inverter-Driven Permanent Magnet Synchronous Machines. In Proceedings of the 2021 IEEE 13th International Symposium on Diagnostics for Electrical Machines, Power Electronics and Drives (SDEMPED), Dallas, TX, USA, 22–25 August 2021; Volume 1, pp. 358–364. [CrossRef]
21. Hang, J.; Ding, S.; Ren, X.; Hu, Q.; Huang, Y.; Hua, W.; Wang, Q. Integration of Interturn Fault Diagnosis and Torque Ripple Minimization Control for Direct-Torque-Controlled SPMSM Drive System. *IEEE Trans. Power Electron.* **2021**, *36*, 11124–11134. [CrossRef]
22. Hang, J.; Wu, H.; Ding, S.; Hua, W.; Wang, Q. A DC-Flux-Injection Method for Fault Diagnosis of High-Resistance Connection in Direct-Torque-Controlled PMSM Drive System. *IEEE Trans. Power Electron.* **2020**, *35*, 3029–3042. [CrossRef]
23. Allafi, I.M.; Foster, S.N. On the Accuracy of Frequency Based Fault Diagnosis for DTC-driven PMSM. In Proceedings of the 2022 International Conference on Electrical Machines (ICEM), Valencia, Spain, 5–8 September 2022; pp. 1628–1634. [CrossRef]
24. Allafi, I.M.; Foster, S.N. Condition Monitoring of Direct Torque Controlled Permanent Magnet Synchronous Machines. In Proceedings of the 2022 IEEE Energy Conversion Congress and Exposition (ECCE), Detroit, MI, USA, 9–13 October 2022; pp. 1–7. [CrossRef]
25. Qiu, X.; Ji, J.; Zhou, D.; Zhao, W.; Chen, Y.; Huang, L. A Modified Flux Observer for Sensorless Direct Torque Control of Dual Three-Phase PMSM Considering Open-Circuit Fault. *IEEE Trans. Power Electron.* **2022**, *37*, 15356–15369. [CrossRef]
26. Wang, Z.; Wang, X.; Cao, J.; Cheng, M.; Hu, Y. Direct Torque Control of T-NPC Inverters-Fed Double-Stator-Winding PMSM Drives With SVM. *IEEE Trans. Power Electron.* **2018**, *33*, 1541–1553. [CrossRef]
27. Wang, X.; Wang, Z.; Cheng, M.; Hu, Y. Remedial Strategies of T-NPC Three-Level Asymmetric Six-Phase PMSM Drives Based on SVM-DTC. *IEEE Trans. Ind. Electron.* **2017**, *64*, 6841–6853. [CrossRef]
28. Strangas, E.G.; Clerc, G.; Razik, H.; Soualhi, A. Applications and Specifics. In *Fault Diagnosis, Prognosis, and Reliability for Electrical Machines and Drives*; Wiley-IEEE Press: Piscataway, NJ, USA, 2022; pp. 125–344. [CrossRef]
29. Zhao, J.; Guan, X.; Li, C.; Mou, Q.; Chen, Z. Comprehensive Evaluation of Inter-Turn Short Circuit Faults in PMSM Used for Electric Vehicles. *IEEE Trans. Intell. Transp. Syst.* **2021**, *22*, 611–621. [CrossRef]
30. Thorsen, O.; Dalva, M. A survey of faults on induction motors in offshore oil industry, petrochemical industry, gas terminals, and oil refineries. *IEEE Trans. Ind. Appl.* **1995**, *31*, 1186–1196. [CrossRef]
31. Hong, J.; Park, S.; Hyun, D.; Kang, T.j.; Lee, S.B.; Kral, C.; Haumer, A. Detection and Classification of Rotor Demagnetization and Eccentricity Faults for PM Synchronous Motors. *IEEE Trans. Ind. Appl.* **2012**, *48*, 923–932. [CrossRef]
32. Espinosa, A.G.; Rosero, J.A.; Cusidó, J.; Romeral, L.; Ortega, J.A. Fault Detection by Means of Hilbert–Huang Transform of the Stator Current in a PMSM With Demagnetization. *IEEE Trans. Energy Convers.* **2010**, *25*, 312–318. [CrossRef]
33. Yao, H.; Yan, Y.; Shi, T.; Zhang, G.; Wang, Z.; Xia, C. A Novel SVPWM Scheme for Field-Oriented Vector-Controlled PMSM Drive System Fed by Cascaded H-Bridge Inverter. *IEEE Trans. Power Electron.* **2021**, *36*, 8988–9000. [CrossRef]
34. Nam, K.H. *AC Motor Control and Electrical Vehicle Applications*, 2nd ed.; CRC PRESS: Boca Raton, FL, USA, 2018.
35. Cintron-Rivera, J.G.; Babel, A.S.; Montalvo-Ortiz, E.E.; Foster, S.N.; Strangas, E.G. A simplified characterization method including saturation effects for permanent magnet Machines. In Proceedings of the 2012 XXth International Conference on Electrical Machines, Marseille, France, 2–5 September 2012; pp. 837–843. [CrossRef]

36. Babel, A.S.; Cintron-Rivera, J.G.; Foster, S.N.; Strangas, E.G. Evaluation of a Parameter Identification Method for Permanent Magnet AC Machines Through Parametric Sensitivity Analysis. *IEEE Trans. Energy Convers.* **2014**, *29*, 240–249. [CrossRef]
37. Zhong, L.; Rahman, M.; Hu, W.; Lim, K. Analysis of direct torque control in permanent magnet synchronous motor drives. *IEEE Trans. Power Electron.* **1997**, *12*, 528–536. [CrossRef]
38. Untaroiu, C.D.; Adam, T.J. Performance-Based Classification of Occupant Posture to Reduce the Risk of Injury in a Collision. *IEEE Trans. Intell. Transp. Syst.* **2013**, *14*, 565–573. [CrossRef]
39. Ibraheem, I. Linear and Quadratic Classifier to Detection of Skin Lesions "Epicutaneus". In Proceedings of the 2011 5th International Conference on Bioinformatics and Biomedical Engineering, Wuhan, China, 10–12 May 2011; pp. 1–5. [CrossRef]
40. Ali, M.Z.; Shabbir, M.N.S.K.; Liang, X.; Zhang, Y.; Hu, T. Machine Learning-Based Fault Diagnosis for Single- and Multi-Faults in Induction Motors Using Measured Stator Currents and Vibration Signals. *IEEE Trans. Ind. Appl.* **2019**, *55*, 2378–2391. [CrossRef]
41. Cortes, C.; Vapnik, V. Support-vector networks. *Mach. Learn.* **1995**, *20*, 273–297. [CrossRef]
42. Cintron-Rivera, J.G.; Foster, S.N.; Strangas, E.G. Mitigation of turn-to-turn faults in fault tolerant permanent magnet synchronous motors. *IEEE Trans. Energy Convers.* **2015**, *30*, 465–475. [CrossRef]
43. Martinez, A.; Kak, A. PCA versus LDA. *IEEE Trans. Pattern Anal. Mach. Intell.* **2001**, *23*, 228–233. [CrossRef]

*energies*

Article

# Abnormal Detection for Running State of Linear Motor Feeding System Based on Deep Neural Networks

**Zeqing Yang [1,2,*], Wenbo Zhang [1], Wei Cui [1], Lingxiao Gao [1,2], Yingshu Chen [1], Qiang Wei [1] and Libing Liu [1,2]**

1 School of Mechanical Engineering, Hebei University of Technology, Tianjin 300130, China
2 Key Laboratory of Hebei Province on Scale-Span Intelligent Equipment Technology, Tianjin 300130, China
* Correspondence: yangzeqing@hebut.edu.cn

**Abstract:** Because the linear motor feeding system always runs in complex working conditions for a long time, its performance and state transition have great randomness. Therefore, abnormal detection is particularly significant for predictive maintenance to promptly discover the running state degradation trend. Aiming at the problem that the abnormal samples of linear motor feed system are few and the samples have time-series features, a method of abnormal operation state detection of a linear motor feed system based on normal sample training was proposed, named GANomaly-LSTM. The method constructs an encoding-decoding-reconstructed encoding network model. Firstly, the time-series features of vibration, current and composite data samples are extracted by the long short-term memory (LSTM) network; Secondly, the three-layer fully connected layer is employed to extract potential feature vectors; Finally, anomaly detection of the system is completed by comparing the potential feature vectors of the two encodings. An experimental platform of the X-Y two-axis linkage linear motor feeding system is built to verify the rationality of the proposed method. Compared with other classical methods such as GANomaly and GAN-AE, the average AUROC index of this method is improved by 17.5% and 9.3%, the average accuracy is enhanced by 11.6% and 15.5%, and the detection time is shortened by 223 ms and 284 ms, respectively. GANomaly-LSTM has successfully proved its superiority for abnormal detection for running state of linear motor feeding systems.

**Keywords:** linear motor feeding system; lack of abnormal samples; deep neural network; anomaly detection; semi-supervised anomaly detection generative adversarial network (GANomaly); long short-term memory (LSTM) network

**Citation:** Yang, Z.; Zhang, W.; Cui, W.; Gao, L.; Chen, Y.; Wei, Q.; Liu, L. Abnormal Detection for Running State of Linear Motor Feeding System Based on Deep Neural Networks. *Energies* **2022**, *15*, 5671. https://doi.org/10.3390/en15155671

Academic Editors: Yuling He, David Gerada, Conggan Ma and Haisen Zhao

Received: 11 July 2022
Accepted: 2 August 2022
Published: 4 August 2022



## 1. Introduction

As a key component of position tracking and positioning control for high-end CNC equipment, robots and precision motion platforms [1], a linear motor feeding system has the characteristics of multi-domain coupling, high integration of functions and dynamic and changeable performance. The complexity of electromechanical thermomagnetic coupling increases the probability of performance degradation, functional failure and malfunction. At the same time, under the combined effect of high-speed operation, mechanical friction, wear, high temperature and corrosion for a long time, unforeseen abnormal states and failures will occur. Therefore, abnormal detection, timely identification of abnormal states and predictive maintenance of the linear motor feeding system are particularly important.

With the increasing complexity of industrial equipment, the traditional way of replacing parts on schedule or judging abnormalities based on human experience can no longer meet the demand. Currently, the methods for anomaly detection of complex mechanical equipment are mainly divided into statistical-based, graph-based and machine learning-based approaches.

Statistical-based approaches need to collect historical data from equipment for statistical analysis to form a large number of normal data samples and abnormal data samples, and detect anomalies by extracting their features. Statistical models for anomaly detection mainly include the Gaussian model, regression model and expectation maximization model. Wang et al. [2] adopted an anomaly detection method based on the Gaussian model, which used the Gaussian model to represent the normal distribution of the data, and then scored the data according to the similarity between the model and the data. While this method can only be applied to data sets with a single attribute, and the distribution law of the data set needs to be known in advance. It is not applicable to the processing of high-dimensional data. Jin et al. [3] developed a bearing anomaly detection and fault prediction method based on an autoregressive model. The abnormality threshold was set by the attributes of Box-Cox transformation and Gaussian distribution. When the health index of the test data was greater than the abnormality threshold, it was judged as abnormal data. However, the threshold set often had errors, and an accurate dividing line between abnormal and normal data cannot be obtained. Liu et al. [4] proposed a novel filter based on an expectation-maximization model to identify anomalies in time-series data. The disadvantage of this model was the absolute dependence on the abnormal threshold. Statistical-based approaches do not re-quire establishing precise system models, and the algorithms are simple and easy to implement. However, this method is unsuitable for dealing with multivariate data, and the lack of fault data samples will lead to inaccurate detection results.

Graph-based approaches abstract entities as vertices and relationships as edges connecting vertices in the graph, providing a powerful means to express the complex relationships between entities. The traditional graph anomaly detection technology mainly obtains the graph statistical information through statistics and probability methods for anomaly detection. The shortcomings of these kind of methods are slow con-tent collection and low efficiency. In recent years, graph-based anomaly detection techniques have gradually developed, mainly used to solve the problem of anomaly detection in complex networks [5,6]. The key to this method is learning the correlation between different graph data. However, the nodes of objects in graph data are connected by edges, resulting in often complex correlations among the nodes. Recently, there is no precedent for their application in linear motor feeding systems.

Machine learning-based approaches include distance-based, deviation-based, density-based, and deep learning-based approaches.

The distance-based methods calculate point or collective anomalies by measuring the distances between data points and adjacent points or between data sequences and adjacent sequences. Objects farther away from others are considered anomalies. Zhang et al. [7] proposed a k-nearest neighbor (KNN) anomaly detection algorithm. It calculated the anomaly score by calculating the average distances between all K adjacent nodes. Li et al. [8] presented a clustering-based anomaly detection approach. The similarity of the time-series was evaluated by using the Euclidean distance function in the original feature space to judge whether the amplitudes are abnormal or not. The advantage of distance-based methods is that they are very efficient and easy to implement when dealing with low-dimensional data. The disadvantage is that they are expensive with regard to calculating distances between multivariable data sets.

The deviation-based approach uses models to make predictions on time-series data. If the deviation between the predicted value and the actual value of a data point exceeds a certain threshold, the data point is judged as abnormal. Li et al. [9] proposed a prediction-based anomaly detection method for time-series. An exception is detected by setting a threshold. Outliers are identified when the deviation between the predicted and actual values is greater than the threshold. Zhou et al. [10] proposed a deviation-based approach to anomaly detection for combined models; whether the sequence is abnormal is judged by the comparison. The abnormal score of a fragment in the sequence is calculated according to the deviation, and then the average score of all fragments is obtained. However, this approach is less effective for datasets with unknown prior knowledge and high-dimensional multivariate data.

The density-based approaches identify outliers by calculating the density of data objects. When the local density of the data object is different from the adjacent area, it will be dis-criminated as an outlier. Classical algorithms based on this method are Local Outlier Factor (LOF) and INFLuenced Outlierness (INFLO). Abdulghafoor et al. [11] proposed a density-based outlier detection method. This method compared the density of the observed object with the surrounding local density, using the LOF as a variable to measure the outliers. When the INFLO [12] algorithm estimates the density distribution of objects, it considers the relationship between the neighborhood and the reverse neighborhood of the object to rank the outliers. The higher the INFLO value of the detected object, the more likely it is to be an outlier. The advantage of the density-based method is that it can detect not only global outliers but also local outliers. The disadvantage is that time correlation on time step is not considered. Therefore, this method cannot be effectively used for anomaly detection of multivariate time-series data.

In recent years, deep learning has been widely adopted in image recognition, object detection, semantic analysis, etc. In order to efficiently and accurately mine effective state information from big mechanical data, deep learning has also been popularly applied in mechanical fault diagnosis [13]. Hoang et al. [14] proposed a convolutional neural network (CNN) that used vibration signals to detect bearing faults. Shao et al. [15] designed a stack transfer autoencoder and used the particle swarm optimization algorithm for fault diagnosis of rotating machinery. Jiang et al. [16] put forward the deep belief networks (DBNs), which directly extracted fault-related features from the original vibration signal and current signal, and the two features for fault diagnosis of a wind turbine gearbox. However, the above deep learning methods require a large amount of historical data in different health states for training. In the operation of linear motor feeding systems, data samples are often only available for normal operating conditions, but not for fault conditions. Due to the insufficient failure samples of the linear motor feed system, the deep learning method is difficult to apply. The reasons for this phenomenon are as follows: (1) The fault situation is far less than the normal situation, and it is difficult to collect fault data; (2) Even if the fault data can be collected, it takes a long time and costs a lot; (3) Some fault data cannot be measured under laboratory conditions. The data generated by the linear motor feeding system is mostly time-series data, with apparent time and sequence features, such as vibration signal and current signal. Collecting abnormal samples for linear motor feeding systems is challenging under current conditions. Therefore, it is an exigent issue to realize the abnormal detection of the running state of the linear motor feeding system in the absence of fault samples.

Recently, the generative adversarial network (GAN) has brought new hope for solving the problem of insufficient samples. GAN is a new network structure that was proposed by Goodfellow [17] in 2014, which is an unsupervised feature learning algorithm based on the idea of adversarial training. The method has been extensively used in anomaly detection because it can use adversarial learning of sample representations for anomaly inference [18]. With the continuous improvement of generative adversarial ideas, many improved generative adversarial networks have been derived, such as the efficient GAN-based anomaly detection (EGBAD) network [19], deep convolutional generative adversarial network (DCGAN) [20], anomaly generative adversarial networks (AnoGAN) [21], et al. These improved GANs can generate training samples by learning the probability distribution of real data, so as to solve the problem of insufficient fault samples in the process of model training. Wang et al. [22] achieved a fault diagnosis approach combining GAN and a stacked denoising autoencoder (SDAE) to generate fault data for the problem of a small number of fault samples in planetary gearboxes. Mao et al. [23] combined GAN with the stacked denoising auto encoder (SDAE) to solve the problem of data imbalance in bearing fault diagnosis. Gao et al. [24] proposed a data augmentation method based on the Wasserstein generative adversarial network with gradient penalty (WGAN-GP) and verified the feasibility of generating fault samples on three datasets. The above studies show that GAN can generate fault samples, dramatically expanding the range and diversity of generated data samples. However, the above models still require a small amount of fault samples, and it is difficult to complete training with a complete lack of them.

Given the above particular situation, Akcay et al. [25] proposed a GANomaly method, which can complete the training of the model without abnormal samples. Subsequently, Akcay et al. [26] further achieved a Skip-GANomaly detection method for images based on GANomaly, which only used normal samples for training. Luo et al. [27] proposed a geological image anomaly detection method based on GANomaly. Liu et al. [28] applied an anomaly detection method based on GANomaly and CNN, which adopted normal time-series data samples for training, and encodes them into two-dimensional images using the Gramian Angular Field (GAF) method. The abnormal detection of vibration signals of long-span bridges was realized, and a good detection effect was achieved. GANomaly is an anomaly detection method commonly used in images. When dealing with time-series data, the gradient disappearance and gradient explosion problems are its traditional limitations, influenced by the choice of activation function and the error back propagation method [29]. The deeper the network layer, the more obvious the problem becomes [30]. Therefore, it is necessary to find an efficacious method.

The long-short-term memory (LSTM) network includes a memory unit, gate structure and attention mechanism, which can effectively solve the aforesaid problems. Methods based on LSTM have excellent anomaly detection capability for time-series data [31]. Li et al. [32] adopted a method based on stacked autoencoders (SAE) and LSTM networks for anomaly detection in vibration signals of rotating machinery. Chen et al. [33] raised an anomaly detection method for time-series data of wind turbines based on LSTM and an auto-encoder (AE) neural network. Vos et al. [34] developed a gear anomaly detection algorithm combining deep learning and LSTM. Ou et al. [35] provided a bearing state anomaly detection method based on the LSTM network. In the training process of this method, only health data was used. Bai et al. [36] used the LSTM network for fault detection of gas turbines. Aiming at the problem that the newly-run gas turbine was difficult with regard to obtaining fault data, this method only used normal data to train the network. Kong et al. [37] developed an anomaly detection method for industrial multidimensional time-series data. This method used a bi-directional LSTM with the attention mechanism in the generator and discriminator of GAN. The results indicated that the method had favorable performance in the task of anomaly detection for industrial time-series data.

On the one hand, GANomaly can complete model training without abnormal samples. On the other hand, LSTM can avoid gradient disappearance and gradient explosion when training time-series data.

Therefore, combining the advantages of GANomaly and the LSTM neural network, an abnormal detection method for the running state of the linear motor feeding system is proposed in this paper. Based on the analysis of the factors affecting the running state of the linear motor feeding system, the abnormal detection network framework is designed by taking vibration signals and current signals as the original data. Firstly, the LSTM network is used to extract the input sample time-series features, then the three-layer fully connected layer is employed to extract potential feature vectors. Secondly, the anomaly score of the input sample is obtained by comparing the difference between the latent feature vectors obtained by the two encodings. The relationship between the abnormal score value and the threshold value is used to judge whether the input sample is abnormal, so as to realize the abnormal detection of the running state of the linear motor feeding system. Finally, the experimental platform of the X-Y two-axis linked linear motor feeding system is built to validate the proposed method experimentally.

The main contributions of this article are listed below:

1. A GANomaly-LSTM method is proposed, which can effectively avert the teasers of gradient disappearance and gradient explosion during the course of training time-series data, and the extracted features can achieve a good clustering effect.
2. This method can realize anomaly detection in the absence of abnormal sample training.

3. The proposed model performs well in anomaly detection of phase-missing current signals and vibration signals, verified under three input conditions.
4. A mass of experiments have been actualized, and the results indicate that the proposed method achieves excellent advantages in effectiveness and performance compared with other classical methods.

The rest of this paper is organized as follows. Section 2 provides a theoretical introduction to the GANomaly-LSTM method. Section 3 builds the experimental platform for the X-Y two-axis linked linear motor feeding system and describes the process of experimental setup, data acquisition and feature extraction. Section 4 provides a detailed analysis of the experimental results. In Section 5, we draw conclusions and discuss directions for future work.

## 2. Anomaly Detection Model of Linear Motor Feeding System

### 2.1. Factors Affecting the Running State of Linear Motor Feeding System

The feeding system of the X-Y two-axis linkage linear motor is employed as the research object. The X-axis is cross orthogonal to the Y-axis, and the Y-axis is located above the X-axis and coupled with the X-axis stator. Factors affecting the running state of the linear motor feeding system mainly include three aspects: (1) Abnormal vibration. Because the linear motor feeding system has no intermediate transmission link, the mechanical damping of the linear motor is small, so the vibration is difficult to be effectively attenuated, which seriously affects the running state of the feeding system. At the same time, the vibration in the processing process is one of the critical factors affecting the machining accuracy, which will reduce not only the surface quality of the workpiece and the dynamic precision of the machine tool but also the productivity [38]; (2) Motor overheating. Severe wear of motor guides will lead to overheating, and increase stator resistance, further leading to abnormal acceleration changes of the linear motor, ultimately affecting the operation state of the feeding system; (3) Excessive load. The vibration will accelerate the wear of power transmission components and overload the transmission and machine tool structures, resulting in the current fluctuation of the linear motor feeding system, thus leading to the change of the operating state [1]. It can be seen that acceleration and current are the two key factors to reflect the evolution of the running state of the linear motor feeding system. Therefore, vibration signals and current signals of the experimental platform under different working conditions are used as the original acquisition time-series data in this paper.

### 2.2. Anomaly Detection Model

#### 2.2.1. The Structure of GANomaly

The GANomaly method adds an adversarial learning strategy to the autoencoder generation model, which is an anomaly detection method that compares the potential features of sample coding. GANomaly determines whether a data sample is abnormal or not based on the difference between the potential feature vector $Z$ and $\hat{Z}$ obtained from the two encoders. GANomaly has strong robustness and anti-noise interference capability, requiring only a small number of abnormal samples or no abnormal samples during learning and training [39].

The GANomaly model framework consists of three parts: a generator, a reconstructed encoder, and a discriminator. The network structure of GANomaly is shown in Figure 1. The encoder $G_E(x)$ and decoder $G_D(Z)$ in Figure 1 are collectively referred to as generators. The input data $x$ passes through the encoder $G_E(x)$ to obtain the latent feature vector $Z$, then passes through the decoder $G_D(Z)$ to obtain the reconstructed data $\hat{x}$. The reconstructed encoder $E(\hat{x})$ encodes the reconstructed data $\hat{x}$ again to get the potential feature vectors $\hat{Z}$ of the reconstructed data. The idea of adversarial learning is introduced in discriminator $D(x, \hat{x})$ to distinguish the differences between the original input data $x$ and the reconstructed data $\hat{x}$. It determines original input data $x$ to be true and reconstructed data $\hat{x}$ to be false. The discriminator is designed as a network with the same structure as the encoder in the generator. Meanwhile, the gap between the reconstructed and original input data is continuously optimized.

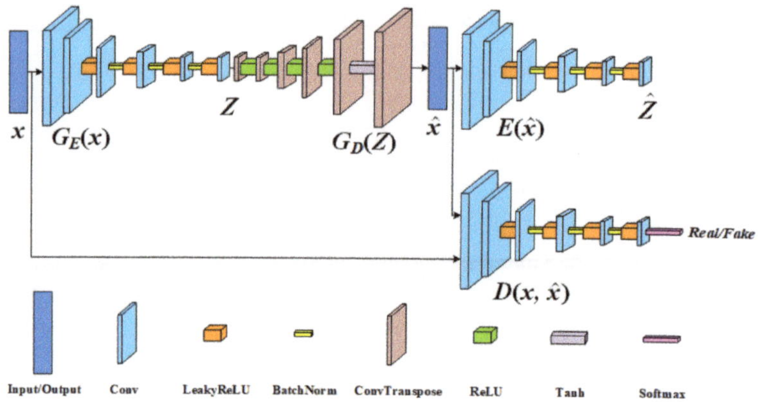


**Figure 1.** Network structure of GANomaly.

In the training stage, the whole model is trained with normal samples. When the model is inputted with an abnormal sample in the testing stage, there is a certain difference between the potential feature vectors obtained by the encoder and the reconstructed encoder because the decoder is trained by normal samples. When the difference is greater than a certain threshold, the input sample is identified as abnormal [40].

2.2.2. The Structure of LSTM

LSTM is a special RNN which can solve the long-term dependency problem well [41]. The standard RNN has only one tanh layer, while the internal structure of LSTM is more complex, consisting of four neural network layers: forgetting gate, input gate, cell state and output gate (as shown in Figure 2) [42]. The advantages and limitations of RNN and LSTM are listed in Table 1.

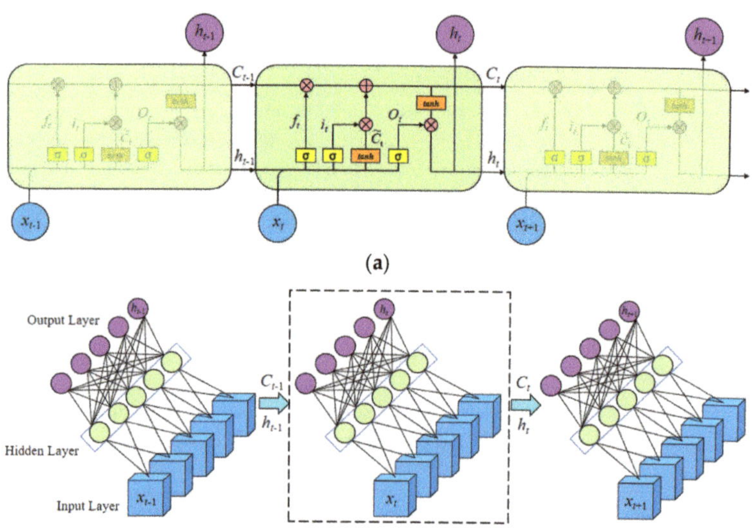


**Figure 2.** Network structure of LSTM: (**a**) Unit structure of LSTM; (**b**) Recurrent structure of LSTM.

**Table 1.** Advantages and limitations of RNN and LSTM.

| | LSTM | RNN |
|---|---|---|
| Advantages | 1. The problem of gradient disappearance and gradient explosion can be overcome when training long sequence data;<br>2. Able to learn long-term dependence;<br>3. Simple to implement. | 1. Train the time-series data;<br>2. The network is simple and easy to operate. |
| Limitations | 1. Because of the large amount of computation in network training, high performance computers are needed. | 1. Cannot process long time-series data;<br>2. There are problems of gradient disappearance and gradient explosion. |

The first layer is "forget gate", and its input is $h_{t-1}$ and $x_t$, mainly focusing on selectively forgetting the information of $h_{t-1}$. Through $\sigma$, a sigmoid activation function, the output $f_t$ is the value within the interval of [0, 1]. "0" represents that the state is inactive and all information is forgotten, while "1" represents the opposite. The sigmoid activation function is shown in Figure 3a. The equation of the forget gate is as follows:

$$f_t = \sigma\left(W_f \left[h_{t-1}, x_t\right] + b_f\right) \tag{1}$$

where $\sigma$ is the sigmoid activation function; $x_t$ is the input at time $t$; $h_{t-1}$ is the output of the time $t-1$; and $W_f$ and $b_f$ are the weight and bias parameters of the forget gate, respectively.

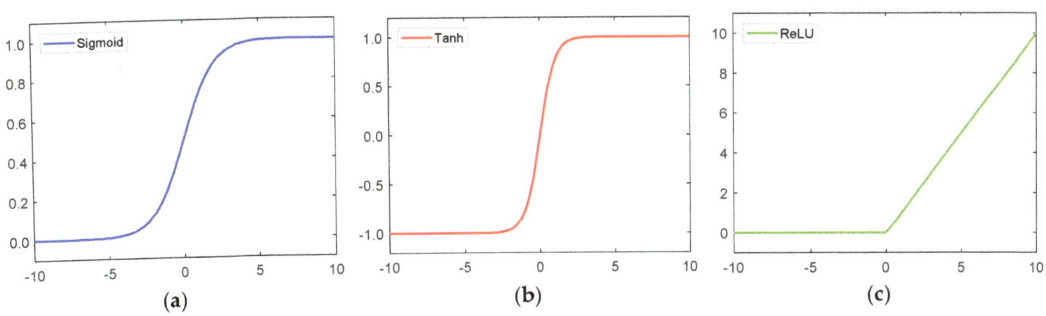


**Figure 3.** Three activation functions: (**a**) Sigmoid; (**b**) Tanh; (**c**) ReLU.

The second layer is "input gate". It determines what information of the cell state needs to retain. The input gate includes the sigmoid layer and tanh layer. The sigmoid layer determines what information needs to be updated at the input gate. The tanh layer creates a matrix to add to the cell state. The tanh activation function is shown in Figure 3b. The input of the input gate is $h_{t-1}$ and $x_t$, so that the equation can be expressed as:

$$i_t = \sigma(W_i \left[h_{t-1}, x_t\right] + b_i) \tag{2}$$

$$\tilde{C}_t = tanh(W_C[h_{t-1}, x_t] + b_C) \tag{3}$$

where $W_i$ is the weight of sigmoid function in the input gate; $b_i$ is the bias parameter of sigmoid function; $W_C$ is the weight of tanh function; $b_C$ is the bias parameter of the tanh function.

The third layer is "cell state". It can update the cell state $C_{t-1}$ to the current cell state $C_t$, which can be expressed as:

$$C_t = f_t C_{t-1} + i_t \widetilde{C}_t \quad (4)$$

The fourth layer is "out gate". It influences the output of the current cell state. The sigmoid function layer determines which parts should be updated. And the tanh function can operate on the cell state $C_t$, so that $tahn(C_t)$ ranges (−1, 1). Multiply the output of the sigmoid function, then the part $h_t$ will be gotten. The output gate equation can be expressed as:

$$o_t = \sigma(W_o[h_{t-1}, x_t] + b_o) \quad (5)$$

$$h_t = o_t \cdot tanh(C_t) \quad (6)$$

where $o_t$ is the output gate; $W_o$ and $b_o$ are the weight and the bias parameter, respectively; $h_t$ is the output of the current cell state and the input of the next state as well.

Besides, Rectified Linear Unit (ReLU) is also a popular activation function, as shown in Figure 3c.

2.2.3. Proposed Method

The Structure of GANomaly-LSTM

Based on the theories of the GANomaly and LSTM network, an anomaly detection method for linear motor feeding system based on GANomaly-LSTM is proposed. As shown in Figure 4, the GANomaly-LSTM network structure comprises three sub-networks: the generation network, the reconstructed encoder network and the discriminant network.

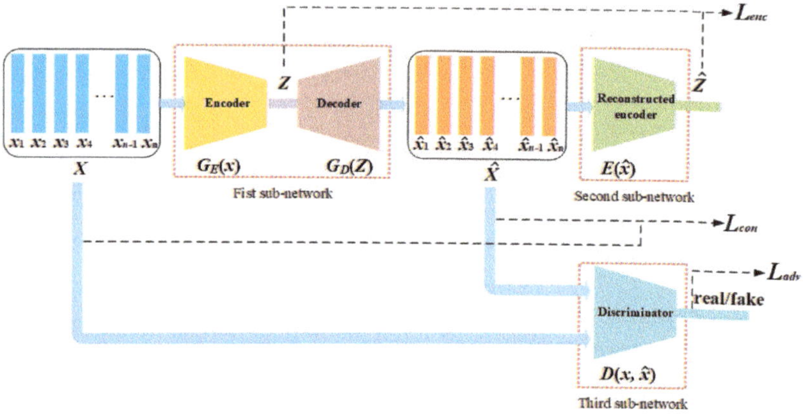


**Figure 4.** The network structure of GANomaly-LSTM.

The first sub-network is the generative network, including encoder $G_E(x)$ and decoder $G_D(Z)$. The encoder $G_E(x)$ structure is expressed in Figure 5. A three-layer LSTM is used to extract time-series features of samples, and then a three-layer fully connected layer is used to extract potential feature vectors. Three batch standardization layers and three rectifying linear unit (ReLU) activation functions are used to optimize the output distribution of the middle layer. The structures of decoder $G_D(Z)$ and encoder $G_E(x)$ are opposite. For the input data $x$, the latent feature vector $Z$ is obtained through the encoder $G_E(x)$, and then the reconstructed data of $\hat{x}$ is obtained through the decoder $G_D(Z)$.

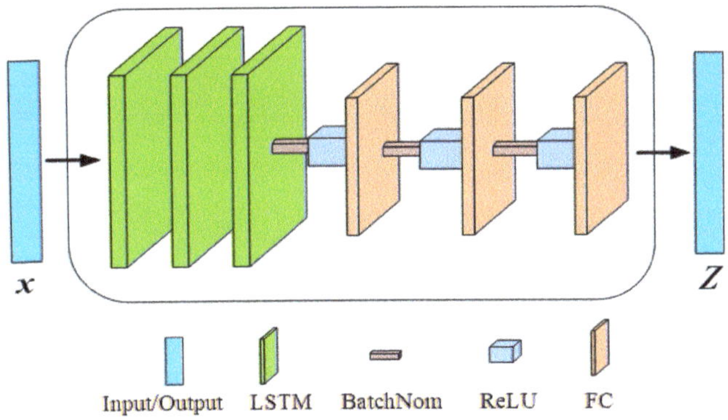


**Figure 5.** Network structure of encoder.

The second subnetwork is the reconstructed encoder network, and by re-encoding the reconstructed data $\hat{x}$, the latent feature vector $\hat{Z}$ of the reconstructed data is obtained. In this subnetwork, the structure of the reconstructed encoder $E(\hat{x})$ is the same as $G_E(x)$.

The third sub-network is the discriminant network, which continuously narrows the gap between the reconstructed data and the original input data by judging the original input data $x$ as true and the reconstructed data $\hat{x}$ as false. Ideally, the reconstructed data is no different from the original input data. By introducing the idea of adversarial training, the generative and discriminant networks play games. On the one hand, it improves the ability of the decoder to recover the input samples, and on the other hand, it enhances the feature extraction ability of the encoder. The discriminator $D(x, \hat{x})$ has the same structure as $G_E(x)$.

Loss Function

In the generative network, the reconstruction error loss is defined as the gap between the original input data and the reconstructed data.

$$L_{con} = |x - \hat{x}| \tag{7}$$

where $L_{con}$ is the reconstruction error loss function of the generative network.

A feature matching error is set in the discriminant network for optimization in the data feature layer.

$$L_{adv} = |f(x) - f(\hat{x})| \tag{8}$$

where $L_{adv}$ is the loss function of the discriminant network, and it specifically refers to the loss of confrontation between the generation network and the discriminant network; $f$ is the transfer function of the model.

Ideally, for normal data, the difference between the latent eigenvector $\hat{Z}$ of the reconstructed data and the latent eigenvector $Z$ is extra small. In order to quantify and optimize this difference in the training phase, the error between latent feature vectors is introduced:

$$L_{enc} = |Z - \hat{Z}| \tag{9}$$

where $L_{enc}$ is the loss function of the latent eigenvector error of the reconstructed data.

For the entire network model, the loss function can be expressed as:

$$L_{total} = \omega_{adv} L_{adv} + \omega_{con} L_{con} + \omega_{enc} L_{enc} \tag{10}$$

where $\omega_{adv}$, $\omega_{con}$ and $\omega_{enc}$ represent the weights of $L_{adv}$, $L_{con}$ and $L_{enc}$, respectively.

The gradient descent method is a popular optimization algorithm in deep learning. Its basic idea is to update the parameters along the opposite direction of the gradient of the loss function about the parameters as the search direction, so that the loss function gradually decreases and finally reaches the minimum value. The gradient descent method can be divided into fixed learning rate optimization algorithms (e.g., SGD, Momentum, and NGA) and adaptive learning rate optimization algorithms (e.g., Adagrad, RMSprop, and Adam) [43].

The Adam optimization algorithm adopts the adaptive learning rate and momentum mechanism to determine the updated direction by considering the previous gradient and the current gradient together, so that the function convergence process is more stable. Moreover, the first-order moment estimation and second-order moment estimation of the gradient can be used to dynamically adjust the learning rates of different parameters to accelerate the convergence rate of the function, so as to obtain the global optimal parameters with less iterations [44]. Therefore, the Adam optimization algorithm is selected to optimize the model parameters in this paper.

Model Validation and Evaluation Criteria

After the training is complete, the model is validated using the test set. Firstly, the generative network can generate reconstructed test samples. In order to obtain the similarity between the potential feature vector $Z$ of the first encoding and the potential feature vector $\hat{Z}$ of the second encoding, the gradient descent method Adam is chosen to update continuously, and then the optimal potential feature vector will be obtained.

$$\overset{min}{\hat{Z}} E(Z,\hat{Z}) = 1 - Simi(Z,\hat{Z}) \tag{11}$$

where *Simi* is the similarity function and $E$ is the error function.

Secondly, anomalies are detected using the gap between the latent feature vector after the first encoding of the test sample and the latent feature vector of the reconstructed data after the second encoding. The anomaly score $A(X)$ of the test sample according to the loss functions $L_{con}$ and $L_{enc}$. is calculated.

$$A(X) = \lambda L_{con} + (1 - \lambda) L_{enc} \tag{12}$$

where $\lambda$ is the weight of the adjustment loss.

Finally, the specific score for judging abnormality is $a(X_i)$, and $A(X)$ is controlled between 0 and 1 by normalization processing. A threshold value $\varphi$ is set, and once the abnormal score $a(X_i) > \varphi$ of the test sample, the sample is judged as an abnormal sample.

$$a(X_i) = \frac{A(X_i) - min(A(X))}{max(A(X)) - min(A(X))} \tag{13}$$

In this paper, AUROC, AUPRC, $F_1$ score and accuracy are chosen as the performance evaluation metrics of the proposed method. According to the actual classification and predicted classification of the test samples, the samples can be divided into four types: true positive (*TP*), true negative (*TN*), false positive (*FP*) and false negative (*FN*). Then the formulas for calculating recall (*R*), precision (*P*), accuracy and $F_1$ score can be expressed as:

$$R = \frac{TP}{TP + FN} \tag{14}$$

$$P = \frac{TP}{TP + FP} \tag{15}$$

$$accuracy = \frac{TP + TN}{TP + FN + FP + TN} \tag{16}$$

$$F_1 = \frac{2RP}{R + P} \tag{17}$$

AUROC is the area under the receiver operating characteristic (ROC) curve with false positive rate (FPR) and true positive rates (TPR), respectively, as its abscissa and ordinate at different threshold conditions. The value of AUROC usually ranges from 0.5 to 1. A larger value indicates better model performance. AUPRC is the area under the precision recall (PR) curve, consisting of the precision and recall rates at different threshold conditions. Accuracy refers to the proportion of correctly predicted samples in all samples. The $F_1$ score comprehensively considers the recall and precision, which is the harmonic mean of the two.

### 2.3. Anomaly Detection Process

The abnormal detection process of the linear motor feeding system proposed in this paper is displayed in Figure 6.

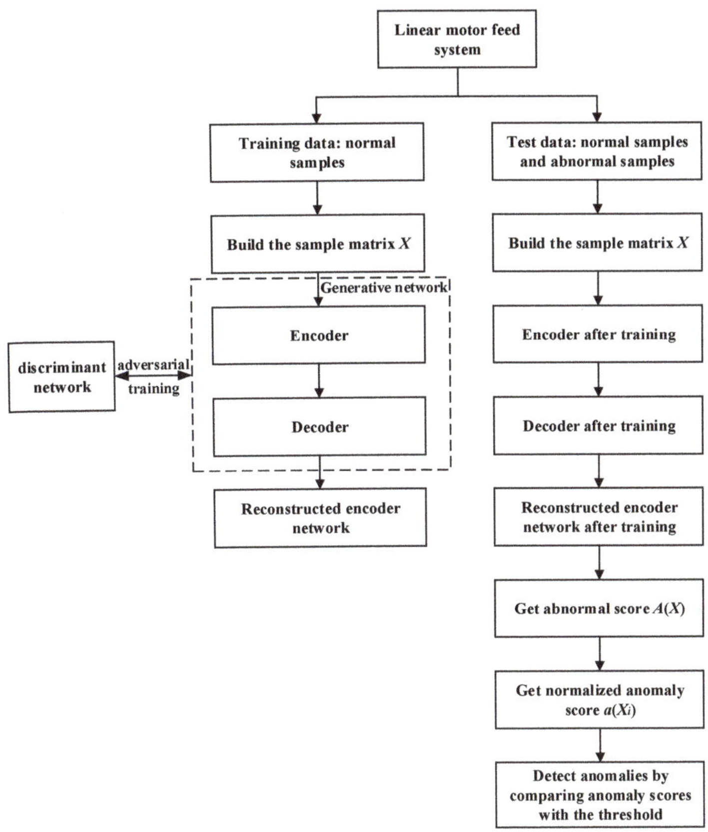


**Figure 6.** Anomaly detection workflow.

Training phase:

- STEP1: The sensor collects relevant time-series signals in real time, and a sample matrix X is constructed for the collected normal samples.

$$X = \begin{bmatrix} x_{11} & x_{12} & \cdots & x_{1n} \\ x_{21} & x_{22} & \cdots & x_{2n} \\ \vdots & \vdots & \ddots & \vdots \\ x_{m1} & x_{m2} & \cdots & x_{mn} \end{bmatrix} \quad (18)$$

where $m$ is the number of samples in a sequence, and $n$ is the number of sensors.

- STEP2: The normal-sample matrix $X$ is inputted into the encoder $G_E(x)$, then the LSTM network extracts the time-series features, later the latent feature vector $Z$ is gained through three-layer fully connected layers, and lastly the reconstructed data $\hat{X}$ is obtained through the decoder $G_D(Z)$.
- STEP3: The discriminant network $D(x, \hat{x})$ discriminates the normal samples $X$ and the reconstructed data samples $\hat{X}$, and continuously narrows the gap between the two during the confrontational training process.
- STEP4: The reconstructed data $\hat{X}$ is inputted into the reconstructed encoder $E(\hat{x})$ network, and then the latent feature vector $\hat{Z}$ is attained.

Test phase:

- STEP1: After the model training, the normal and abnormal samples are used to construct the sample matrix $X$ for testing. At this time, the discriminant network is no longer used. In the testing stage, network model parameters are fixed and outputted by training stage. The potential feature vector $Z$ is obtained through $G_E(x)$, then the reconstructed data $\hat{X}$ is gained through $G_D(Z)$, and finally, the potential feature vector of the reconstructed data $\hat{Z}$ is garnered through $E(\hat{x})$.
- STEP2: The anomaly score $A(X)$ of the input sample $X$ is computed according to the loss functions $L_{con}$ and $L_{enc}$. The final anomaly score $a(X_i)$ is obtained by normalizing $A(X)$.
- STEP3: It is determined whether the input sample is abnormal or not according to the relationship between the abnormal score $a(X_i)$ and a certain threshold $\varphi$. If $a(X_i) > \varphi$, the input sample will be classified as an abnormal sample; otherwise it is a normal sample.

## 3. Experimental Setup and Feature Extraction

### 3.1. Construction of Experimental Platform and Data Collection

In order to verify the validity of the proposed method, an experimental platform for the linear motor feeding system is built, as presented in Figure 7 and Table 2. The platform consists of an X-Y two-axis linkage linear motor feeding system, a YD623 tri-axial accelerometer, a TL-1 signal conditioner, a USB-1608FS data acquisition card, a linear motor feeding system controller, and a computer processing center. The changeable working conditions of the experimental platform include feed speed, displacement and load, etc. The vibration signals in three directions and three-phase current signals are collected by setting different working conditions.

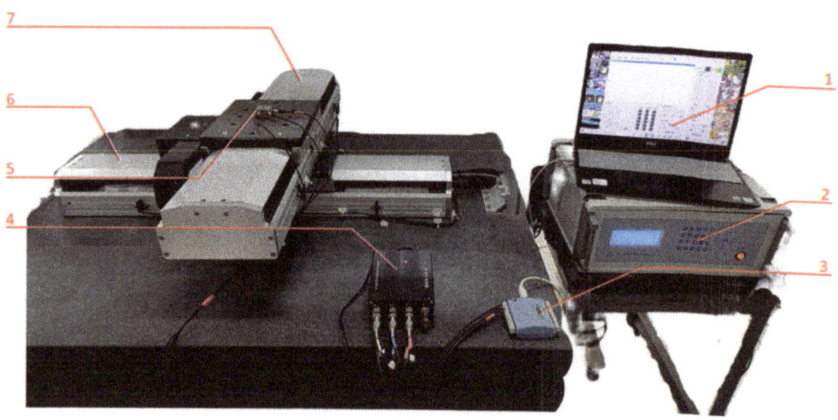


**Figure 7.** X-Y two-axis linkage linear motor feeding system experimental platform.

**Table 2.** Numbers and names of acquisition devices.

| Number | Equipment Name |
|---|---|
| 1 | Computer processing center |
| 2 | Linear motor feeding system controller |
| 3 | Data acquisition card |
| 4 | Signal conditioner |
| 5 | Accelerometer |
| 6 | X-axis feeding system |
| 7 | Y-axis feeding system |

In order to obtain the vibration signals and current signals of the linear motor feeding system, the YD623 tri-axial accelerometer is fixed directly above the linear motor feeding system, and the vibration signals are collected by the USB-1608FS data acquisition card. The current sensor is located inside the system and connected to the computer via the appropriate data line. The current signals are measured by the software Composer. In this experiment, the sampling frequency is set to 50 kHz, and the sampling interval is 1 min. During the experiment, vibration signals in three directions are collected, and only the signals in the vertical direction are used in the data processing. The vibration signals of 2029 sampling points in the vertical direction are collected each time, and a total of 1,623,200 data points are collected in 800 groups. The three-phase current signals of 1533 sampling points are collected each time, and a total of 1,226,400 data points are collected in 800 groups.

During the experiment, vibration signals collected each time are stored in CSV files, and current signals are stored in MAT files. According to experimental conditions, each CSV file and MAT file are stored in corresponding folders, respectively.

### 3.2. Design of Experimental Working Conditions and Description of Data Samples

In order to obtain comprehensive and diverse experimental samples, the vibration signals and current signals of the linear motor feeding system under different operation commands are collected in this experiment. The collected data is divided into 16 categories according to the execution command parameters, and each category contains several time-acceleration sequence data and time-current data. The experimental working conditions for collecting the vibration and current signals in this experiment are shown in Table 3. Figure 8a,b shows the vertical vibration signals and three-phase current signals which are collected in working condition 5 under a displacement interval of How 420 mm.

In this experiment, 800 samples of each of the vibration signal and current signal are collected. The dataset is divided into a 70% training set and 30% test set [45,46]. There are 560 samples in the training set, all of which are normal samples. The number of samples in the test set is 240, which are divided into two parts equally. Part of them are directly used as a normal sample for testing, and the other part are replaced with an abnormal sample. However, the life cycle of the linear motor feeding system is very long, and the probability of abnormality is small. Therefore, running to an abnormal state is an extremely time-consuming experiment in order to perform abnormal testing to obtain abnormal data. To overcome this difficulty, we introduce abnormal samples containing random amplitude shock noise in random positions of vibration signals. The abnormal samples of the vibration signal test set are shown in Figure 8c (the arrow position is the position where the noise is introduced). In order to obtain abnormal samples of the current signals, we collected the current signals of the missing phase. The abnormal samples of the current signals test set are shown in Figure 8d.

**Table 3.** Experimental working condition settings.

| Number | Load (kg) | X/Y Axis | Displacement Interval (mm) | Speed (mm/s) |
|---|---|---|---|---|
| 1 | No load | X axis | 60, 180, 300, 420 | 60 |
| 2 | No load | X axis | 60, 180, 300, 420 | 80 |
| 3 | No load | X axis | 60, 180, 300, 420 | 100 |
| 4 | No load | X axis | 60, 180, 300, 420 | 120 |
| 5 | No load | Y axis | 60, 180, 300, 420 | 60 |
| 6 | No load | Y axis | 60, 180, 300, 420 | 80 |
| 7 | No load | Y axis | 60, 180, 300, 420 | 100 |
| 8 | No load | Y axis | 60, 180, 300, 420 | 120 |
| 9 | No load | X-Y axis linkage | 60, 180, 300, 420 | 60 |
| 10 | No load | X-Y axis linkage | 60, 180, 300, 420 | 80 |
| 11 | No load | X-Y axis linkage | 60, 180, 300, 420 | 100 |
| 12 | No load | X-Y axis linkage | 60, 180, 300, 420 | 120 |
| 13 | Load 10 kg | X axis | 60, 180, 300, 420 | 60 |
| 14 | Load 10 kg | X axis | 60, 180, 300, 420 | 80 |
| 15 | Load 10 kg | X axis | 60, 180, 300, 420 | 100 |
| 16 | Load 10 kg | X axis | 60, 180, 300, 420 | 120 |
| 17 | Load 10 kg | Y axis | 60, 180, 300, 420 | 60 |
| 18 | Load 10 kg | Y axis | 60, 180, 300, 420 | 80 |
| 19 | Load 10 kg | Y axis | 60, 180, 300, 420 | 100 |
| 20 | Load 10 kg | Y axis | 60, 180, 300, 420 | 120 |
| 21 | Load 10 kg | X-Y axis linkage | 60, 180, 300, 420 | 60 |
| 22 | Load 10 kg | X-Y axis linkage | 60, 180, 300, 420 | 80 |
| 23 | Load 10 kg | X-Y axis linkage | 60, 180, 300, 420 | 100 |
| 24 | Load 10 kg | X-Y axis linkage | 60, 180, 300, 420 | 120 |

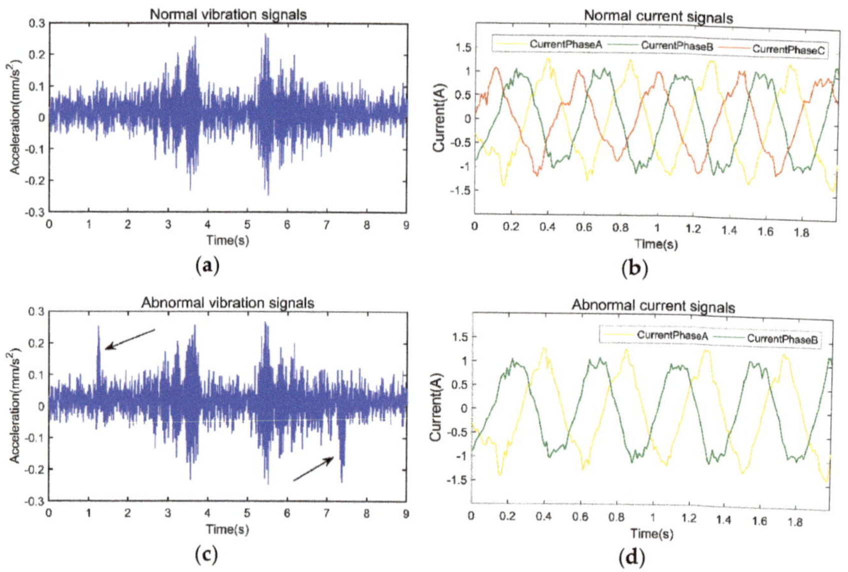


**Figure 8.** Normal and abnormal time domain vibration and current signals in working condition 5 under (displacement interval 420 mm): (**a**) Normal vibration signals; (**b**) Normal current signals; (**c**) Abnormal vibration signals (the arrow position is the position where the noise is introduced); (**d**) Abnormal current signals.

### 3.3. Experimental Environment and Model Parameters

This paper adopts PyTorch, a deep learning open-source framework, to construct the neural network model, complete the training, and test the model. PyTorch is widely used in computer vision, natural language processing and other fields.

The hardware environment of the computing deviceincludes a 4-core CPU Intel Xeon E3-123lv3, with a clock speed of 3.4 GHz; the GPU is AMD Radeon VII, with a memory capacity of 16 GB, and a memory speed of 4 Gbps and 3840 cores.

The Adam optimization algorithm is used in the training process, and the first-order moment weight $\beta_1 = 0.6$, while the second-order moment weight $\beta_2 = 0.999$. Since the GPU used in this experiment has enough video memory, the batch size used is 128, which can speed up the operation speed and convergence speed, and reduce the parameter jitter during training. Other model parameters are shown in Table 4.

**Table 4.** The parameters of the model.

| Parameters | $\omega_{adv}$ | $\omega_{con}$ | $\omega_{enc}$ | $\lambda$ | Learning Rate |
|---|---|---|---|---|---|
| Value | 1 | 1 | 0.5 | 0.8 | 0.0001 |

### 3.4. Signal Feature Extraction

In order to achieve a better abnormality detection effect of the linear motor feeding system, the extracted features must be appropriately selected. There is a lot of high noise and redundant information in the original data collected by sensors, which causes the dimension problem. The method based on deep learning can solve the above issues very well. An LSTM network is used to extract three effective time domain features from the original vibration and current signals, namely, standard deviation, skewness and kurtosis. After repeated experiments, it is found that these three statistical features are susceptible to the abnormal changes in the vibration signals and current signals. Two hundred and forty feature datasets are extracted from the vibration and current signal test sets, respectively. Figure 9 shows three time-domain feature scatter plots of vibration signals and current signals in working condition 5 under a displacement interval of 420 mm. The results show that both the sample features of the vibration signals and the current signals can achieve good classification. However, the sample feature aggregation effect of the current signals is better than that of the vibration signal, and the classification effect is more pronounced.

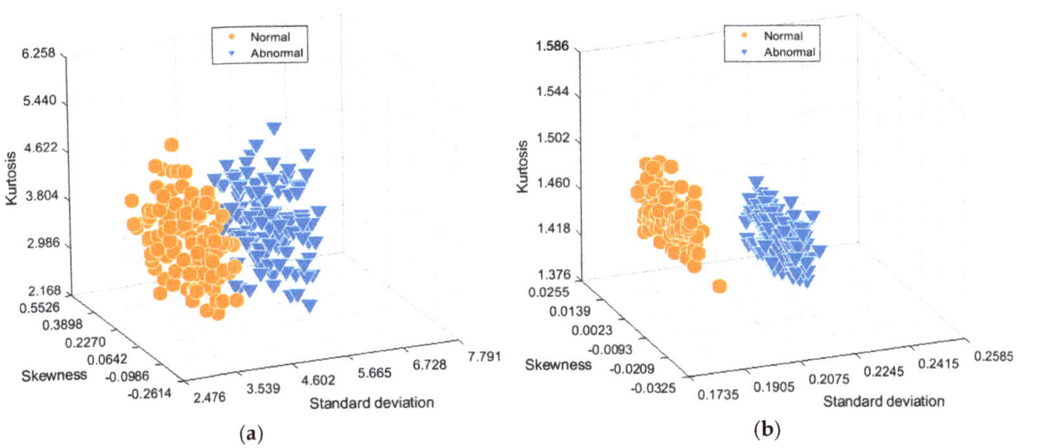


**Figure 9.** The distribution scatters diagram of the three-time domain features in working condition 5 under displacement interval 420 mm: (**a**) Feature distribution of vibration signals; (**b**) Feature distribution of the current signals.

## 4. Analysis of Experimental Results

To verify the performance of the proposed method, the performances under the following three different input conditions are compared.

- Case1: Vibration data sample;
- Case2: Current data sample;
- Case3: Vibration and Current Composite Data Sample.

Figure 10 exhibits the ROC curves of our method under different input conditions. It is obvious that the AUROC indicators under the three input conditions are above 90%, especially in Case2, which reaches 98.5%, the maximum value, compared to 94.6% and 97.7% in the other two cases, an improvement of 3.9% and 0.8%, respectively. In order to further certify the accuracy of the comparison, we introduce the AUPRC indicator for comparison. The experimental results are shown in Table 5, and the change trends of the two indicators are generally consistent. For the AUPRC, it reaches 98.2% in Case2, compared to 94.8% and 96.9% in the other two cases, an improvement of 3.4% and 1.3%, respectively. Figure 11 is the histogram of normal and abnormal scores when the method in this paper uses current data samples alone. It can be seen that our model can distinguish normal and abnormal classes well. To further compare the accuracy of the models under the three input conditions, we calculated the precision score, recall score and $F_1$ score under the three input conditions. As shown in Figure 11, the dividing line between normal and abnormal scores is about 0.5, that is, the threshold $\varphi = 0.5$. Therefore, our model distinguishes samples with scores less than 0.5 as normal samples and otherwise abnormal. We set the threshold at 0.44, when the recall rate of abnormal samples is 1, to compare the scores under three input conditions. As can be seen from Table 6, the proposed method achieves the maximum values on the three metrics of Case2, which are 96.5%, 100%, and 98.2%, respectively.

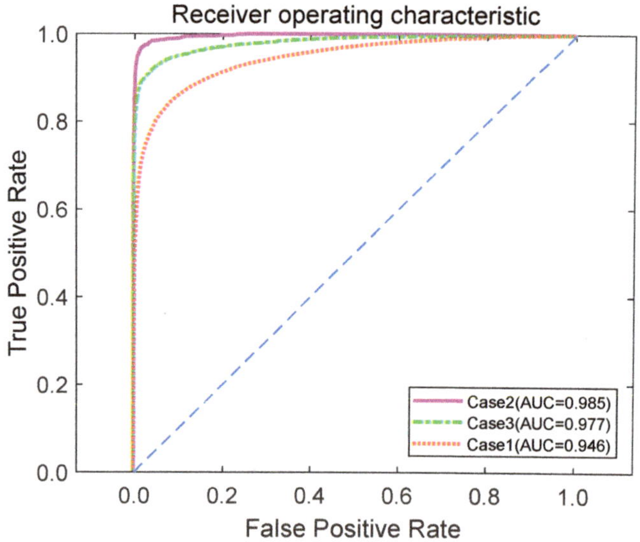


**Figure 10.** ROC curves of our method under different input conditions.

**Table 5.** Experimental results of our method under three input conditions.

| Case | AUROC | AUPRC |
|---|---|---|
| Case 1 | 0.946 | 0.948 |
| Case 2 | 0.985 | 0.982 |
| Case 3 | 0.977 | 0.969 |

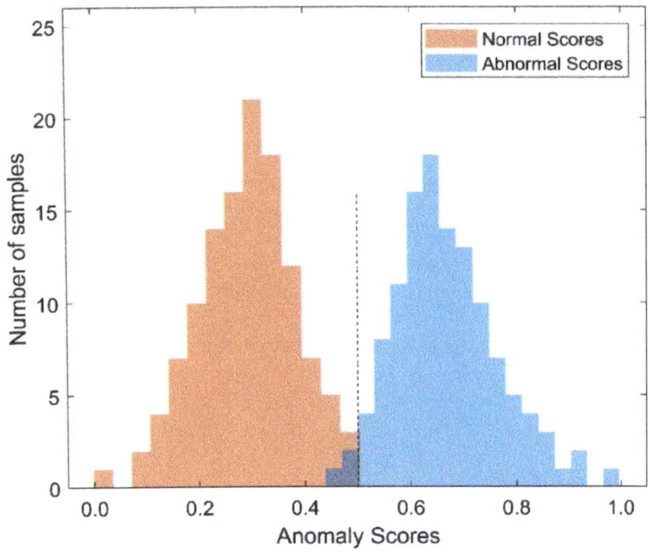


**Figure 11.** Histograms of normal and abnormal scores for samples of current data tested with our method.

**Table 6.** Evaluation results of our method under three input conditions (the threshold is set to 0.44).

| Case | Precision | Recall | $F_1$ |
|---|---|---|---|
| Case 1 | 0.921 | 0.953 | 0.937 |
| Case 2 | 0.965 | 1.000 | 0.982 |
| Case 3 | 0.943 | 0.984 | 0.963 |

The aforementioned experimental results reflect that the proposed method is a valid and feasible anomaly detection method for time-series data, and remarkable results are achieved under the above three input conditions. That is, the GANomaly-LSTM method has excellent effects on the identification of current phase missing and abnormal vibration of the linear motor feeding system.

To further test and verify the performance of the proposed method, we compared the performances of the three methods under three input conditions. The other two methods are GANomaly and GAN-AE. In order to ensure the effectiveness of the method comparison, the parameter settings of these two methods are exactly the same as the proposed method. The reasons for choosing these two methods to compare with the proposed method are: (1) The network structures of these two methods are similar to that of the proposed method, which is based on the extension of the GAN and adopts the structure of encoding-decoding-encoding; (2) These two methods and the proposed method only use normal samples during training, and use normal and abnormal samples during testing; (3) All three detect anomalies by the difference between the latent features after two encodings. Therefore, it is essential to choose these two methods to compare with our method.

During training, Case1 uses 560 normal samples of vibration signals, Case2 uses 560 normal samples of current signals, and Case3 uses the training set of the first two. In the testing process, Case1 uses the vibration signals of 120 normal samples and 120 abnormal samples, Case2 uses the current signals of 120 normal samples and 120 abnormal samples, and Case3 uses the test sets of the former two at the same time. Figure 12 shows the ROC curves of the three methods under three input conditions. It can be seen that the AUROC metric of the proposed method under the three input conditions is significantly higher than the other two methods. It is shown in Table 7 that our method achieves the optimal accuracy in the AUROC indicator under the three input conditions, and achieves good performance in the task of abnormal detection of the running state of the linear motor feeding system, showing tremendous advantages. Specifically, the AUROC indicators of GANomaly are 87.3%, 88.1% and 87.5%, respectively, while GAN-AE performs poorly, at only 72.8%, 83.2% and 82.2%, respectively. That is, the AUROC metrics of the proposed method are 94.6%, 98.5% and 97.7%, which are 7.3%, 10.4% and 10.2% higher than GANomaly. Figure 13 compares the experimental results of the three methods under three input conditions in the form of histograms. It can be distinctly shown that all three methods achieve the best results under the condition of Case2. Under Case2, we further compare the average accuracy and detection time of the three methods, shown in Table 8, the average accuracies of the proposed method, GANomaly, and GAN-AE achieves 98%, 86.4% and 82.5%; the detection times are 134 ms, 357 ms and 418 ms, respectively. Our method has an average accuracy improvement of 11.6% and 15.5%, and a detection time shortened by 223 ms and 284 ms, respectively, compared with the classical two methods.

**Table 7.** AUROC metrics for three methods under three input conditions.

| AUROC | Case1 | Case2 | Case3 |
|---|---|---|---|
| GAN-AE | 0.728 | 0.832 | 0.822 |
| GANomaly | 0.873 | 0.881 | 0.875 |
| Our method | 0.946 | 0.985 | 0.977 |

**Table 8.** Comparison of the average accuracy and detection time of the three methods when using current data samples alone.

| Case | Average Accuracy | Detection Time (ms) |
|---|---|---|
| GAN-AE | 0.825 | 418 |
| GANomaly | 0.864 | 357 |
| Our method | 0.980 | 134 |

The above experimental results demonstrate that the proposed method achieves a more pronounced improvement in detection accuracy, a shorter detection time and more significant results. In general, the proposed method is highly effective for abnormal detection of the running state of linear motor feeding systems, meeting the application requirements in industrial production. This method solves the task of anomaly detection in the absence of abnormal sample training, and has profound significance for improving industrial production efficiency and equipment predictive maintenance.

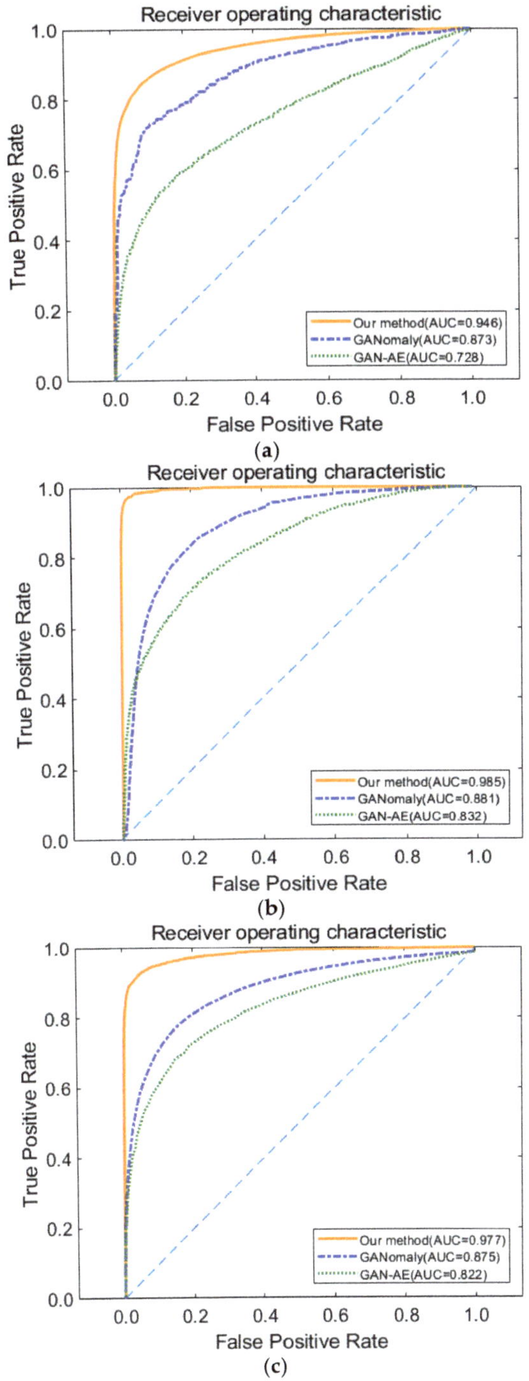


**Figure 12.** ROC curves of three methods under three input conditions: (**a**) Case1; (**b**) Case2; (**c**) Case3.

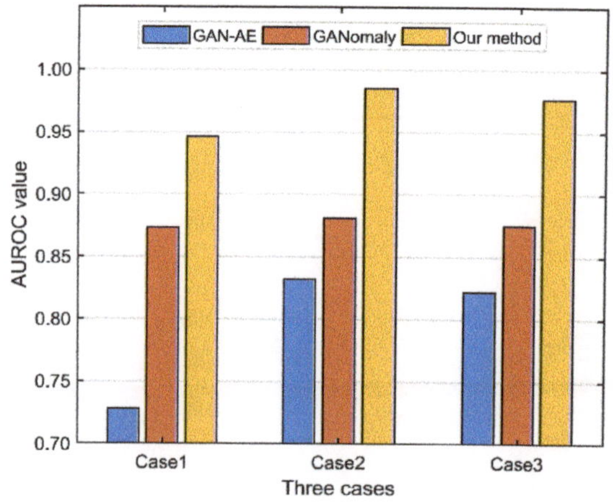


**Figure 13.** Histograms of experimental results for three methods under three input conditions.

## 5. Conclusions

This paper proposes an anomaly detection method based on GANomaly-LSTM for the running status of the linear motor feeding system. This method solves the problem of anomaly detection in the linear motor feeding systems when there is no abnormal sample and the sample timing is significant. This method compares the anomaly detection performance of the model under three input conditions (vibration data sample, current data sample, vibration and current combination data sample). The AUROC indexes are 94.6%, 98.5% and 97.7%, and the F1 scores are 93.7%, 98.2% and 96.3%, respectively. The consequences show that the proposed method has favorable detection performances. In addition, compared with GANomaly and GAN-AE, the proposed method improved the average AUROC indicator by 9.3% and 17.5%, respectively, thus greatly enhancing the anomaly detection accuracy. In sum, the GANomaly-LSTM method has excellent effects on the identification of the current phase missing and abnormal vibration of the linear motor feeding system. Due to the variety of abnormal conditions in the linear motor feeding system and varying types of abnormal samples, the limitation of our method lies in the need for retraining and retesting different types of samples. In addition, the applicability of the method for other anomalies such as current single phase short circuit and voltage anomalies needs to be further investigated. For future works, we expect to investigate more data types such as torque and power along with the current and vibration signals of the linear motor feeding system, and to compare the effects of different sample sizes on detection accuracy to achieve more efficient and comprehensive anomaly detection. In future research, we hope to realize further fault location based on anomaly detection to achieve more accurate predictive maintenance.

**Author Contributions:** Conceptualization, Z.Y.; methodology, W.Z.; software, W.Z.; validation, W.Z., W.C. and Y.C.; resources, Z.Y., L.G., Q.W. and L.L.; data curation, Z.Y. and W.Z.; writing—original draft preparation, W.Z.; writing—review and editing, Z.Y., W.C., L.G., Y.C., Q.W. and L.L.; funding acquisition, Z.Y. and L.L. All authors have read and agreed to the published version of the manuscript.

**Funding:** This work is partially supported by National Natural Science Foundation of China (Grant number 52175461, 11632004 and U1864208); Intelligent Manufacturing Project of Tianjin (Grant number 20201199); Fund for the High-level Talents Funding Project of Hebei Province (Grant number B2021003027); Key Program of Research and Development of Hebei Province (Grant number 202030507040009); Innovative Research Groups of Natural Science Foundation of Hebei Province (Grant number A2020202002); Top Young Talents Project of Hebei Province, China (Grant number 210014); Diversified investment fund projects of Tianjin applied basic research (Grant number 21JCZDJC00710).

**Institutional Review Board Statement:** Not applicable.

**Informed Consent Statement:** Not applicable.

**Data Availability Statement:** Not applicable.

**Conflicts of Interest:** The authors declare no conflict of interest.

## References


1. Altintas, Y.; Verl, A.; Brecher, C.; Uriarte, L.; Pritschow, G. Machine tool feed drives. *CIRP Ann. Manuf. Technol.* **2010**, *60*, 779–796. [CrossRef]
2. Wang, B.; Mao, Z. Outlier detection based on Gaussian process with application to industrial processes. *Appl. Soft. Comput.* **2019**, *76*, 505–516. [CrossRef]
3. Jin, X.; Sun, Y.; Que, Z.; Wang, Y.; Zhou, X. Anomaly detection and fault prognosis for bearings. *IEEE Trans. Instrum. Meas.* **2016**, *65*, 2046–2054. [CrossRef]
4. Liu, J.; Liu, L.; Du, J.; Sang, J. TLE outlier detection based on expectation maximization algorithm. *Adv. Space Res.* **2021**, *68*, 2695–2712. [CrossRef]
5. Su, J.; Dong, Y.; Yan, M.; Qian, J.; Xin, Y. Research progress of anomaly detection for complex networks. *Control Decis.* **2021**, *36*, 1293–1310.
6. Li, Z.; Jin, X.; Zhuang, C.; Sun, Z. A Survey of Graph-Oriented Anomaly Detection Research. *J. Softw.* **2021**, *32*, 167–193. [CrossRef]
7. Zhang, S.; Li, X.; Zong, M.; Zhu, X.; Wang, R. Efficient kNN Classification with Different Number of Nearest Neighbors. *IEEE Trans. Neural Netw. Learn. Syst.* **2018**, *29*, 1774–1785. [CrossRef] [PubMed]
8. Li, J.; Izakian, H.; Pedrycz, W.; Jamal, I. Clustering-based anomaly detection in multivariate time series data. *Appl. Soft. Comput.* **2021**, *100*, 106919. [CrossRef]
9. Li, T.; Comer, M.L.; Delp, E.J.; Desai, S.R.; Mathieson, J.L.; Foster, R.H.; Chan, M.W. Anomaly Scoring for Prediction-Based Anomaly Detection in Time Series. In Proceedings of the 2020 IEEE Aerospace Conference, Big Sky, MT, USA, 7–14 March 2020; pp. 1–7.
10. Zhou, Y.; Ren, H.; Li, Z.; Wu, N.; Al-Ahmari, A.M. Anomaly detection via a combination model in time series data. *Appl. Intell.* **2021**, *51*, 4874–4887. [CrossRef]
11. Abdulghafoor, S.A.; Mohamed, L.A. A local density-based outlier detection method for high dimension data. *Int. J. Nonlinear Anal. Appl.* **2022**, *13*, 1683–1699.
12. Jin, W.; Tung, A.K.H.; Han, J.; Wang, W. Ranking Outliers using Symmetric Neighborhood Relationship. In *Advances in Knowledge Discovery and Data Mining*; Ng, W.K., Kitsuregawa, M., Li, J., Chang, K., Eds.; Lecture Notes in Computer Science; Springer International Publishing: Berlin/Heidelberg, Germany, 2006; Volume 3918, pp. 577–593, ISBN 978-3-540-33206-0.
13. Zhao, R.; Yan, R.; Chen, Z. Deep learning and its applications to machine health monitoring. *Mech. Syst. Signal Proc.* **2019**, *115*, 213–237. [CrossRef]
14. Hoang, D.T.; Kang, H.J. Rolling element bearing fault diagnosis using convolutional neural network and vibration image. *Cogn. Syst. Res.* **2019**, *53*, 42–50. [CrossRef]
15. Shao, H.; Ding, Z.; Cheng, J.; Jiang, H. Intelligent fault diagnosis among different rotating machines using novel stacked transfer auto-encoder optimized by PSO. *ISA Trans.* **2020**, *105*, 308–319.
16. Jiang, G.; Zhao, J.; Jia, C.; He, Q.; Xie, P.; Meng, Z. Intelligent fault diagnosis of gearbox based on vibration and current signals: A multimodal deep learning approach. In Proceedings of the 2019 Prognostics and System Health Management Conference, Qingdao, China, 25–27 October 2019; pp. 1–6.
17. Goodfellow, I.J.; Pouget-Abadie, J.; Mirza, M.; Xu, B.; Warde-Farley, D.; Ozair, S.; Courville, A.; Bengio, Y. Generative adversarial networks. Available online: https://arxiv.org/abs/1406.2661 (accessed on 20 June 2022).
18. Xia, X.; Pan, X.; Li, N.; He, X.; Ma, L.; Zhang, X.; Ding, N. GAN-based Anomaly Detection: A Review. *Neurocomputing* **2022**, *493*, 497–535. [CrossRef]
19. Zenati, H.; Foo, C.S.; Lecouat, B.; Manek, G.; Chandrasekhar, V.R. Efficient Ganbased Anomaly Detection. Available online: https://arxiv.org/abs/1802.06222 (accessed on 20 June 2022).
20. Radford, A.; Metz, L.; Chintala, S. Unsupervised representation learning with deep convolutional generative adversarial networks. Available online: https://arxiv.org/abs/1511.06434 (accessed on 20 June 2022).
21. Schlegl, T.; Seeböck, P.; Waldstein, S.M.; Schmidt-Erfurth, U.; Langs, G. Unsupervised anomaly detection with generative adversarial networks to guide marker discovery. In *Information Processing in Medical Imaging*; Niethammer, M., Styner, M., Aylward, S., Zhu, H., Ogue, L., Yap, P., Shen, D., Eds.; Lecture Notes in Computer Science; Springer International Publishing: Cham, Switzerland, 2017; Volume 10265, pp. 146–157, ISBN 978-3-319-59049-3.
22. Wang, Z.; Wang, J.; Wang, Y. An intelligent diagnosis scheme based on generative adversarial learning deep neural networks and its application to planetary gearbox fault pattern recognition. *Neurocomputing* **2018**, *310*, 213–222. [CrossRef]
23. Mao, W.; Liu, Y.; Ding, L.; Yuan, L. Imbalanced Fault Diagnosis of Rolling Bearing Based on Generative Adversarial Network: A Comparative Study. *IEEE Access* **2019**, *7*, 9515–9530. [CrossRef]

24. Gao, X.; Deng, F.; Yue, X. Data augmentation in fault diagnosis based on the Wasserstein generative adversarial network with gradient penalty. *Neurocomputing* **2020**, *396*, 487–494. [CrossRef]
25. Akçay, S.; Atapour-Abarghouei, A.; Breckon, T.P. GANomaly: Semi-supervised anomaly detection via adversarial training. In *Computer Vision-ACCV 2018*; Jawahar, C., Li, H., Mori, G., Schindler, K., Eds.; Lecture Notes in Computer Science; Springer International Publishing: Cham, Switzerland, 2019; Volume 11363, pp. 622–637, ISBN 978-3-030-20892-9.
26. Akçay, S.; Atapour-Abarghouei, A.; Breckon, T.P. Skip-GANomaly: Skip connected and adversarially trained encoder-decoder anomaly detection. In Proceedings of the 2019 International Joint Conference on Neural Networks (IJCNN), Budapest, Hungary, 14–19 July 2019; pp. 1–8.
27. Luo, Z.; Zuo, R.; Xiong, Y.; Wang, X. Detection of geochemical anomalies related to mineralization using the GANomaly network. *Appl. Geochem.* **2021**, *131*, 105043. [CrossRef]
28. Liu, G.; Niu, Y.; Zhao, W.; Duan, Y.; Shu, J. Data anomaly detection for structural health monitoring using a combination network of GANomaly and CNN. *Smart. Struct. Syst.* **2022**, *29*, 53–62.
29. Hong, Z.; Yang, Z.; Yang, C.; Liao, S.; Sun, Y.; Xing, Y. Triple-GAN with Fixed Memory Step Gradient Descent Method and Xwish Activation Function. In Proceedings of the 2020 39th Chinese Control Conference (CCC), Shenyang, China, 27–29 July 2020; pp. 3188–3193.
30. Duan, S.; Chen, Z.; Wu, Q.M.J.; Cai, L.; Lu, D. Multi-scale gradients self-attention residual learning for face photo-sketch transformation. *IEEE Trans. Inf. Forensic Secur.* **2020**, *16*, 1218–1230. [CrossRef]
31. Lindemann, B.; Maschler, B.; Sahlab, N.; Weyrich, M. A survey on anomaly detection for technical systems using LSTM networks. *Comput. Ind.* **2021**, *131*, 103498. [CrossRef]
32. Li, Z.; Li, J.; Wang, Y.; Wang, K. A deep learning approach for anomaly detection based on SAE and LSTM in mechanical equipment. *Int. J. Adv. Manuf. Technol.* **2019**, *103*, 499–510. [CrossRef]
33. Chen, H.; Liu, H.; Chu, X.; Liu, Q.; Xue, D. Anomaly detection and critical SCADA parameters identification for wind turbines based on LSTM-AE neural network. *Renew. Energy* **2021**, *172*, 829–840. [CrossRef]
34. Vos, K.; Peng, Z.; Jenkins, C.; Shahriar, M.R.; Borghesani, P.; Wang, W. Vibration-based anomaly detection using LSTM/SVM approaches. *Mech. Syst. Signal Proc.* **2022**, *169*, 108752. [CrossRef]
35. Ou, X.; Wen, G.; Huang, X.; Su, Y.; Chen, X.; Lin, H. A deep sequence multi-distribution adversarial model for bearing abnormal condition detection. *Measurement* **2021**, *182*, 109529. [CrossRef]
36. Bai, M.; Liu, J.; Ma, Y.; Zhao, X.; Long, Z.; Yu, D. Long Short-Term Memory Network-Based Normal Pattern Group for Fault Detection of Three-Shaft Marine Gas Turbine. *Energies* **2021**, *14*, 13. [CrossRef]
37. Kong, F.; Li, J.; Jiang, B.; Wang, H.; Song, H. Integrated generative model for industrial anomaly detection via bi-directional LSTM and attention mechanism. *IEEE Trans. Ind. Inform.* **2021**. [CrossRef]
38. Yang, H.; Wang, Z.; Zhang, T.; Du, F. A review on vibration analysis and control of machine tool feed drive systems. *Int. J. Adv. Manuf. Technol.* **2020**, *107*, 1–23. [CrossRef]
39. Yan, K. Chiller fault detection and diagnosis with anomaly detective generative adversarial network. *Build. Environ.* **2021**, *201*, 107982. [CrossRef]
40. Du, J.; Guo, L.; Song, L.; Liang, H.; Chen, T. Anomaly detection of aerospace facilities using GANomaly. In Proceedings of the 2020 5th International Conference on Multimedia Systems and Signal Processing, Wuhan, China, 8–10 May 2020; pp. 40–44.
41. Wang, F.; Liu, X.; Deng, G.; Yu, X.; Li, H.; Han, Q. Remaining life prediction method for rolling bearing based on the long short-term memory network. *Neural Process. Lett.* **2019**, *50*, 2437–2454. [CrossRef]
42. Huang, R.; Wei, C.; Wang, B.; Yang, J.; Xu, X.; Wu, S.; Huang, S. Well performance prediction based on Long Short-Term Memory (LSTM) neural network. *J. Pet. Sci. Eng.* **2022**, *208*, 109686. [CrossRef]
43. Ruder, S. An Overview of Gradient Descent Optimization Algorithms. Available online: https://arxiv.org/abs/1609.04747 (accessed on 30 July 2022).
44. Kingma, D.P.; Ba, J. Adam: A method for Stochastic Optimization. Available online: https://arxiv.org/abs/1412.6980 (accessed on 30 July 2022).
45. Chen, Z.; Gryllias, K.; Li, W. Intelligent Fault Diagnosis for Rotary Machinery Using Transferable Convolutional Neural Network. *IEEE Trans. Ind. Inform.* **2020**, *16*, 339–349. [CrossRef]
46. Ravikumar, K.N.; Yadav, A.; Kumar, H.; Gangadharan, K.V.; Narasimhadhan, A.V. Gearbox fault diagnosis based on Multi-Scale deep residual learning and stacked LSTM model. *Measurement* **2021**, *186*, 110099. [CrossRef]

*energies*

MDPI

*Article*

# In-Situ Efficiency Estimation of Induction Motors Based on Quantum Particle Swarm Optimization-Trust Region Algorithm (QPSO-TRA)

**Mahamadou Negue Diarra, Yifan Yao, Zhaoxuan Li, Mouhamed Niasse, Yonggang Li and Haisen Zhao ***

School of Electrical and Electronic Engineering, North China Electric Power University, Beijing 102206, China; diarraneguess82@yahoo.fr (M.N.D.); yaoyifan2000@ncepu.edu.cn (Y.Y.); 120202201393@ncepu.edu.cn (Z.L.); mhniasse@ncepu.edu.cn (M.N.); 51350586@ncepu.edu.cn (Y.L.)

* Correspondence: zhaohisen@ncepu.edu.cn

**Abstract:** The accuracy estimation of induction motors' efficiency is beneficial and crucial in the industry for energy savings. The requirement for in situ machine efficiency estimation techniques is increasing in importance because it is the precondition to making the energy-saving scheme. Currently, the torque and speed identification method is widely applied in online efficiency estimation for motor systems. However, the higher precision parameters, such as stator resistance $R_s$ and equivalent resistance of iron losses $R_{fe}$, which are the key to the efficiency estimation process with the air gap torque method, are of cardinal importance in the estimation process. Moreover, the computation burden is also a severe problem for the real-time data process. To solve these problems, as for the torque and speed-identification-based efficiency estimation method, this paper presents a lower time burden method based on Quantum Particle Swarm Optimization-Trust Region Algorithm (QPSO-TRA). The contribution of the proposed method is to transform the disadvantages of former algorithms to develop a reliable hybrid algorithm to identify the crucial parameters, namely, $R_s$ and $R_{fe}$. Sensorless speed identification based on the rotor slot harmonic frequency (RSHF) method is adopted for speed determination. This hybrid algorithm reduces the computation burden by about 1/3 compared to the classical genetic algorithm (GA). The proposed method was validated by testing a 5.5 kW motor in the laboratory and a 10 MW induction motor in the field.

**Keywords:** in situ efficiency; induction motors; quantum particle swarm optimization; trust region algorithm; QPSO-TRA; rotor slot harmonics frequencies

**Citation:** Diarra, M.N.; Yao, Y.; Li, Z.; Niasse, M.; Li, Y.; Zhao, H. In-Situ Efficiency Estimation of Induction Motors Based on Quantum Particle Swarm Optimization-Trust Region Algorithm (QPSO-TRA). *Energies* **2022**, *15*, 4905. https://doi.org/10.3390/en15134905

Academic Editor: Armando Pires

Received: 27 May 2022
Accepted: 28 June 2022
Published: 5 July 2022



## 1. Introduction

The increasing demand for energy combined with rising energy costs has led to a desire to increase energy utilization efficiency. Induction motors (IMs) are the most commonly used motors in the industry. They are essential components in the chains of drive systems. The efficiency estimation strategies of all induction motors require good knowledge of the machine parameters to ensure an accurate estimation. The problem of estimating the parameters of induction motors and their efficiency is considered the most critical and complex matter in industries, in in situ situations. However, the higher accuracy of IM parameters is of capital importance in all industrial processes since it directly affects the performance of the control systems. One of the main reasons for monitoring a motor's efficiency is to ensure that energy conversion occurs with the least energy supply. Another reason is that the nominal parameters obtained from manufacturers might not be the same as those under actual motor operating conditions. This is due to several factors such as aging, rewinding, and voltage unbalance. Hence, it is necessary to estimate these parameters in in situ for any in situ induction motors efficiency estimation. As for the speed, it can be affected by bearing faults. Estimating the speed in in situ has several advantages. It can be used to check bearing faults and motor efficiency estimation online.

Therefore, estimating induction motors' parameters and speed in in situ can be helpful for all industrialists because it can give information about the motors' healthy state and avoid extra cost for speed measuring devices. While effort is being made to design and manufacture more efficient induction motors (IMs), monitoring a machine's operational efficiency helps reduce energy costs by ensuring that it is running in its optimal efficiency range. Numerous works have been published for estimating the efficiency of the electrical machines in situ, under the loaded condition without disturbing their operation [1,2]. To assess energy efficiency and enhance the overall performance of the industrial processes, it is essential to identify energy loss and monitor the efficiency in real time [3].

Generally, torque and speed measurements are made to estimate the induction motor's efficiency. However, if an efficiency estimation of an induction motor is required for an in situ situation, whose operation cannot be disrupted because of the ongoing critical industrial process, torque determination becomes problematic.

The procedure of determining the parameters of an induction motor is well-known for estimating efficiency. The parameters of the induction motor's per-phase equivalent model can be determined by the no-load and locked rotor test or the IEEE Standard 112 impedance test method. No-load and locked rotor tests are not practicable in in situ [4,5]. In in situ situations, the Genetic Algorithm (GA) is mainly used for parameter identification, demanding more central processing unit (CPU) and computational time [6]. To sum up, the methods for efficiency estimation can be categorized as follows.

(1) Slip methods;
(2) Current methods;
(3) Segregated loss method;
(4) Equivalent circuit-based method;
(5) Nonintrusive air gap torque (NAGT) methods;
(6) Air gap torque (AGT) methods;
(7) Shaft torque method.

Among the above methods, each of the following methods for efficiency estimation has its advantages and disadvantages according to the application domain. Some of them cannot be employed for in situ efficiency estimation. However, the current and slip methods are straightforward and nonintrusive, but they are lacking in high accuracy [7], which is unsuitable for any efficiency assessment. Despite having very high precision, the segregated loss method [8] is intrusive for use in in situ conditions. To limit the intrusiveness, empirical results are used, which could diminish the accuracy of those procedures [9]. The equivalent circuit method requires no-load and locked rotor tests or variable voltages and frequency supplies, making them unsuitable for in situ applications. The air-gap torque (AGT) and the shaft torque method consider harmonic distortion because it uses the operating voltage and current signals to calculate this torque [10], which makes it appreciable for in situ efficiency estimation.

The most challenging task in efficiency estimation is the determination of the output power. The online efficiency estimation method can be classified into the equivalent circuit and the torque and speed methods from the above analysis. The first method must determine all the losses and subtract them from the input power. The second one must determine the shaft torque through the air gap torque, and the speed is estimated as described in Section 2.1 to determine the output power. For the second method, the core losses resistance $R_{fe}$ and the stator resistance $R_s$ must first be identified online for air gap torque calculation, these are dependent on the operating condition. Furthermore, the problem in the current identification method of these parameters is that the time and the computation burden of the algorithm used are high enough, or the algorithm used for parameters identification is not accurate enough. The AGT and the shaft torque method are adopted in this paper to calculate the output power.

The contribution of this paper is firstly to propose a new algorithm to identify the motor parameters simultaneously. The computation time and burden can be reduced significantly since the global search is avoided. In the presented method, a hybrid algorithm derived from PSO and TRA is used, the Quantum Particle Swarm Optimization-Trust Region Algorithm QPSO-TRA, to improve the deficiencies of PSO and TRA. The time burden of the proposed hybrid QPSO-TRA is slightly larger than that of the standard PSO, but the identification is more accurate than the standard PSO.

## 2. Efficiency Estimation Method

This study adopts an efficiency estimation method based on torque and speed identification. The output torque and rotor speed should first be determined by air gap torque and rotor slot harmonic methods [11]. Then, the output power $P_2$ can be obtained. The input power $P_1$ is calculated by measured current and voltage. Finally, efficiency can be calculated by $P_2/P_1$. To obtain $P_2$, several processes are involved, which have to be accurately investigated to avoid calculation mistakes.

### 2.1. Torque and Speed Determination Methods

During the operation of the induction motors, despite many internal excitations that can rise to system complexity, the external environment, such as unsymmetrical voltage sags or external load variation, can also be unstable, etc. The electromechanical behavior of an induction motors change according to the actual load changes from the start-up no-load to the full-load condition. In addition, considering that the electromechanical power is the transmitted power from the stator to the rotor through the air gap, it includes the actual load effect since it depends on the actual current and voltage and the core loss power $P_{fe}$ as can be seen in (2) because the motors' running current is affected by the load changes. The AGT method was proposed in [12] for in situ efficiency estimation. It is calculated by using the voltage, current, and stator resistance $R_s$ and the core losses $P_{fe}$ data. However, the authors in [12] did not consider the change of the resistance under the operating temperature, which make it unsuitable for real-time calculation. Because of variable current in the winding under different load conditions, the temperature changes affect the motor's winding resistance. The temperature of a machine at the actual loading condition is required to adjust or correct the stator and rotor windings resistances. An initial temperature is determined from the input current at each measured load to overcome this problem, as shown in Equation (1), or directly measured with a temperature sensor. For the best calculation of the air gap torque, the resistance is corrected according to real-time temperature in this study, whereas there is the presence of $R_{s,cor}$ in Equation (2) instead of $R_s$. The AGT is calculated as in Equation (2).

$$T_{load,Est} = \frac{I_{load}}{I_{rated}}(T_{rated} - T_{ambien}) + T_{ambien} \tag{1}$$

where $I_{rated}$ is the rated load current stated on the machine's nameplate, $I_{load}$ is the actual load current, $T_{load,Est}$ is the actual temperature, $T_{ambien}$ is the ambient temperature, and $T_{rated}$ is the rated load temperature

$$T_{ag} = \frac{p*\sqrt{3}}{6}\{(2*i_a + i_c) * \int [v_{ca} - R_{s,cor}(i_c - i_a)dt] - (i_b - i_a) * \int [v_{ab} - R_{s,cor}(2*i_a + i_b)dt]\} - \frac{60P_{fe}}{2\pi n_1} \tag{2}$$

where $p$ is the number of poles; $i_a$ and $i_c$ are line currents; $v_{ab}$ and $v_{ca}$ are line voltages; $R_{s,cor}$ is the stator resistance at the operating temperature, and $P_{fe}$ is the core losses. $P_{fe}$ is calculated using the identified core resistance $R_{fe}$.

Speed determination is crucial in motors' efficiency estimation. This technique is designed to work without interrupting the industrial process. Measuring the speed of a built IM is a relatively easy task. Still, to measure it, one needs an extra device (speed sensor) which can be expensive and can increase the manufacturing costs for the manufacturer. Therefore, we need to estimate the speed to avoid these additional costs. One phase current is analyzed to extract the rotor harmonics high frequency and then estimate the IM speed. In this method, one does not need to calculate the losses between the input and output power separately to have efficiency. The output power is directly calculated by using the torque and speed as in (15) and (16). To do so, we need to use $R_{s,cor}$ and $R_{fe}$ in the air gap torque calculation process. This method can avoid extra production costs and considers the motor's temperature change. The different techniques for rotor slot harmonic frequency (RSHF) are summarized below.

An enhanced approach to speed determination utilizing the motor's current harmonics, which arise for stator core form, rotor shaft misalignment, bearing position, and rotor bar resistance variation, was conducted in [13]. The current harmonics can be expressed by Equation (3):

$$f_{sh} = f_1\left[(\lambda Z + n_d)\frac{(1-s)}{p} + \delta\right] \tag{3}$$

where $f_{sh}$ is the current harmonic frequency; $f_1$ is the supply frequency; $\lambda = 0, 1, 2, 3, \ldots$, $Z$ is the number of rotor slots; $n_d = 0 \pm 1, \ldots$ is the order of rotor eccentricity; $s$ is the slip ratio; $p$ is the number of pole pairs; $\delta = \pm 1, \pm 3, \pm 5$, is the air gap magnetomotive force (MMF) harmonic order.

The fast Fourier transform (FFT) technique successfully improved the speed detection of inverter-fed induction motors via the stator current signal [14]. The generation of rotor slot harmonics (RSHs) in machines is primarily due to the irregular allocation of rotor currents in a finite number of slots as well as the permeance fluctuation in the air gap related to slot opening [15]. In [16], harmonics were used to detect the rotor slot number. It is possible to study the speed and the slip estimation in induction motors using rotor slot harmonics. Thus, the slip can be expressed by Equation (4):

$$s = 1 - \frac{p}{Z}\left(\frac{f_{sh}}{f_1} - \delta\right) \tag{4}$$

From Equation (4), it is possible to determine the rotational speed on the shaft of the induction motor. For that, the expression of the slip is required to be $n = n_s(1 - s)$, where $n_s$ is the synchronous speed and $s$ is the slip; $n_s$ is given by $n_s = 60 f_1/p$. The rotational speed is thus given by Equation (5):

$$n = \frac{60}{Z}(f_{sh} - \delta f_1) \tag{5}$$

where $f_1$ is supply frequency, and the rotor slot harmonics frequency $f_{sh}$ is determined by applying digital signal processing techniques.

### 2.2. Core Losses Determination

Core losses are caused by time-varying magnetic flux. The core losses can be separated into two components, the hysteresis and eddy current effects, as in Equation (6).

$$P_{fe} = P_h + P_e = K_h \omega_e \psi^2 + K_e \omega_e^2 \psi^2 \tag{6}$$

where $P_h$ and $P_e$ are the hysteresis and eddy current components of core losses, respectively, $\Psi$ is the stator winding flux linkage, $\omega_e$ is the angular frequency, and $K_h$ and $K_e$ are the coefficients of hysteresis and eddy current loss, respectively.

According to Equations (7) and (8), core losses depend only on the angular frequency and the flux linkage if the variation of the coefficients $K_h$ and $K_e$ are neglected. In fact, $K_h$ and $K_e$ rely on the core material's lamination and its thickness. Their variation is slow compared to the angular frequency and flux linkage variation. However, knowing that the core losses are linked to the winding flux linkage and based on the classical electric machine theory, which is assumed to be a sinusoidally varying field, it is evident that the winding flux linkage is proportional to the air gap flux density.

$$P_{fe} = \psi^2(K_h \omega_e + K_e \omega_e^2) \tag{7}$$

$$P_{fe} = \omega_e^2 \psi^2 / R_{fe} \tag{8}$$

The standard way to determine the core losses is the no-load test. This method is not practicable in the field due to the ongoing production process. After analyzing the above phenomena, this paper proposes a new algorithm to determine the core losses resistance. The core resistance $R_{fe}$ is identified by the new algorithm QPSO-TRA and then uses the voltage to determine the core losses as shown in Equation (9). After $R_{fe}$ identification, this method is robust and straightforward because it considers the voltage fluctuation.

$$P_{core} = 3\frac{V_{ph}^2}{R_{fe}} \tag{9}$$

### 2.3. *Stray Load Loss Determination*

According to the IEEE Standard 112-2004 [8], stray load loss at rated load can be assumed, as shown in Table 1.

**Table 1.** Assumed values for stray load loss [8].

| Machines Rating (kW) | Stray Load Loss Percent |
|---|---|
| 1–90 | 1.8% |
| 91–375 | 1.5% |
| 376–1850 | 1.2% |
| 1851 and greater | 0.9% |

Moreover, it can also be calculated using the International Standard IEC 60034-2-1 [7], as in Equation (10).

$$P_{sll} = P_{in,fl}\left[0.025 - 0.005\log_{10}\left(\frac{P_{out}}{1\ \mathrm{kW}}\right)\right] \tag{10}$$

where $P_{sll}$ is the full-load stray load loss, $P_{in,fl}$ is the full-load input power, and $P_{out}$ is the full-load output power.

In [17], a new stray load loss formula has been proposed. It also uses the input power, but it does not use the output power, unlike in Formula (10). The increase in input power due to the unbalance and voltage distortion can be considered. The proposed procedure is, as in Equation (11):

$$P_{sll} = 0.011 * P_{in,fl} \tag{11}$$

where $P_{sll}$ is the stray load loss, and $P_{in,fl}$ is the input power under full-load conditions.

Figure 1 compares three stray load loss determination methods, namely, IEEE method, IEC method, and the improved method in [14]. It is imperative to emphasize that all of the foregoing similarities are only valid if, and only if, the motor under test and the data motor have an identical power rating value. It is found that compared with the tested stray losses, the method in [14] is more accurate. Therefore, this method is adopted to estimate the stray losses in this study.

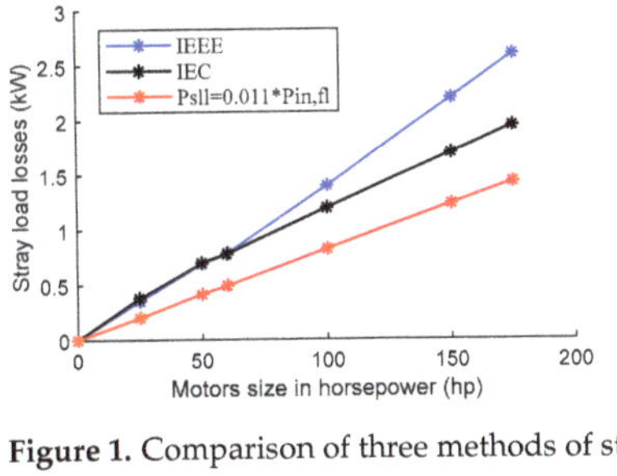


**Figure 1.** Comparison of three methods of stray load loss determination.

### 2.4. Friction and Winding ($P_{FW}$) Losses Determination

An estimation strategy for friction and windage losses has been developed to eliminate this intrusive requirement. This new strategy is based on full-load input power and depends on the motor pole pair number [18], as follows:

(a) If the number of poles is 2, then the friction and windage losses will be calculated using the Equation (12):

$$P_{FW} = 2.5\% * P_{in,fl} \tag{12}$$

(b) If the number of poles is 4, the friction and windage losses will be calculated by Equation (13).

$$P_{FW} = 1.2\% * P_{in,fl} \tag{13}$$

(c) If the number of poles is 6, the friction and windage losses can be calculated by Equation (14).

$$P_{FW} = 1.0\% * P_{in,fl} \tag{14}$$

### 2.5. Output Torque and Efficiency Determination

From the identified electromagnetic torque, the shaft output torque can be obtained by subtracting stray load loss and friction windage loss as in Equation (15) [19]:

$$T_{shaft} = T_{ag} - \frac{P_{sll} + P_{FW}}{\omega_r} \tag{15}$$

where $T_{shaft}$ is the shaft torque, $T_{ag}$ is the air gap torque, and $\omega_r$ is the rotor speed.

Then, output and input power can be calculated by Equations (16) and (17).

$$P_{output} = T_{shaft} * \omega_r \tag{16}$$

$$P_{in} = \sqrt{3} * U * I * \cos\varphi \tag{17}$$

where $U$ is the phase-to-phase voltage, $I$ is the stator current, and $\cos\varphi$ is the power factor.

The motor's efficiency is calculated as in Equation (18).

$$\eta_{est} = \frac{P_{\text{out}}}{P_{\text{in}}} = \frac{T_{\text{shaft}} * \omega_r}{P_{\text{in}}} \tag{18}$$

### 2.6. Problems in Current Efficiency Determination

The main problem with the above method is accurately identifying the parameters $R_s$ and $R_{fe}$, which are used in the torque estimation process. In most recent papers, the core loss resistance $R_{fe}$ is ignored, affecting the online efficiency because of the loss caused by this resistance since the AGT will not be accurately determined for the output torque calculation. Several versions of the GA and optimization-based algorithms were used to help to obtain the parameters. The problems of parameters identification technique accuracy are summarized as follows:

(1) $R_s$ and $R_{fe}$ identification with higher precision is vital in all efficiency estimation processes. However, both of them are variants; the first one is a function of the operating condition and the second one is a function of the voltage condition. Therefore, identifying them in real time is essential for efficiency estimation.

(2) A Genetic Algorithm (GA)'s computation time and burden are high. In [4], it is stated that three GAs were constructed to extract the necessary parameters; some versions of GA use binary coding [14], which demands significant time and computational space. The standard PSO is a global search algorithm. Despite having a fast and

straightforward convergence, one of the drawbacks of this algorithm is its global research abilities. If the best particle is a local one in a PSO system, it cannot be accurately identified. In fact, in PSO, the particles will congregate near the global particles if the number of iterations increases.

A new algorithm is proposed to solve these problems. This algorithm does not need to be used three times or coded in binary like some versions of GA, which significantly reduces the computation time and burden, and it is not a global search algorithm. The conception of any optimization algorithm is not a simple task. In this study, the proposed algorithm is more complicated than the standard PSO algorithm in conception and implementation. Still, from the point of view of accuracy, it is better than the standard PSO and some other PSO-based algorithms since it makes full use of the research space. The details of this new algorithm are given in Section 3.

## 3. Quantum Particles Swarm Optimization-Trust Region Algorithm (QPSO-TRA)

QPSO-TRA is inspired by the PSO algorithm, other PSO-based algorithms, and the TRA algorithm. QPSO-TRA has been inspired by the advantages and disadvantages of some standard former algorithms. It studies the weaknesses of these algorithms and transforms them into benefits for a better investigation of the problems to be solved.

### *3.1. Standard Particle Swarm Optimization PSO*

The mechanism of standard PSO is motivated by the complex social movement shown by biological species such as flocks of birds, schools of fish, swarms of bees, and sometimes social behaviors of the human being. PSO is a relatively simple algorithm and converges fast [20].

In a PSO system, each particle represents a potential solution to the problem and it updates its location by following two optima. One is its personal best location named $P_{\text{best}}$, the best position found by it so far. The other is the global best position $P_{\text{gbest}}$, namely, the best position located by its neighborhood particles so far. The movement of every particle is regulated by the efficacy of its previous location and that of their neighbors. The velocity and the location are formulated as in Equations (19) and (20):

$$v_i(t+1) = \omega * v_i(t) + c_1 * r_1 * (P_{\text{best}_i}(t) - x_i(t)) + c_2 * r_2 * (P_{P_{\text{gbest}_i}}(t) - x_i(t)) \quad (19)$$

$$X_i(t+1) = x_i(t) + v_i(t+1) \quad (20)$$

The inertia weight can be updated by Equation (21):

$$\omega(t) = \omega_{\max} - \frac{t(\omega_{\max} - \omega_{\min})}{iter_{\max}} \quad (21)$$

The velocity value is chosen to be $0.1 \leq k \leq 0.5$ and is set by the user [21]. However, the problem with the PSO algorithm is that it is a global research algorithm. It quickly and easily falls into a local optimum.

### *3.2. Quantum Particles Swarm Optimization (QPSO)*

Inspired by trajectory analysis of the PSO and quantum mechanics, the Quantum Particle Swarm Optimization (QPSO) is an algorithm for problem optimization. The QPSO can find the optimal solution in search space and has the advantage of fewer control parameters, simplicity in software programming, and a relatively fast convergence rate. QPSO is better than standard PSO because, in the standard PSO algorithm, the particles are restricted to a small space for every iteration of the swarm. This restriction will weaken the global search abilities, affect the optimization accuracy, and lead to premature convergence [22]. To improve the search abilities and optimization efficiency and to avoid premature convergence, a quantum is introduced into the PSO. This quantum performs the mutations of particles to increase the particles' diversity, and their movement is governed by quantum displacement.

The Gaussian sample is based on the personal best position *Pi* and the global personal best position *Pg*. The position of each particle is updated by Equation (22). However, if the global best particle is a local optimum of the problem to be optimized, it is more challenging to break out from this local optimum. Thus, to have a complete and satisfactory use of research space, QPSO must join a local optimal research algorithm such as the "Trust Region Algorithm (TRA)" to enhance its performance.

$$x_{ij} = (t+1) = G\left(\mu_{ij}(t), \sigma_{ij}^2(t)\right) \tag{22}$$

where $x_{ij}(t) = \left(p_{ij}(t) + p_{gj}(t)\right)/2, \sigma_{ij}^2(t) = \left|p_{ij}(t) - p_{gj}(t)\right|$ are, respectively, the mean and standard deviation of the Gaussian distribution.

### 3.3. *Trust Region Algorithm (TRA)*

TRA is a well-known iterative approach for solving unbounded optimization problems. It uses the method developed by Newton and offers a high degree of convergence and stability [23]. For a given problem, TRA divides the research space into subspaces for a better investigation. At the start, a small region is initialized at each iteration as the trust region centering on the current iteration point. This operation is continuously conducted until all the research space is investigated.

However, multimodal and multivariate problems which necessitate optimization are sometimes trapped in local optima, making it challenging to obtain the optimal global solution. This trap is due to the fact that the following searching direction of TRA is only based on the capacity of the locally developed objective function. Therefore, to overcome this problem, the TRA needs to be associated with a global optimization algorithm.

### 3.4. *Hybrid Algorithm Formed by QPSO and TRA (QPSO-TRA)*

Because QPSO is a global optimization algorithm, if the global best particle is a local optimum of the problem to be optimized, it is challenging for QPSO to break out from this local optimum. As for the TRA, it frequently becomes stuck in local optima. A hybrid algorithm is formed by combining QPSO and TRA to avoid the early convergence at the last step of the QPSO algorithm and improve the convergence speed. The advantage of TRA and its research abilities is applied to QPSO. This hybrid algorithm helps to increase or enhance the particle diversity and the global search abilities of the QPSO algorithm. The combined, joined, or hybrid algorithm is then called QPSO-TRA. QPSO-TRA achieves quick and efficient convergence performance at a low computational cost compared to GA. Its computation time is slightly higher than PSO, but it is more accurate than PSO. In the hybrid algorithm, each personal best position is updated according to Equation (23):

$$\left\{ \begin{array}{c} p_q(t+1) = x_q(t+1), \\ p_i(t+1) = \left\{ \begin{array}{l} p_i(t), fit(p_i(t)) \le fit(x_i(t+1)) \\ x_i(t+1), fit(p_i(t)) > fit(x_i(t+1)) \end{array} \right. \quad (i \ne q) \end{array} \right. \tag{23}$$

Let Equation (23) be a function (*F*) and $\{p_g(t)\}_{t=0}^{\infty}$ a series or succession generated by the algorithm. Then $\lim_{n\to\infty} P(p_g(t)) \in R_\varepsilon = 1$, where $R_\varepsilon$ is the optimal region, $P(p_g(t)) \in R_\varepsilon$ is the probability at stage α, and the point $p_g(t)$ generated by the algorithm is in $R_\varepsilon$.

It should be considered that $\{p_g(t)\}$ is a sequence generated by the QPSO-TRA algorithm, where $p_g(t)$ is the current best position of the swarm on time $(t)$. As the function *F* is defined as in Equation (24), from the limit, it can be seen that the algorithm converges well.

$$\left\{ F(p_g(t), x_i(t)) = \left\{ \begin{array}{l} p_g(t), fit(p_g(t)) \le fit(x_i(t)) \\ x_i(t), fit(p_g(t)) > fit(x_i(t)) \end{array} \right. \right. \tag{24}$$

QPSO-TRA satisfies the convergence theories from the above statements, so it is an optimization algorithm. Moreover, from the theorem, QPSO-TRA is an optimum search algorithm that benefits from different algorithms.

The different steps of implementing the algorithm are detailed in Figure 2 and explained in the next section. As constraints, the values of each particle's lower and upper bounds are set. The lower bound (lb) and upper bound (ub) are set for the motor's parameters. One can set as lb = [0.00 0.00 0.00 0.00 0.00 0.00], ub = [2.4.00 1.00 1.20 2853 155.00 5.50] for the six parameters of the motors, namely, stator resistance, stator reactance, rotor reactance, core resistance, magnetizing reactance, and rotor resistance, respectively. The lower and upper bounds depend on the size of the motors. Once the parameters are obtained, they are used in the efficiency estimation process.

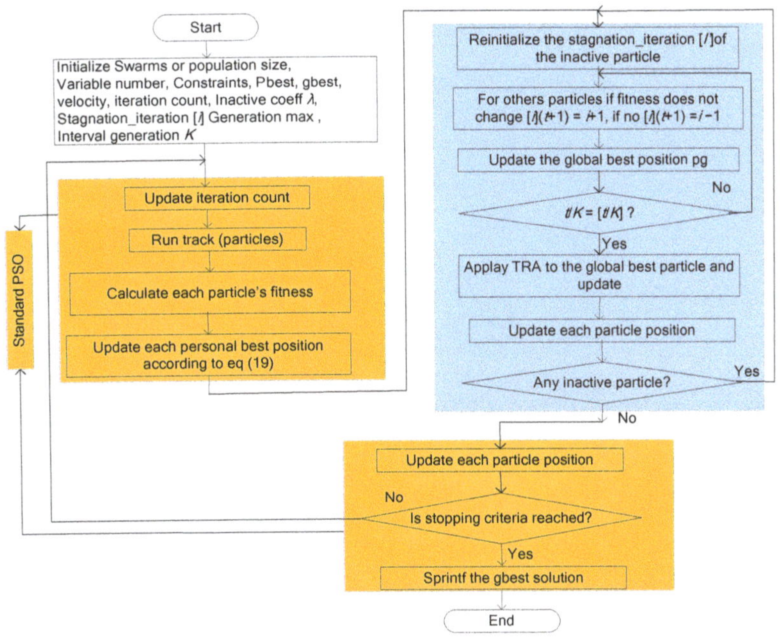


**Figure 2.** Flowchart of QPSO-TRA.

## 4. Experimental Validation

### *4.1. Laboratory Test*

A Matlab program was made for the implementation to identify the parameters of a 5.5 kW, 380 V, 11.7 A, 1450 r/min, and design B induction motor using the proposed algorithm in the laboratory. Figure 3 depicts the test rig of 5.5 kW motor. The test was performed with a power analyzer, a torque transducer and visualizer, a data acquisition system, a DC generator as load, and an auto-transformer used for different voltage conditions. The motor was run at full-load condition to obtain its stable temperature, and the hot temperature was measured. The equivalent circuit parameters method was associated with the algorithm to estimate the motor parameters. The optimization of parameters identification of the induction motor was mathematically expressed as follows: find $X = (x_1, x_2, \ldots, x_n)$ such that $F(X)$ is a minimum. The following are the identification variables for the optimization $(X)$: stator leakage reactance $X_1$; core resistance $R_{fe}$; core magnetization reactance $X_m$, and rotor resistance $R_2$. The rotor leakage reactance $X_2$ is calculated with the value of stator leakage reactance $X_1$, as shown in Table 2. Four parameters out of six will be determined using the QPSO-TRA algorithm, as cited above.

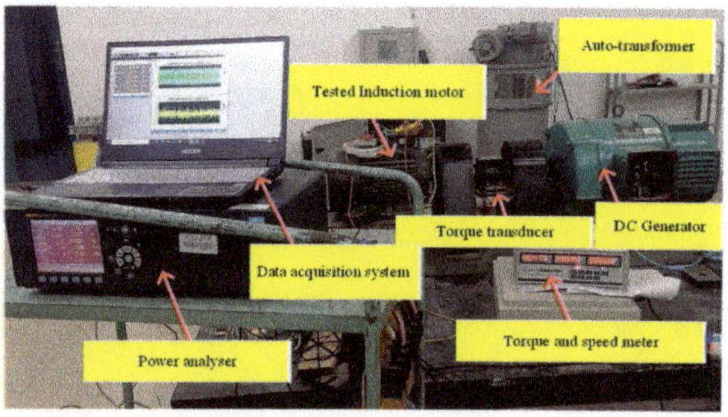


**Figure 3.** Test rig of 5.5 kW motor.

**Table 2.** Ratio of $X_1/X_2$ according to the design class [8].

| Design Class | $X_1/X_2$ |
|---|---|
| A | 1.00 |
| B | 0.67 |
| C | 0.43 |
| D | 1.00 |
| Wound rotor | 1.00 |

Those four parameters represent the four variable particles in the QPSO-TRA algorithm. The algorithm needs one initial value, the value of stator resistance $R_s$. This value is measured at standstill and is corrected with temperature rise during the identification process. In addition, the values of $R_{s,cor}$, and $X_2$ were monitored during all the iterations to respect the algorithm's lower and upper bounds constraints. A stagnation_iteration coefficient [*i*] is used to monitor each particle's fitness value to record the particle's location change. If the fitness value of a particle $X_i$ does not improve during an iteration, the stagnation_iteration [*i*] is extended by one, otherwise it is dropped by one. To consider a particle to be dormant or inactive, its stagnation_iteration [*i*] must respond to $[i] \geq \lambda$ and then has to be reinitialized to give it a chance to be a new candidate for the research. Reinitialization will assist the dormant particle in escaping from the local optimum and enhancing its search skills. The dormant coefficient is chosen by the user based on the research space, and also its value is determined by the problem's severity. The dormant coefficient λ was set to 0.95 in our case.

In the iteration process of the QPSO, the TRA technique is performed on the current global best particle at each iteration *k*. Considering the computing severity of TRA, this combined approach harnesses the experience and knowledge of each particle swarm during the iteration. In the implementation process, a beginning point $\alpha_0$ and trust region radius $r_0$ are first specified. Then, using the iterative procedures, a series of points $\{\alpha_a\}$ is constructed to search for the ideal solution. At each iteration, a tiny region is created called the trust region, centered on the current iteration point. By dealing with a subproblem inside this region, a trial step k is determined. Next, an evaluation function is utilized to evaluate if the trial step should be accepted, and the trust region radius for the following iteration is determined. If the testing step is accepted, $\alpha_{(a+1)} = \alpha_a - \beta_a$; otherwise, $\alpha_{(a+1)} = \alpha_a$. If the trial stage is successful, the new radius will be enlarged or maintained. The trust region exclusively uses the problem's information and region knowledge. The ideal combination is depicted in Equation (25). Because the original QPSO-TRA method is an optimization technique for obtaining optimum values, it leads the particle to deviate from the subregional minimum value and returns the optimal values when the iteration ends, where the fitness represents the objective function, $\{X_i\}$ is the function of parameters to be identified, $\beta = [K,\alpha]$, $K$ is an integer in the interval [1,6], and $\alpha = 50$ is the population size for the optimization:

$$\begin{cases} fitness = \min_{\beta=[K,\alpha]}\{X_i\} \\ K = [1,6] \\ \alpha = 250 \end{cases} \tag{25}$$

As constraints, the values of each particle's lower and upper bounds are set. The lower bound (lb) and upper bound (ub) are set for the motor's parameters. In our case, the lower and upper bounds were set to lb = [0.00 0.00 0.00 0.00 0.00 0.00] and ub = [2.4.00 1.00 1.20 2853 155.00 5.50] for the six parameters of the motors, namely, stator resistance, stator reactance, rotor reactance, core resistance, magnetizing reactance, and rotor resistance, respectively. These lower and upper bounds depend on the size of the motors. Then, the algorithm runs until the stopping criterion is reached. After their identification by QPSO-TRA, $R_{s,cor}$ is used for air gap torque determination, and $R_{fe}$ is used for core losses determination in the proposed efficiency estimation algorithm. The remaining parameters can be used to judge the motor winding healthy state.

The current signal is collected using a data acquisition system for the motor speed detection technique, as shown in Figure 4a. FFT was used to obtain the current frequency spectrum. All the frequency components in the spectrum cannot be directly used. A range of harmonic frequencies must be defined as a criterion for harmonics frequency determination. The number of the spectrum component is reduced by determining lower and upper threshold values, a range of probable frequencies for harmonics investigation. The high harmonics frequencies determine this range in no-load $f_{sh1}$ and $f_{sh2}$ in the full-load conditions. For example, suppose $n_1$, $n_2$, and $z$ are, respectively, the synchronous speed (no-load speed), the rated speed, and the rotor slot number. In that case, the harmonic frequency is determined using the formulas $f_{sh1} = \frac{n_1 z}{60} \pm f_0$ and $f_{sh2} = \frac{n_2 z}{60} \pm f_0$, where $f_0$ is the supply frequency. The frequency components that are not in the range are eliminated to reduce or limit the research space. Then, those frequencies that are in the range are investigated to calculate the probable rotor slot harmonic frequency (RSHF). In our case, the determination range was 680–800 Hz. The frequency that has high amplitude is the identified slot harmonic frequency. Once the RSHF is obtained, Equation (5) is used to determine the motor speed. Figure 4b shows that the identified slot harmonic frequency is 734 Hz, and the speed is 1466 r/min. Figure 5a illustrates the estimated versus measured speed according to load ratio; whiles Figure 5b depicts estimated versus measured efficiency according to load ratio, which show a good agreement with the measured results.

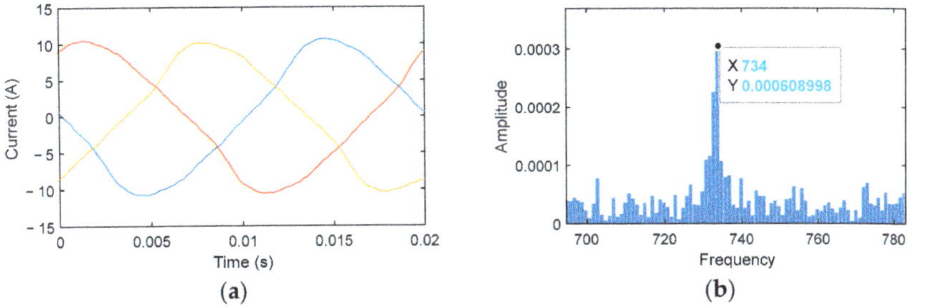


**Figure 4.** Measured stator current and rotor harmonic frequency: (**a**). measured phase-A currents; (**b**). extracted rotor slot harmonics frequency (734 Hz) for 5.5 kW motor, with slot number Z = 28 and pole pair number $p = 2$; the corresponding speed is 1465.71 r/min after recorded data analysis.

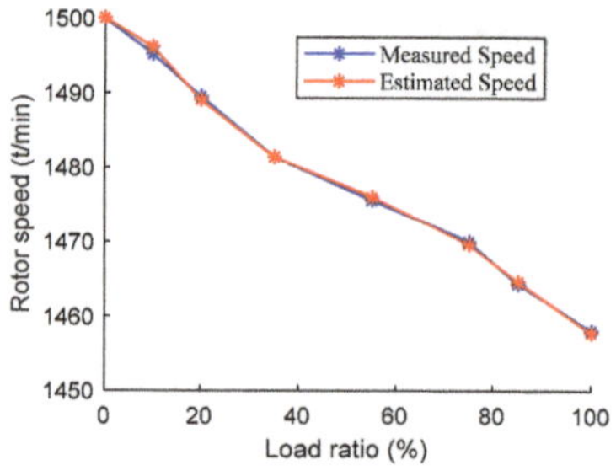

(a)

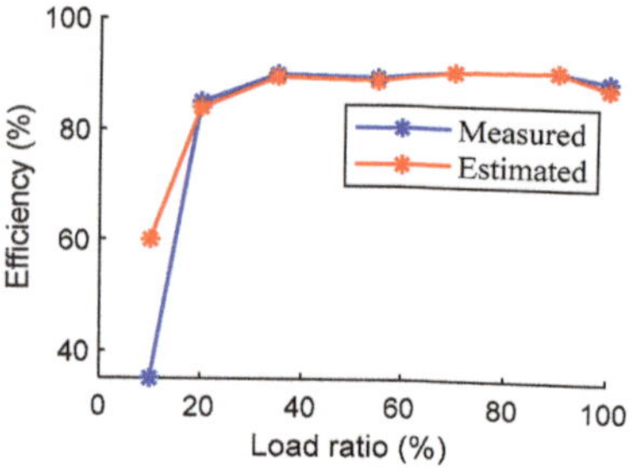

(b)

**Figure 5.** Measured versus estimated: (**a**). estimated versus measured speed according to load ratio; (**b**). estimated versus measured efficiency according to load ratio.

Table 3 shows the extracted parameters and compares PSO and QPSO-TRA algorithms for motor parameters determination. The speed has also been measured using a contactless method and both results were compared to evaluate the speed detection technique. After the full load, partial loads such as 75%, 50%, and 25% were also tested. Table 4 depicts the motor measured efficiencies and their associated speeds. The full charge and partial loads results are shown in Table 4. The relative errors of both efficiencies and speed can also be seen. Table 4 also illustrates the estimated efficiencies with PSO and QPSO-TRA and the estimated speeds versus measured speeds of the partial loads in the laboratory. The relative error deviations of the estimated efficiencies from the measured values are presented. For comparison purposes, PSO and QPSO-TRA results are shown in Table 4 for a laboratory test.

**Table 3.** Comparison of determined parameters of 5.5 kW motor.

| Parameters | PSO | QPSO-TRA |
|---|---|---|
| $R_1$ | 2.43 | 2.43 |
| $R_2$ | 1.92 | 1.72 |
| $R_{fe}$ | 2803.87 | 2687.675 |
| $X_m$ | 114.15 | 150.003 |
| $X_1$ | 0.88 | 0.85 |
| $X_2$ | 1.31 | 1.27 |

**Table 4.** 5.5 kW motor estimated efficiencies and speeds versus measured in the laboratory.

| Load (%) | Speeds (r/min) | | Relative Errors | Efficiencies (%) | | | Relative Errors PSO/QPSO-TRA |
|---|---|---|---|---|---|---|---|
| | Measured | Estimated | | Measured | Estimated | | |
| | | | | | PSO | QPSO-TRA | |
| 100 | 1468.04 | 1465.71 | 0.0158 | 88.63 | 91.05 | 89.32 | 0.1900 |
| 75 | 1475.08 | 1473.88 | 0.0081 | 91.41 | 90.94 | 91.65 | −0.0780 |
| 50 | 1481.4 | 1482.5 | −0.0007 | 90.04 | 89.55 | 89.95 | −0.0446 |
| 25 | 1489.5 | 1490.8 | −0.0008 | 83.86 | 89.33 | 83.53 | 0.6492 |

It is evident that the error increases with the load factor of 25%. At 25%, the motor is under light load condition. At this point, the losses are more considerable than if the load point is close to the rated load and the output power is small, which is not recommended because it is not beneficial.

### 4.2. Field Application

The proposed algorithm has been applied on an 11,000 kW, 10 kV feeding pump induction motor in a power plant, the rated current and speed are 710 A and 1491 rpm, respectively. Figure 6 shows the measured voltage and current waveforms of the above motor. In the field test, a self-developed monitoring system is installed on the motor terminal, and the running interface is to complete the corresponding data monitoring and analysis. Besides of the voltage and current waveforms, the monitored data also include voltage, current, torque, speed, power factor, efficiency, voltage deviation, unbalance and harmonic components and so on. Here, when monitoring and recording a large amount of data, only a typical example of data is given in Figure 6 and Table 5. It can be seen that.

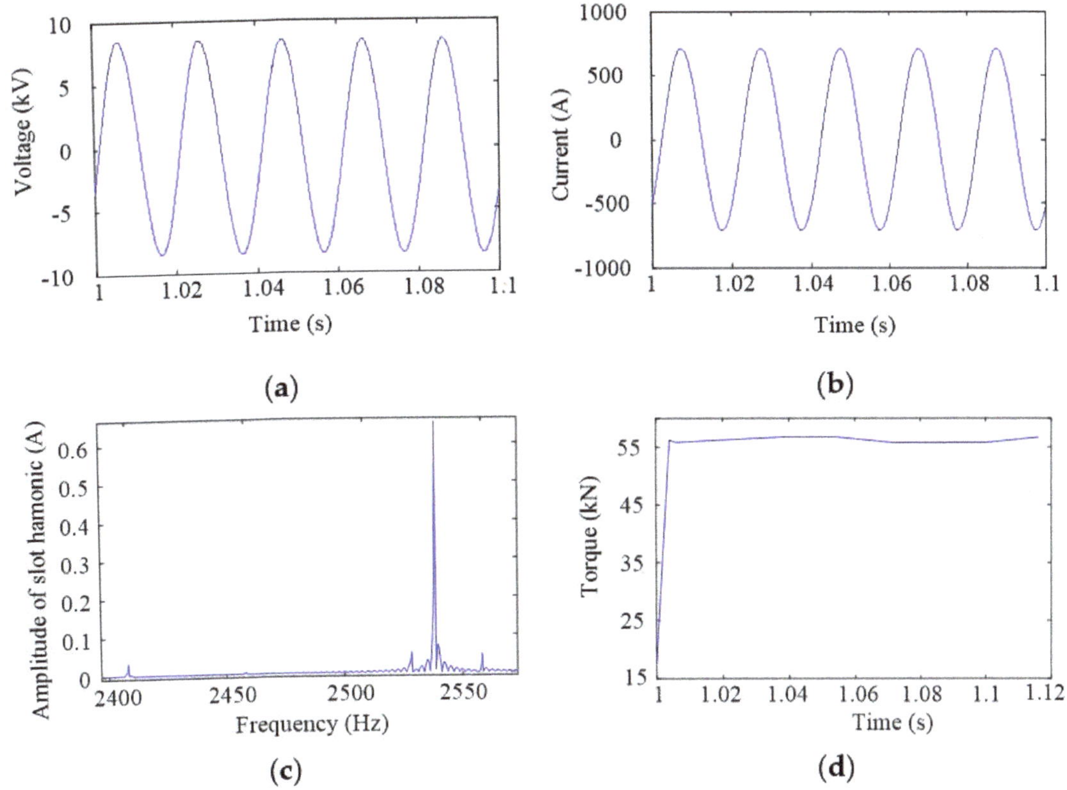


**Figure 6.** Measured waveforms with the developed monitoring system and estimated air-gap torque and slot harmonic obtained by the presented method. (**a**). Voltage. (**b**) Current. (**c**) Air-gap torque. (**d**) Rotor slot harmonic.

**Table 5.** Extracted parameters by QPSO-TRA and PSO.

| **Parameters** | **10 MW Motor** | |
|---|---|---|
| | **QPSO-TRA** | **PSO** |
| $R_1$ | 0.223 | 0.351 |
| $R_2$ | 0.172 | 0.272 |
| $R_{fe}$ | 28,687.675 | 29,381.178 |
| $X_m$ | 2350.003 | 2715.078 |
| $X_1$ | 0.115 | 0.231 |
| $X_2$ | 0.115 | 0.231 |

(1) Figure 6a,b are the measured voltage and current waveforms, and Figure 6c,d are identified rotor slot harmonics and electromagnetic torque. The input power can be

calculated by measured voltage and current waveforms, namely, 8922.27 kW. Furthermore, it is difficult to measure the shaft torque directly in the field, therefore, the output power are derived by the measured speed and pressure and flow of pumped water, namely, 8565.92 kW, respectively. And motor efficiency under this condition is 96.01%.

(2) From Figure 6c,d, the identified rotor slot harmonic is 2540 Hz, and the related rotor speed is 1494.5 rpm, and average electromagnetic torque under this condition based on Equation (2) is 55 kNm. It can be calculated that the output power is 8603.33 kW. And the identified efficiencies are 95.63% and 96.43%, with PSO and QPSO-TRA, respectively.

(3) Table 5 shows the extracted parameters of the motor using QPSO-TRA and PSO algorithms, respectively. Table 6 shows the motor data and the estimation results. It can be concluded that the estimated results show that the relative error is about 0.83% in the field test, which indicates that the proposed method can meet the efficiency estimation requirement of motor systems.

**Table 6.** Measured versus estimated efficiencies of the motors by PSO and QPSO-TRA methods.

| Measured Efficiency (%) | Estimated Efficiency by PSO (%) | Estimated Efficiency QPSO-TRA (%) | Relative Errors PSO/QPSO-TRA |
|---|---|---|---|
| 96.01 | 95.63 | 96.43 | 0.83 |

To sum up, the proposed algorithm demonstrates an sufficient accuracy level when the predicted or estimated efficiencies are compared to their related measured values.

## 5. Conclusions

This study proposes an approach for in situ efficiency estimation of induction motors based on the QPSO-TRA algorithm. With the proposed method, the stator resistance $R_s$ and the core losses resistance $R_{fe}$ can be accurately identified online. The main conclusions are as follows.

(1) A new algorithm to identify the motor parameters is designed to deal with the problem mentioned early for induction motor parameters identification. This new algorithm is called QPSO-TRA. After comparing the extracted parameters with QPSO-TRA and standard PSO, it is found that QPSO-TRA can achieve a better result than the standard PSO, as can be seen in Tables 3 and 5.

(2) With the proposed hybrid algorithm, global research is avoided when investigating the full research space. The computation time and burden are reduced to 1/3 compared to the Genetic Algorithm and also the algorithm is faster, reducing the computational time, compared to simple TRA.

(3) Experiments in the laboratory and the field were performed to analyze the performance of the proposed efficiency estimation algorithm by testing some induction motors of various types and rating power in the laboratory rather than in the field. The relative errors in Tables 4 and 6 are less than 2%, which is a good indicator to show the high accuracy of the proposed method. Finally, the high accuracy level showed by the proposed technique validated the efficiency estimation technique.

**Author Contributions:** Methodology—M.N.D. and Y.Y.; Conceptualization—Y.L. and H.Z.; Validation, M.N.; Data analysis, M.N.D., Y.Y., and H.Z.; writing—original draft preparation, M.N.D., Y.Y. and Z.L.; writing—review and editing, H.Z.; supervision, Y.L. and H.Z. All authors have read and agreed to the published version of the manuscript.

**Funding:** This research was funded by the National Natural Science Foundation of China under Grant No. 52177041.

**Institutional Review Board Statement:** Not applicable.

**Informed Consent Statement:** Not applicable.

**Data Availability Statement:** Not applicable.

**Acknowledgments:** The authors would like to thank SOACAP-MALI (Société Africaine de Chaussures et Articles en Plastiques: African Company for Shoes and Plastic Products) for their support which helped to improve the outcome feasibility of the research.

**Conflicts of Interest:** The authors declare no conflict of interest.

## References

1. Gajjar, C.S.; Kinyua, J.M.; Khan, M.A.; Barendse, P.S. Analysis of a Nonintrusive Efficiency Estimation Technique for Induction Machines Compared to the IEEE 112B and IEC 34-2-1 Standards. *IEEE Trans. Ind. Appl.* **2015**, *51*, 4541–4553. [CrossRef]
2. Esen, G.K.; Ozdemir, E. A New Field Test Method for Determining Energy Efficiency of Induction Motor. *IEEE Trans. Instrum. Meas.* **2017**, *66*, 3170–3179. [CrossRef]
3. Santos, V.S.; Eras, J.J.C.; Gutierrez, A.S.; Ulloa, M.J.C. Assessment of the energy efficiency estimation methods on induction motors considering real-time monitoring. *Energy Convers. Manag.* **2019**, *136*, 237–2470. [CrossRef]
4. Al-Badri, M.; Pillay, P.; Angers, P. A Novel In Situ Efficiency Estimation Algorithm for Three-Phase IM Using GA, IEEE Method F1 Calculations, and Pretested Motor Data. *IEEE Trans. Energy Convers.* **2015**, *30*, 1092–1102. [CrossRef]
5. Hsu, J.S.; Sorenson, P.L. Field Assessment of Induction Motor Efficiency through Air-Gap Torque. *IEEE Trans. Energy Convers.* **1996**, *11*, 489–494. [CrossRef]
6. Bijan, M.G.; Al-Badri, M.; Pillay, P.; Angers, P. Induction Machine Parameter Range Constraints in Genetic Algorithm Based Efficiency Estimation Techniques. *IEEE Trans. Ind. Appl.* **2018**, *54*, 4186–4197. [CrossRef]
7. Siraki, G.; Pillay, P. An in situ efficiency estimation technique for induction machines working with unbalanced supplies. *IEEE Trans. Energy Convers.* **2012**, *27*, 85–95. [CrossRef]
8. *Std 112-2004 (Revision IEEE Std 112-1996)*; IEEE Standard Test Procedure for Polyphase Induction Motors and Generators 2004. IEEE Press: Piscataway, NJ, USA, 2018; pp. 1–83.
9. Hsu, J.; Scoggins, B. Field test of motor efficiency and load changes through air-gap torque. *IEEE Trans. Energy Convers.* **1995**, *10*, 477–483. [CrossRef]
10. Santos, V.S.; Felipe, P.R.V.; Sarduy, J.R.G.; Lemozy, N.A.; Jurado, A.; Quispe, E.C. Procedure for Determining Induction Motor Efficiency Working Under Distorted Grid Voltages. *IEEE Trans. Energy Convers.* **2015**, *30*, 331–339. [CrossRef]
11. Siraki, A.G.; Pillay, P.; Angers, P. Full Load Efficiency Estimation of Refurbished Induction Machines from No-Load Testing. *IEEE Trans. Energy Convers.* **2013**, *28*, 317–326. [CrossRef]
12. Siraki, G.; Pillay, P. Comparison of two methods for full-load in situ induction motor efficiency estimation from field testing in the presence of over/undervoltages and unbalanced supplies. *IEEE Trans. Ind. Appl.* **2012**, *48*, 1911–1921. [CrossRef]
13. Desai, C.; Pillay, P. Back EMF, torque-angle, and core loss characterization of a variable-flux permanent-magnet machine. *IEEE Trans. Transp. Electrif.* **2019**, *5*, 371–384. [CrossRef]
14. Zhang, Z.; Deng, Z.; Sun, Q.; Peng, C.; Gu, Y.; Pang, G. Analytical Modeling and Experimental Validation of Rotor Harmonic Eddy-Current Loss in High-Speed Surface-Mounted Permanent Magnet Motors. *IEEE Trans. Magn.* **2018**, *55*, 100811. [CrossRef]
15. Climente-Alarcon, V.; Antonino-Daviu, J.A.; Haavisto, A.; Arkkio, A. Diagnosis of Induction Motors Under Varying Speed Operation by Principal Slot Harmonic Tracking. *IEEE Trans. Ind. Appl.* **2015**, *51*, 3591–3599. [CrossRef]
16. Silva, W.L.; Lima, A.M.N.; Oliveira, A. Speed Estimation of an Induction Motor Operating in the Nonstationary Mode by Using Rotor Slot Harmonics. *IEEE Trans. Instrum. Meas.* **2014**, *64*, 984–994. [CrossRef]
17. Al-Badri, M.; Pillay, P.; Angers, P. A Novel in Situ Efficiency Estimation Algorithm for Three-Phase Induction Motors Operating with Distorted Unbalanced Voltages. *IEEE Trans. Ind. Appl.* **2017**, *53*, 5338–5347. [CrossRef]
18. Stopa, M.M.; Resende, M.R.; Luiz, A.S.A.; Justino, J.C.G.; Rodrigues, G.G.; Filho, B.J.C. A Simple Torque Estimator for In-Service Efficiency Determination of Inverter-Fed Induction Motors. *IEEE Trans. Ind. Appl.* **2020**, *56*, 2087–2096. [CrossRef]
19. Salomon, C.P.; Sant'Ana, W.C.; da Silva, L.E.B.; Lambert-Torres, G.; Bonaldi, E.L.; de Oliveira, L.E.L.; da Silva, J.G.B. Induction Motor Efficiency Evaluation Using a New Concept of Stator Resistance. *IEEE Trans. Instrum. Meas.* **2015**, *64*, 2908–2917. [CrossRef]
20. Bramerdorfer, G.; Tapia, J.A.; Pyrhonen, J.J.; Cavagnino, A. Modern Electrical Machine Design Optimization: Techniques, Trends, and Best Practices. *IEEE Trans. Ind. Electron.* **2018**, *65*, 7672–7684. [CrossRef]
21. Wang, P.-P.; Chen, X.-X.; Zhang, Y.; Hu, Y.-J.; Miao, C.-X. IBPSO-Based MUSIC Algorithm for Broken Rotor Bars Fault Detection of Induction Motors. *Chin. J. Mech. Eng.* **2018**, *31*, 80. [CrossRef]
22. Liu, F.; Gao, J.; Liu, H. A Fault Diagnosis Solution of Rolling Bearing Based on MEEMD and QPSO-LSSVM. *IEEE Access* **2020**, *8*, 101476–101488. [CrossRef]
23. Hu, C.; Reeves, S.J. Trust Region Methods for the Estimation of a Complex Exponential Decay Model in MRI With a Single-Shot or Multi-Shot Trajectory. *IEEE Trans. Image Process.* **2015**, *24*, 3694–3706. [PubMed]

energies

MDPI

Article

# Optimization Method Based on Hybrid Surrogate Model for Pulse-Jet Cleaning Performance of Bag Filter

**Shirong Sun [1], Libing Liu [1,2], Zeqing Yang [1,2,*], Wei Cui [1], Chenghao Yang [1,*], Yanrui Zhang [3] and Yingshu Chen [1]**

1 School of Mechanical Engineering, Hebei University of Technology, Tianjin 300130, China; dreamingsun_sunsr@foxmail.com (S.S.); tjxiaobing@163.com (L.L.); cw211512@foxmail.com (W.C.); rosecys@163.com (Y.C.)
2 Key Laboratory of Hebei Province on Scale-Span Intelligent Equipment Technology, Tianjin 300130, China
3 Experimental Training Center, Hebei University of Technology, Tianjin 300401, China; 2008021@hebut.edu.cn
* Correspondence: yangzeqing@hebut.edu.cn (Z.Y.); c.yang@hebut.edu.cn (C.Y.)

**Abstract:** The pulse-jet cleaning process is a critical part of the bag filter workflow. The dust-cleaning effect has a significant impact on the operating stability of bag filters. Aiming at the multi-parameter optimization problem involved in the pulse-jet cleaning process of bag filters, the construction method of hybrid surrogate models based on second-order polynomial response surface models (PRSMs), radial basis functions (RBFs), and Kriging sub-surrogate models is investigated. With four sub-surrogate model hybrid modes, the corresponding hybrid surrogate models, namely PR-HSM, PK-HSM, RK-HSM, and PRK-HSM, are constructed for the multi-parameter optimization involved in the pulse-jet cleaning process of bag filters, and their objective function is the average pressure on the inner side wall of the filter bag at 1 m from the bag bottom. The genetic algorithm is applied to search for the optimal parameter combination of the pulse-jet cleaning process. The results of simulation experiments and optimization calculations show that compared with the sub-surrogate model PRSM, the evaluation indices RMSE, $R^2$, and RAAE of the hybrid surrogate model RK-HSM are 9.91%, 4.41%, and 15.60% better, respectively, which greatly enhances the reliability and practicability of the hybrid surrogate model. After using the RK-HSM, the optimized average pressure $F$ on the inner side wall of the filter bag at 1 m from the bag bottom is −1205.1605 Pa, which is 1321.4543 Pa higher than the average pressure value under the initial parameter condition set by experience, and 58.4012 Pa to 515.2836 Pa higher than using the three sub-surrogate models, verifying its usefulness.

**Keywords:** hybrid surrogate model; performance optimization; Kriging; RBF; genetic algorithm; pulse-jet cleaning

**Citation:** Sun, S.; Liu, L.; Yang, Z.; Cui, W.; Yang, C.; Zhang, Y.; Chen, Y. Optimization Method Based on Hybrid Surrogate Model for Pulse-Jet Cleaning Performance of Bag Filter. *Energies* **2023**, *16*, 4652. https://doi.org/10.3390/en16124652

Academic Editors: Enrique Romero-Cadaval, David Gerada, Yuling He, Conggan Ma and Haisen Zhao

Received: 3 February 2023
Revised: 17 May 2023
Accepted: 8 June 2023
Published: 12 June 2023



## 1. Introduction

Industrial dust and industrial soot emissions are the main pollution sources affecting the quality of the atmospheric environment [1]. As effective facilities for controlling dust, especially fine particles, bag filters have the characteristics of high efficiency and good economy and are widely used for dust treatment in coal, electric power, steel, cement, and waste incineration industries [2,3]. The pulse-jet cleaning process is an important part of the bag filter workflow, and its performance is directly related to the effect of energy saving and emission reduction. By adopting advanced methods, optimized cleaning parameters can be obtained, thus improving the overall operating efficiency of bag filters.

In order to optimize the pulse-jet cleaning parameters of bag filters, relevant scholars obtained better cleaning parameters by constructing a digital simulation and optimization model for bag filters, which can improve cleaning performance. Kang et al. [2] adopted a numerical simulation for a slit-nozzle optimization design to remove the caked-on dust of the dead zone. Park et al. [4] relied on numerical simulation to study the influence of filter bag length on the distribution of filtration velocity and proposed a scheme to improve the uniformity of filtration velocity. Fan et al. [5] conducted a numerical simulation of the bag filter dust-cleaning process based on fluid dynamics to study the effects of factors such as pulse-jet pressure, nozzle diameter, pulse-jet height, and filter bag length on the flow field distribution inside the bag, established a quadratic polynomial prediction model with four influencing factors, and modified the models to provide theoretical guidance for the optimization of the bag filter dust-cleaning system.

Complex and computationally expensive simulations and physical experiments are often required to quantify the economic and engineering performance of complex products such as bag filters. The surrogate model approach has been widely recognized to avoid the computational burden of directly adopting the finite element model for parameter optimization.

The surrogate model refers to the mapping between the design variables and the optimization objective based on the existing design point data with low computational effort and a short calculation period, while the calculation results are similar to the numerical analysis or physical test results. The focus of surrogate modeling is the construction of approximate models to evaluate system performance and obtain the relationship between the inputs and outputs, so as to obtain the influence of each specific variable on the optimization target, which can bring a more refined design experience to the designers [6]. Common surrogate model methods include the polynomial response surface method (PRSM), radial basis function (RBF), and Kriging.

Xu et al. [7] applied the PRSM surrogate model to the improvement of bentonite in shield construction. The results displayed the optimal parameters for strongly weathered granite strata at different shield advance speeds, thus providing a technical guarantee for efficient construction. Thakre et al. [8] used the PRSM surrogate model in the parameter sensitivity analysis of the selective laser sintering process, which provided conditions for realizing the robust performance of additive manufacturing processes.

Sun et al. [9] introduced the global optimization technique based on the RBF surrogate model into the correction and optimization of key parameters of honeycomb sandwich plates in spacecraft, which reduced the dynamics analysis error. Xu et al. [10] used the RBF surrogate model to optimize the parameters of the annular jet pump to improve its hydraulic performance in submarine trenching and dredging.

Wang et al. [11] employed the Kriging surrogate model to optimize the design of drum brake stability, and it was found that the introduction of Kriging could greatly improve the solution efficiency. Li et al. [12] adopted the Kriging model for the design parameter optimization of a flat-type permanent magnet linear synchronous motor for improving average thrust while reducing thrust fluctuation.

Different surrogate models have their own applicable scenarios [13]. The PRSM is suitable for solving low-order problems in low-dimensional spaces, Kriging is appropriate for solving complex nonlinear problems in low-dimensional spaces and slight nonlinear problems in high-dimensional spaces, and RBF is fit for solving high-order nonlinear problems in high-dimensional spaces. Moreover, the optimization results of a single surrogate model for different problems are highly variable and unpredictable, which brings some risks to engineering optimization.

To address the above issues, a hybrid surrogate model approach based on multiple surrogate models can reduce the risk of using a single surrogate model for optimization while improving the accuracy of the model. Li et al. [14] combined three surrogate models to construct a hybrid surrogate model for the structural design of permanent magnet drives, and the optimization design effect was significantly improved. Xie et al. [15] applied the hybrid surrogate model to the comprehensive dynamic performance optimization of rail vehicles to obtain more reasonable suspension parameters, resulting in significant improvements in both the smoothness and stability of the vehicle operation. He et al. [16] applied the hybrid surrogate model to the shape and structural optimization design of the multi-bubble pressure cabin in an underwater vehicle, which enabled the pressure cabin to have a stronger load capacity and a significantly lower weight.

Compared with different single surrogate models, optimization approaches based on the hybrid surrogate model can combine the superior characteristics of sub-surrogate models to make the model more accurate, thus improving the accuracy of prediction and optimization results. The pulse-jet cleaning process of the bag filter is a complex, nonlinear, dynamic process. In this paper, a more suitable hybrid surrogate model approach will be selected to optimize the pulse-jet cleaning-related parameters to improve the performance of the bag filter.

## 2. Analysis of the Bag Filter Dust-Cleaning Optimization Problem

The dust-cleaning system of the pulse-jet bag filter is mainly composed of the compressed air cylinder, the clean air chamber, the tube sheet, several filter bags, and the corresponding pulse-jet valves, injection pipes, nozzles, etc. The dust-cleaning process starts with the initiation of the pulse-jet valves, and the pulse-jet time is only roughly 0.1 s to 0.5 s. The schematic diagram of the pulse-jet cleaning state is shown in Figure 1 [17]. When the pulse-jet valve is activated, the high-speed pulse airflow immediately enters the filter bag from the compressed air cylinder through the injection pipe and nozzle. At the same time, it will also cause a low-pressure area at the entrance of the filter bag, thus inducing a secondary airflow into the bag filter, while the secondary airflow rate is several times the initial pulse-jet compressed air [18]. As a large amount of airflow enters the filter bag, the pressure inside the filter bag rises sharply.

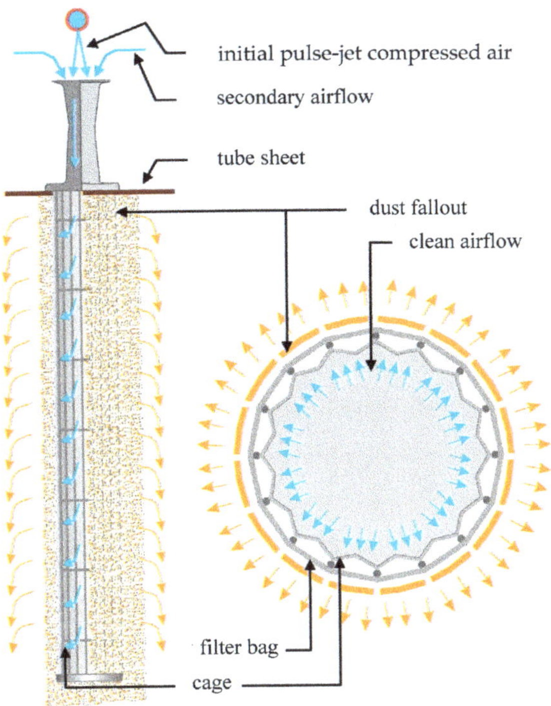


**Figure 1.** The schematic diagram of the pulse-jet cleaning state.

The strong airflow impact will make the filter bag vibrate, while the pressure difference between the inside and the outside of the filter bag will make the bag, along with its surface dust cake, move outward in a radial direction. Under the action of tension, the radial velocity of the filter bag will gradually drop to zero, and then the filter bag will shrink inward in a radial direction [19]. Because the dust cake attached to the outer wall of the filter bag is not subject to the tension, the dust cake will overcome the adhesion with the filter bag under the action of inertia and fall into the dust hopper under the action of gravity, thus completing the dust-cleaning process.

In the process of pulse-jet cleaning, there are many factors that affect the dust-cleaning performance [18] such as the air cylinder volume, pulse-jet pressure, pulse-jet height, pulse-jet time, structure of nozzles, filter material type, filter bag diameter and length, clean air chamber volume, and so on.

An optimization model should be established for the pulse-jet system, and then the multiple relevant design parameters can be optimized simultaneously to obtain the optimum dust-cleaning performance. In the study of the optimization of dust-cleaning performance, four parameters which have a significant impact on the dust-cleaning performance, as well as being easily accessible in engineering, were selected. They are the pulse-jet pressure (*P*), filter bag length (*L*), filter bag diameter (*D*), and nozzle diameter (*d*). According to the actual conditions of engineering, the design intervals of the variables are shown in Table 1.

**Table 1.** The design intervals of the variables.

| Variable | Initial Empirical Value | Lower Limit Value | Upper Limit Value |
|---|---|---|---|
| $P$/kPa | 250 | 100 | 400 |
| $L$/m | 8 | 6 | 10 |
| $D$/mm | 145 | 130 | 160 |
| $d$/mm | 16 | 8 | 20 |

In previous studies on the optimization of bag filter dust-cleaning performance, dust detachment from the filter bag was mainly evaluated by the sidewall peak pressure. In the process of dust cleaning, the sidewall pressure of the filter bag is unevenly distributed in the axial direction, and the pressure is relatively small in the bottom part of the filter bag, so the effect of dust cleaning is also poor. In other words, when the conditions at the bottom of the filter bag meet the cleaning requirements, the entire bag's cleaning performance will also be guaranteed. When considering the force required, the dust cake and the filter bag should be considered as a whole. In practical engineering applications, it is difficult to accurately obtain the outer pressure of the filter bag due to the complex characteristics of the dust cake. Therefore, the average pressure on the inner side wall of the filter bag at 1 m from the bag bottom is selected as the optimization target, which is recorded as *F*. That is, the maximum *F* is the optimization goal of the cleaning performance. The multi-parameter optimization model for pulse-jet cleaning of the bag filter can be expressed as:

$$\begin{gathered} find P, L, D, d \\ \max . F \\ s.t. 100\,\mathrm{kPa} \leq P \leq 400\,\mathrm{kPa} \\ 6\,\mathrm{m} \leq L \leq 10\,\mathrm{m} \\ 130\,\mathrm{mm} \leq D \leq 160\,\mathrm{mm} \\ 8\,\mathrm{mm} \leq d \leq 20\,\mathrm{mm} \end{gathered} \tag{1}$$

## 3. Hybrid Surrogate Model for Multi-Parameter Optimization

The surrogate model transforms the black box problem into a smooth and continuous explicit mathematical problem, which greatly reduces the computational complexity while satisfying the accuracy requirements. Different surrogate models have different characteristics and apply to different optimization problems. If they can be combined to build on their strengths and avoid their weaknesses, the optimization results may be better. Therefore, the hybrid surrogate model method is proposed to expect better optimization results.

Set as follows, $\mathbf{x} = (x_1, \cdots, x_n)$ is the n-dimensional input variable, $y$ is the output response, and the corresponding response data set is $\mathbf{Y} = \left(y^{(1)}, \cdots, y^{(N)}\right)^{\mathrm{T}}$ for the training sample data set $\mathbf{X} = \left(\mathbf{x}^{(1)}, \cdots, \mathbf{x}^{(N)}\right)^{\mathrm{T}}$ of size $N$. The functional relationship between $\mathbf{x}$ and $y$ can be expressed as:

$$y = \hat{y}(\mathbf{x}) + \varepsilon \tag{2}$$

where $\hat{y}(\mathbf{x})$ is the surrogate model output function, and $\varepsilon$ is the error when approximating $y$ with $\hat{y}(\mathbf{x})$.

The essence of the hybrid surrogate model is a weighted linear superposition of different single surrogate models called sub-surrogate models [20], which can be expressed as:

$$\begin{gathered} \hat{y}_{\mathrm{HSM}}(\mathbf{x}) = \sum_{i=1}^{m} \alpha_i \hat{y}_i(\mathbf{x}) \\ \sum_{i=1}^{m} \alpha_i = 1 \end{gathered} \tag{3}$$

where $\hat{y}_{\mathrm{HSM}}(\mathbf{x})$ is the output of the hybrid surrogate model, $m$ is the number of sub-surrogate models, $\hat{y}_i(\mathbf{x})$ is the output of the $i$th sub-surrogate model, and $a_i$ is the weight of $\hat{y}_i(\mathbf{x})$.

Using two sub-surrogate models cannot provide enough sub-surrogate model hybrid modes, and using four or more sub-surrogate models will greatly increase the amount of calculation. In order to make better use of the advantages of each sub-surrogate model and reduce the influence of its defects on the hybrid surrogate model, there are three sub-surrogate models selected by the proposed hybrid surrogate model method, which are the PRSM model, the RBF model, and the Kriging model.

### 3.1. Sub-Surrogate Model

#### 3.1.1. PRSM

PRSM was first proposed by the mathematicians Box and Wilson in 1951. The process of PRSM is to construct a mapping relationship between the design variables and the target response through least squares regression based on the given input and output values. It can be expressed as:

$$\begin{aligned} y &= \boldsymbol{\beta}^{\mathrm{T}} \cdot \boldsymbol{f}(\mathbf{x}) \\ &= \beta_1 f_1(\mathbf{x}) + \beta_2 f_2(\mathbf{x}) + \cdots + \beta_k f_k(\mathbf{x}) \end{aligned} \tag{4}$$

where $f_i(\mathbf{x})(i = 1, 2, \cdots, k)$ is the $i$th polynomial, $\beta_i$ is the corresponding polynomial coefficient to be solved, $\boldsymbol{f}(\mathbf{x})$ is the polynomial matrix, and $\boldsymbol{\beta}$ is the coefficient matrix.

The most common usage of PRSM in engineering applications is the second-order PRSM [21]. Its general formula can be expressed as:

$$\hat{y}_{\mathrm{PRSM}} = \beta_0 + \sum_{i=1}^{n} \beta_i x_i + \sum_{i=1}^{n} \beta_{ii} x_i^2 + \sum_{i=1}^{n} \sum_{i<j}^{n} \beta_{ij} x_i x_j \tag{5}$$

where $\hat{y}_{\mathrm{PRSM}}$ is the estimated value of the response, and $x_i$ is the $i$th component of the n-dimensional variable $\mathbf{x}$. $\beta_0$, $\beta_i$, $\beta_{ii}$, and $\beta_{ij}$ are the coefficients of the objective function to be solved, respectively, which can be arranged in a certain order to form the column vector $\boldsymbol{\beta}$.

The polynomial response surface method uses polynomials with explicit expressions to replace complex simulation analysis models, which are easy to construct. The models constructed by PRSM have good continuity and derivability to reduce the influence of numerical noise. At the same time, the influence of each variable on the output target response can be judged by the coefficient of each component [22]. The engineering practice shows that the PRSM can effectively reduce the number of simulation analyses. However, if there are too many design variables, the number of sample points will increase accordingly, and the computational burden will increase significantly.

3.1.2. RBF

RBF is a method that uses discrete multivariate data to fit unknown functions. The basic principle is to linearly weight the model constructed with a typical radial function as the basis function, transforming the multi-dimensional problem into a one-dimensional problem with the Euclidean distance between the known sample points and the unknown point to be measured as the independent variable [23]. The function value corresponding to the unknown sample point $\mathbf{x}$ based on RBF can be expressed as:

$$\hat{y}_{\mathrm{RBF}} = \sum_{i=1}^{N} \lambda_i \varphi\left(\left\|\mathbf{x} - \mathbf{x}^{(i)}\right\|\right) = \boldsymbol{\lambda}^{\mathrm{T}} \boldsymbol{\varphi} \tag{6}$$

where $\|\cdot\|$ is the Euclidean distance, and $\lambda_i$ is the weight coefficient of the radial function for the $i$th sample point $\mathbf{x}^{(i)}$.

The response vector [24] can be expressed as:

$$\mathbf{y} = \boldsymbol{A} \cdot \boldsymbol{\lambda}$$
$$\boldsymbol{A} = \begin{pmatrix} \varphi\left(\left\|\mathbf{x}^{(1)} - \mathbf{x}^{(1)}\right\|\right) & \cdots & \varphi\left(\left\|\mathbf{x}^{(1)} - \mathbf{x}^{(N)}\right\|\right) \\ \vdots & \ddots & \vdots \\ \varphi\left(\left\|\mathbf{x}^{(N)} - \mathbf{x}^{(1)}\right\|\right) & \cdots & \varphi\left(\left\|\mathbf{x}^{(N)} - \mathbf{x}^{(N)}\right\|\right) \end{pmatrix}_{N\times N} \tag{7}$$

where $\varphi(r)$ is the radial function and $r$ is the Euclidean distance. When the sample points do not coincide and $\boldsymbol{A}$ is positive definite [25], the above equation has a unique solution $\boldsymbol{\lambda} = \boldsymbol{A}^{-1} \cdot \boldsymbol{y}$. After that, the function value at the unknown sample point can be predicted.

Among these common radial functions, the multiquadric (MQ) function $\varphi(r,c) = \left(r^2 + c^2\right)^{0.5}$ shows good effects in terms of accuracy, stability, and computational efficiency [26]. The RBF is an interpolation-type surrogate model, and the approximation accuracy is strongly influenced by the shape coefficient $c(c > 0)$. The optimal value of $c$ depends on the distribution of sample points and approximate targets.

The RBF has the advantages of simple structure, good flexibility, high computational efficiency, and fast convergence, making it suitable for solving problems with multiple variables. However, the model is sensitive to numerical noise, so it is slightly insufficient for fitting strongly non-linear responses [27].

3.1.3. Kriging

Kriging is an unbiased estimation model with minimum estimation variance [28]. This model adds a random process on the basis of the global regression model, which not only considers the influence of the distance relationship between the sample points on the output, but also the influence of the spatial distribution and location relationship between the sample points on the output [29]. It can be generally expressed as:

$$y = F(\boldsymbol{\beta}, \mathbf{x}) + z(\mathbf{x}) = \boldsymbol{f}^{\mathrm{T}}(\mathbf{x})\boldsymbol{\beta} + z(\mathbf{x}) \tag{8}$$

where $F(\boldsymbol{\beta}, \mathbf{x})$ is the global regression model to reflect the overall trend of the system response in the design space, $\boldsymbol{f}(\mathbf{x})$ is the polynomial function vector, $\boldsymbol{\beta}$ is the regression coefficient vector, and $z(\mathbf{x})$ is a stochastic process with a mean of zero, a variance of $\sigma^2$, and a non-zero covariance, which reflects the approximation of local deviation. The covariance satisfies:

$$\mathrm{Cov}\left[z\left(\mathbf{x}^{(i)}\right), z\left(\mathbf{x}^{(j)}\right)\right] = \sigma^2 R\left(\mathbf{x}^{(i)}, \mathbf{x}^{(j)}\right) \tag{9}$$

where $R\left(\mathbf{x}^{(i)},\mathbf{x}^{(j)}\right)$ is a spatial correlation function, indicating the spatial correlation between sample points $\mathbf{x}^{(i)}$ and $\mathbf{x}^{(j)}$, which plays a decisive role in the accuracy of the simulation [30].

$$R\left(\mathbf{x}^{(i)},\mathbf{x}^{(j)}\right)=\prod_{k=1}^{n}R_k(\theta_k,d_k) \tag{10}$$

where $R_k(\theta_k,d_k)$ is the kernel function of the correlation function $R$; $d_k$ is the difference between the $k$th component of the sample points $x_k^{(i)}$ and $x_k^{(j)}$, namely $d_k=x_k^{(i)}-x_k^{(j)}$; $n$ is the dimension of the sample points, that is, the number of design variables; and $\theta_k$ is the model parameter of the kernel function in the $k$th direction.

The GAUSS model has the characteristics of being smooth and differentiable everywhere [31]; therefore, it is usually adopted in engineering applications to construct the correlation function model.

$$R\left(\mathbf{x}^{(i)},\mathbf{x}^{(j)}\right)=\exp\left(-\sum_{k=1}^{n}\theta_k\left|\mathbf{x}_k^{(i)}-\mathbf{x}_k^{(j)}\right|^2\right) \tag{11}$$

The Kriging model needs to minimize the mean square error of prediction $\varphi(\mathbf{x})$ under the condition of unbiased estimation.

$$\begin{gathered}\mathrm{E}(\hat{y}_{\mathrm{KRG}})=\mathrm{E}(y)\\ \varphi(\mathbf{x})=\mathrm{MSE}[\hat{y}_{\mathrm{KRG}}]=E\left[(\hat{y}_{\mathrm{KRG}}-y)^2\right]\end{gathered} \tag{12}$$

The Kriging model is based on the information of known sample points, fully considering the spatial correlation characteristics of variables. Moreover, the model has both local and global statistical properties. These characteristics make Kriging advantageous in solving problems of high nonlinearity to achieve desirable results [32].

*3.2. Method for Determining the Weighting Factor*

There are two key aspects of the modeling process of the hybrid surrogate model: one is to choose suitable single surrogate models as its sub-surrogate models, and the other is to determine the hybrid strategy, that is, to determine the weight coefficients through an effective calculation method.

The inverse proportional averaging method regards the accuracy indexes of each sub-surrogate model as irrelevant and has no systematic deviation [33]. The proportion of each sub-surrogate model in the hybrid surrogate model will be calculated according to this method.

$$\omega_i=\frac{E_i^{-1}}{\sum\limits_{i=1}^{m}E_i^{-1}} \tag{13}$$

where $E_i$ is the approximate capability evaluation value of the $i$th sub-surrogate model. In this paper, the root mean square error (RMSE) is picked as $E$.

*3.3. Hybrid Surrogate Model Optimization Algorithm*

The optimization algorithm proposed in this paper is a multi-parameter optimization strategy based on a hybrid surrogate model. The initial sample set is obtained through simulation experiments, and the hybrid surrogate model will be constructed. The accuracy of the hybrid surrogate model should be inspected in combination with the actual problem, and if it meets the requirements, the global optimal solution of the model will be obtained by the genetic algorithm; otherwise, the hybrid agent model needs to be updated by selecting a suitable addition strategy until the model accuracy meets the requirements. The process of optimization based on the hybrid surrogate model is displayed in Figure 2.

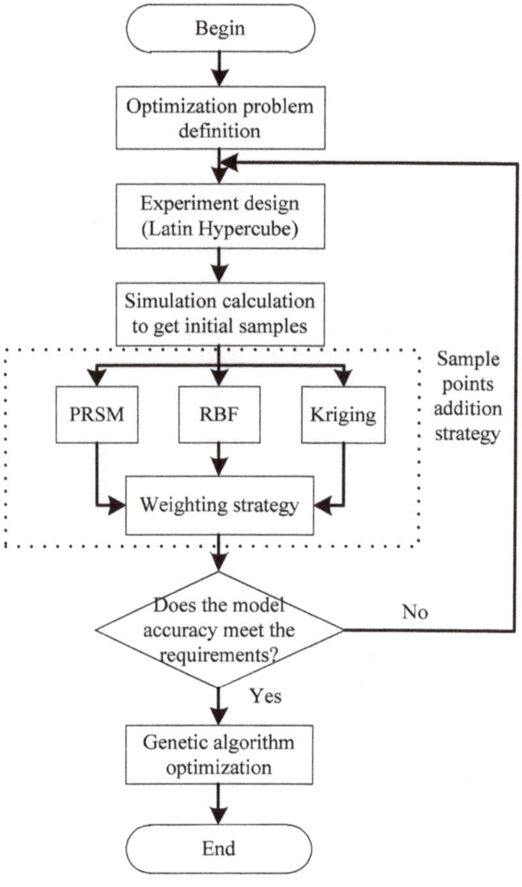


**Figure 2.** The process of optimization based on the hybrid surrogate model.

## 4. Optimization of Dust-Cleaning Performance of Bag Filter

### *4.1. Simulation and Modeling*

This study relies on the bag filter laboratory, as shown in Figure 3a, and the digital model of the bag filter, as shown in Figure 3b, is constructed based on physical equipment. In order to facilitate the study, a single filter bag is taken as the object, and some simplifications are applied. The finite element analysis model is shown in Figure 3c.

In this paper, the main settings of the numerical simulation are as follows:

1. The pressure inlet boundary condition is adopted at the inlet of the injection pipe. Considering the change of outlet pressure during the opening and closing process of the pulse-jet valve, the user-defined function (UDF) is adopted to define the function of pulse-jet pressure changing with time. According to the physical experiment, the curve of pulse-jet pressure changing with time is shown in Figure 4.
2. The boundary condition of the filter bag is set to the porous-jump medium.
3. The pulse-jet cleaning process satisfies the law of conservation of energy, the law of conservation of mass, and the law of conservation of momentum.

**Figure 3.** The simplification of bag filter: (**a**) the bag filter laboratory; (**b**) the digital model; (**c**) the finite element analysis model.

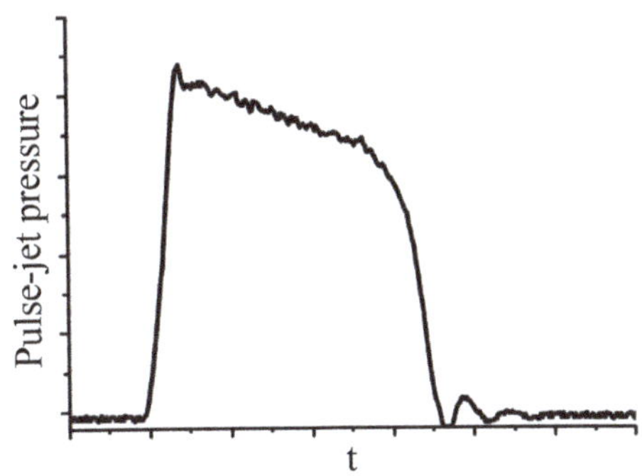


**Figure 4.** The curve of pulse-jet pressure changing with time.

Too few sample points make it difficult to guarantee the surrogate model's accuracy, while too many sample points will increase the computation cost. To ensure the space-filling of sample points and improve the computational efficiency, the Latin hypercube sampling method [34] based on the maximizing minimum distance criterion is employed to perform the experimental design, and 70 sets of sample points are obtained to construct the surrogate models. To confirm the correctness of the proxy model, 30 sets of validation sample points are obtained using the same method to test the accuracy. The distribution of modeling sample points and validation sample points is shown in Figure 5.

According to the experimental design conditions of 100 sets of sample points, the ANSYS is used to simulate the pulse-jet dust-cleaning process of the bag filter, as shown in Figure 6.

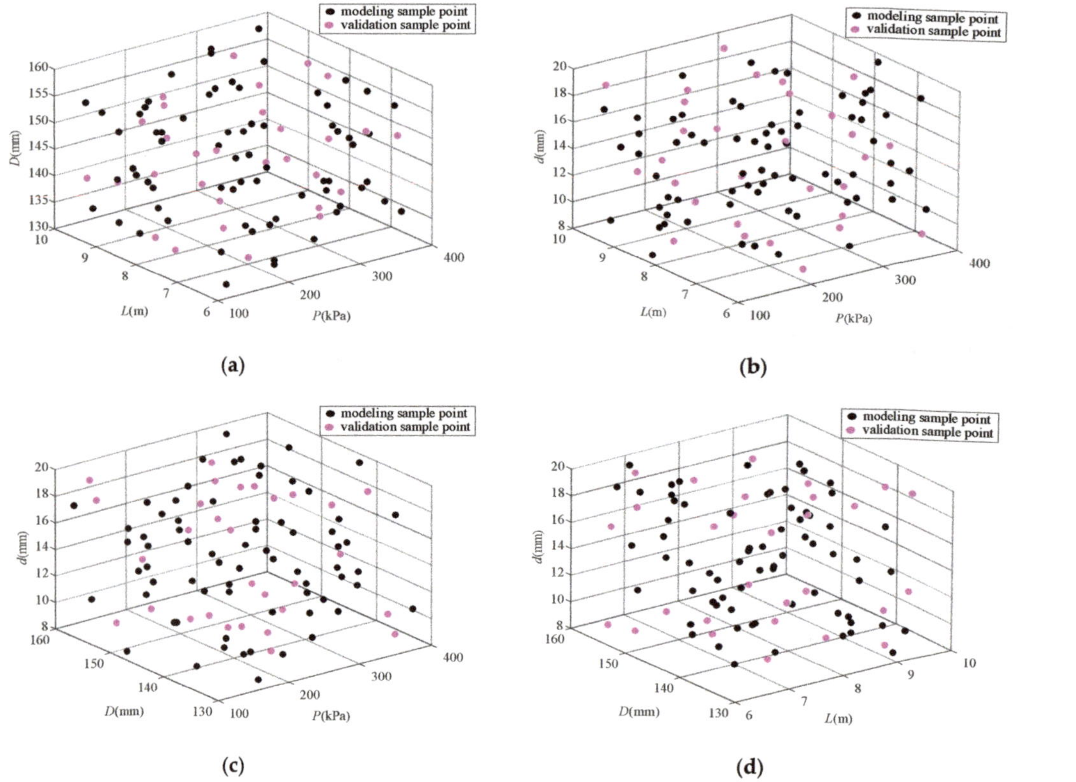


**Figure 5.** The distribution of modeling sample points and validation sample points: (**a**) *P-L-D*, (**b**) *P-L-d*, (**c**) *P-D-d*, (**d**) *L-D-d*.

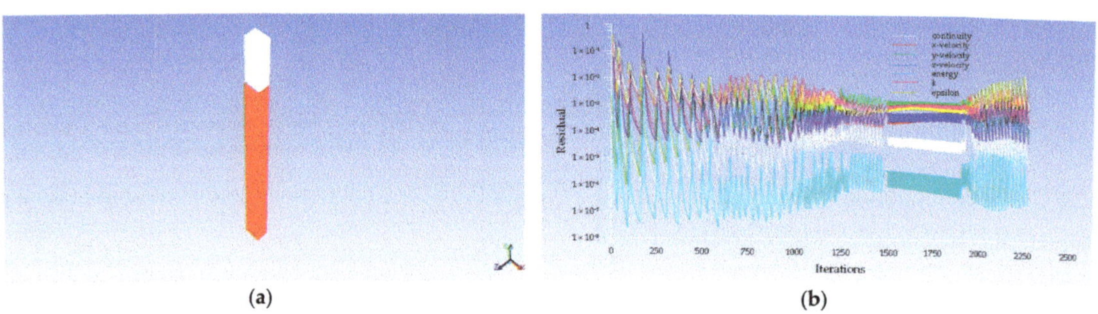


**Figure 6.** The simulation process: (**a**) the mesh; (**b**) the iterations and residuals image of the simulation process.

The surrogate models will be constructed based on the data of the 70 sample points. Meanwhile, considering the large range of data variation in each dimension, it is necessary to normalize the input variables to reduce their impact on the accuracy of the surrogate models [35].

$$x_{\text{new}} = \frac{x_i - x_{i\min}}{x_{i\max} - x_{i\min}} \tag{14}$$

where $x_{new}$ is the normalized design variable, $x_i$ is the initial design variable, and $x_{i\max}$ and $x_{i\min}$ are the minimum and maximum values of the initial design variable. It can be obtained:

$$\begin{aligned} x_P &= \frac{P-100}{300} \\ x_L &= \frac{L-6}{4} \\ x_D &= \frac{D-130}{30} \\ x_d &= \frac{d-8}{12} \end{aligned} \tag{15}$$

where $x_P$, $x_L$, $x_D$, and $x_d$ are the normalized values of $P$, $L$, $D$, and $d$, respectively.

According to the normalization and simulation results of the modeling sample points, the sub-surrogate models can be constructed.

1. Second-order PRSM sub-surrogate model

$$\begin{aligned} F_{\text{PRSM}}(\mathbf{x}) &= 0.0430x_P^2 - 1.2778x_L^2 - 0.1340x_D^2 - 3.3243x_d^2 - 0.2987x_P x_L \\ &-0.3057x_P x_D - 3.5838x_P x_d - 0.0723x_L x_D - 0.2685x_L x_d - 0.4268x_D x_d \\ &+1.1371x_P + 1.7622x_L + 0.4638x_D + 3.6502x_d - 2.9011 \end{aligned} \tag{16}$$

2. RBF sub-surrogate model

In this paper, the MQ function is employed as the kernel function of the RBF sub-surrogate model, which can be expressed as:

$$F_{\text{RBF}}(\mathbf{x}) = \sum_{i=1}^{70} \lambda_i \left[ (x_P - x_{iP})^2 + (x_L - x_{iL})^2 + (x_D - x_{iD})^2 + (x_d - x_{id})^2 + c^2 \right]^{\frac{1}{2}} \tag{17}$$

where $x_{iP}$, $x_{iL}$, $x_{iD}$, and $x_{id}$ denote the four-dimensional components of the $i$th modeling sample point, respectively. The solution can be obtained: $c = 0.3457$, $\lambda = [-13.8543, 70.6401, -9.6336, 7.0372, \cdots, 13.9376, 13.0330]$.

3. Kriging sub-surrogate model

In this paper, the quadratic function $g(\mathbf{x})$ is used as the global approximation function in the Kriging sub-surrogate model.

$$F_{\text{KRG}}(\mathbf{x}) = g(\mathbf{x}) + z(\mathbf{x}) \tag{18}$$

### 4.2. Accuracy Evaluation

In order to ensure the validity of sub-surrogate models, error analysis is required. Commonly used indicators for evaluating the model accuracy are root mean square error (RMSE), coefficient of certainty ($R^2$) [36], and relative average absolute error (RAAE) [37].

$$\begin{aligned} \text{RMSE} &= \sqrt{\frac{\sum_{i=1}^{N}(y_i-\hat{y}_i)^2}{N}} \\ \text{R}^2 &= 1 - \frac{\sum_{i=1}^{N}(y_i-\hat{y}_i)^2}{\sum_{i=1}^{N}(y_i-\overline{y}_i)^2} \\ \text{RAAE} &= \frac{\sum_{i=1}^{N}|y_i-\hat{y}_i|}{N \cdot \sqrt{\frac{1}{N}\sum_{i=1}^{N}(y_i-\overline{y}_i)^2}} \end{aligned} \tag{19}$$

According to the formulas, the accuracy indexes of three sub-surrogate models are shown in Table 2. It can be seen that the accuracy of the RBF sub-surrogate model is better than the other two.

**Table 2.** Model accuracy of three sub-surrogate models.

| Sub-Surrogate Model | RMSE | $R^2$ | RAAE |
|---|---|---|---|
| Second-order PRSM | 0.358211 | 0.810372 | 0.353255 |
| RBF | 0.327596 | 0.841400 | 0.309310 |
| Kriging | 0.342097 | 0.827049 | 0.331309 |

### 4.3. Construction of Hybrid Surrogate Model

The basic form of the hybrid surrogate model for pulse-jet cleaning performance can be expressed as:

$$F_{HSM}(\mathbf{x}) = \omega_{PRSM}F_{PRSM} + \omega_{RBF}F_{RBF} + \omega_{KRG}F_{KRG} \tag{20}$$

Three sub-surrogate models can construct multiple hybrid surrogate models. When the number of sub-surrogate models is 2, there are three hybrid surrogate models. When the number of sub-surrogate models is 3, there is one hybrid surrogate model.

1. Hybrid surrogate model constructed by the second-order PRSM and RBF (PR-HSM)

According to formula 13, the weights of the second-order PRSM model and the RBF model in the hybrid surrogate model RH-HSM are $\omega_{PR\text{-}PRSM} = 0.477680$ and $\omega_{PR\text{-}RBF} = 0.522320$.

$$F_{PR\text{-}HSM}(\mathbf{x}) = 0.477680F_{PRSM}(\mathbf{x}) + 0.522320F_{RBF}(\mathbf{x}) \tag{21}$$

2. Hybrid surrogate model constructed by the second-order PRSM and Kriging (PK-HSM)

According to formula 13, the weights of the second-order PRSM model and the Kriging model in the hybrid surrogate model PK-HSM are $\omega_{PK\text{-}PRSM} = 0.488495$ and $\omega_{PK\text{-}KRG} = 0.511505$.

$$F_{PK\text{-}HSM}(\mathbf{x}) = 0.488495F_{PRSM}(\mathbf{x}) + 0.511505F_{KRG}(\mathbf{x}) \tag{22}$$

3. Hybrid surrogate model constructed by the RBF and Kriging (RK-HSM)

According to formula 13, the weights of the RBF model and the Kriging model in the hybrid surrogate model RK-HSM are $\omega_{RK\text{-}RBF} = 0.510827$ and $\omega_{RK\text{-}KRG} = 0.489173$.

$$F_{RK\text{-}HSM}(\mathbf{x}) = 0.510827F_{PRSM}(\mathbf{x}) + 0.489173F_{KRG}(\mathbf{x}) \tag{23}$$

4. Hybrid surrogate model constructed by three sub-surrogate models (PRK-HSM)

According to formula (13), the weights of the second-order PRSM model, the RBF model, and the Kriging model in the hybrid surrogate model PRK-HSM are $\omega_{PRK\text{-}PRSM} = 0.318415$, $\omega_{PRK\text{-}RBF} = 0.348172$, and $\omega_{PRK\text{-}KRG} = 0.333413$.

$$F_{PRK\text{-}HSM}(\mathbf{x}) = 0.318415F_{PRSM}(\mathbf{x}) + 0.348172F_{RBF}(\mathbf{x}) + 0.333413F_{KRG}(\mathbf{x}) \tag{24}$$

The accuracy indexes of four hybrid surrogate models are shown in Table 3. It can be seen from Table 3 that the accuracy of the hybrid surrogate model RK-HSM is better than the other three hybrid surrogate models.

The accuracy of the hybrid surrogate model is affected by the accuracy of its sub-surrogate model. Comparing the accuracy of three sub-surrogate models and four hybrid surrogate models, it can be seen that hybrid surrogate models are better, among which the hybrid surrogate model RK-HSM performs best. The model accuracy indexes RMSE, $R^2$, and RAAE of RK-HSM are improved by 9.91%, 4.41%, and 15.60%, respectively, compared with the sub-surrogate model PRSM. With RMSE as the overall evaluation metric, the performance of the seven surrogate models is ranked as RK-HSM > RBF > PR-HSM > PRK-HSM > Kriging > PK-HSM > Second-order PRSM; with $R^2$ as the overall evaluation metric, the ranking is similarly RK-HSM > RBF > PR-HSM > PRK-HSM > Kriging > PK-HSM > Second-order PRSM; and, with RAAE as the overall evaluation metric, the ranking is RK-HSM > PR-HSM >

PRK-HSM > RBF > Kriging > PK-HSM > Second-order PRSM. The accuracy metrics of the sub-surrogate model RBF and Kriging are better than some constructed hybrid agent models, which is related to the construction principle of the surrogate model.

**Table 3.** Model accuracy of four hybrid surrogate models.

| Hybrid Surrogate Model | RMSE | $R^2$ | RAAE |
|---|---|---|---|
| PR-HSM | 0.329314 | 0.839732 | 0.303630 |
| PK-HSM | 0.348637 | 0.820373 | 0.342030 |
| RK-HSM | 0.322719 | 0.846087 | 0.298155 |
| PRK-HSM | 0.330431 | 0.838644 | 0.308248 |

Applying the best-performing hybrid surrogate model RK-HSM in this study, the values of the average pressure $F_{RK\text{-}HSM}$ on the inner side wall of the filter bag at 1 m from the bag bottom during the pulse-jet dust-cleaning process under different parameter combinations can be obtained, as shown in Figure 7, which can be used to study the influence of different parameters on the pulse-jet dust-cleaning process. The parameter conditions corresponding to the images in Figure 7 are shown in Table 4.

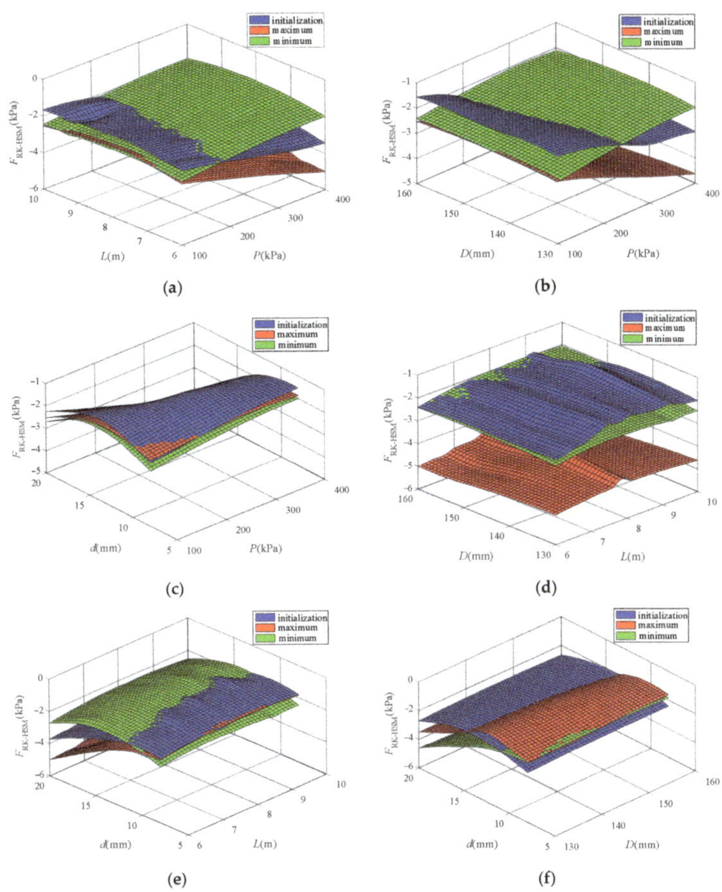


**Figure 7.** The variation of $F_{RK\text{-}HSM}$ with parameters: (**a**) *P*-*L*-$F_{RK\text{-}HSM}$; (**b**) *P*-*D*-$F_{RK\text{-}HSM}$; (**c**) *P*-*d*-$F_{RK\text{-}HSM}$; (**d**) *L*-*D*-$F_{RK\text{-}HSM}$; (**e**) *L*-*d*-$F_{RK\text{-}HSM}$; and (**f**) *D*-*d*-$F_{RK\text{-}HSM}$.

Table 4. The parameter conditions corresponding to the images.

| Image | Sub-Image | P/kPa | L/m | D/mm | d/mm |
|---|---|---|---|---|---|
| (a) | initialization<br>maximum<br>minimum | 100~400 | 6~10 | 145<br>160<br>130 | 16<br>20<br>8 |
| (b) | initialization<br>maximum<br>minimum | 100~400 | 8<br>10<br>6 | 130~160 | 16<br>20<br>8 |
| (c) | initialization<br>maximum<br>minimum | 100~400 | 8<br>10<br>6 | 145<br>160<br>130 | 8~20 |
| (d) | initialization<br>maximum<br>minimum | 250<br>400<br>100 | 6~10 | 130~160 | 16<br>20<br>8 |
| (e) | initialization<br>maximum<br>minimum | 250<br>400<br>100 | 6~10 | 145<br>160<br>130 | 8~20 |
| (f) | initialization<br>maximum<br>minimum | 250<br>400<br>100 | 8<br>10<br>6 | 130~160 | 8~20 |

As can be seen from Figure 7, the output response $F_{\text{RK-HSM}}$ is comprehensively affected by the multiple parameters. The change rules of the average pressure on the inner side wall of the filter bag at 1 m from the bag bottom $F_{\text{RK-HSM}}$ can be roughly understood through these images. However, as these response surfaces are interlaced, it is difficult to describe the variation completely and accurately. In order to obtain the optimal parameter combination which results in the maximum $F$, the genetic algorithm is chosen in this paper.

### 4.4. Genetic Algorithm for Seeking Optimal Parameters

A genetic algorithm [38] is a bionic algorithm conceived to search for the optimal solution based on the principle of biological evolution. As a global search heuristic algorithm, the genetic algorithm is suitable for solving complex optimization problems with strong robustness [39]. In this paper, combining the characteristics of the genetic algorithm and the common experience of parameter setting, the operational parameters of the genetic algorithm are set as shown in Table 5.

Table 5. Genetic algorithm parameters.

| Genetic Algorithm Parameter | Set Value |
|---|---|
| Number of populations | 200 |
| Termination algebra | 150 |
| Crossover probabilities | 0.8 |
| Probability of variation | 0.05 |

Under the initial experience setting condition ($P_0$ = 250 kPa, $L_0$ = 8 m, $D_0$ = 145 mm, and $d_0$ = 16 mm), the simulation result of $F$ is −2526.6148 Pa. For the seven surrogate models constructed above, including three sub-surrogate models and four hybrid surrogate models, the genetic algorithm is applied to seek the optimal combination of the four variable parameters ($P$, $L$, $D$, and $d$) involved in the pulse-jet dust-cleaning process to maximize $F$. The results are shown in Table 6.

**Table 6.** Genetic algorithm-optimization-seeking results.

| Surrogate Model Configuration | | Optimized Parameter Combination | | | | Simulation Result/Pa | Pressure Increase/% |
|---|---|---|---|---|---|---|---|
| | | *P*/kPa | *L*/m | *D*/mm | *d*/mm | | |
| Sub-Surrogate model | Second-order PRSM | 399.9669 | 8.2384 | 143.0993 | 8.0004 | −1723.4441 | 31.79 |
| | RBF | 225.3032 | 9.1380 | 147.1087 | 12.7751 | −1430.5546 | 43.38 |
| | Kriging | 267.3377 | 8.0878 | 131.7959 | 12.1181 | −1263.5617 | 49.99 |
| hybrid surrogate model | PR-HSM | 108.1891 | 8.6201 | 135.7152 | 14.4433 | −1333.7011 | 47.21 |
| | PK-HSM | 100.0094 | 8.7184 | 140.9819 | 14.5812 | −1532.9513 | 39.33 |
| | **RK-HSM** | **265.7430** | **8.0866** | **130.7656** | **12.11737** | **−1205.1605** | **52.30** |
| | PRK-HSM | 100.0550 | 8.7151 | 135.9512 | 14.6379 | −1331.2843 | 47.31 |

As can be seen from Table 6, the adoption of the genetic algorithm based on surrogate models can increase the simulation result of average pressure $F$ in the pulse-jet dust-cleaning process by more than 30% compared with the result under the initial empirical parameter condition ($P_0$ = 250 kPa, $L_0$ = 8 m, $D_0$ = 145 mm, and $d_0$ = 16 mm). In addition, the comprehensive performance of the hybrid surrogate models is better than that of the sub-surrogate models. Among these surrogate models, the hybrid surrogate model RK-HSM performs more outstandingly, increasing the $F$ by 52.30% to −1205.1605 Pa, which significantly improves the dust-cleaning effect of the bag filter. In summary, the hybrid surrogate model RK-HSM performs superiorly in both model accuracy and genetic algorithm seeking, that is, the hybrid surrogate model RK-HSM is more suitable for solving the optimization of bag filter dust-cleaning performance.

## 5. Conclusions

This paper proposes a hybrid surrogate model construction method based on the second-order PRSM, the RBF, and the Kriging as sub-surrogate models. Based on different hybrid modes, four hybrid surrogate models are constructed, namely PR-HSM, PK-HSM, RK-HSM, and PRK-HSM. On this basis, combined with the genetic algorithm-parameter-seeking strategy, the hybrid surrogate model optimization methods are provided for optimizing the pulse-jet cleaning process of industrial bag filters.

For the multi-parameter optimization of the bag filter pulse-jet cleaning process, the accuracy indexes RMSE, $R^2$, and RAAE of the hybrid surrogate model RK-HSM are improved by 9.91%, 4.41%, and 15.60%, respectively, compared with the sub-surrogate model PRSM, which greatly enhances the reliability and practicality of the surrogate model method. What's more, the accuracy metrics of the sub-surrogate models RBF and Kriging are better than some constructed hybrid agent models, which is related to the construction principle of the surrogate model. Therefore, the choice of the surrogate model should be appropriate to the problem to be solved, rather than aiming at the complexity of the method.

Aiming at maximizing the average pressure $F$ on the inner side wall of the filter bag at 1 m from the bag bottom, the genetic algorithm is used to search for the optimal parameter combination, which includes the pulse-jet pressure ($P$), filter bag length ($L$), filter bag diameter ($D$), and nozzle diameter ($d$). The simulation results show that among the seven surrogate models, including three sub-surrogate models and four hybrid surrogate models, the hybrid surrogate model RK-HSM performs more prominently, increasing the $F$ by 52.30% to −1205.1605 Pa, compared with the results under the initial empirical parameter condition. The application of the optimization method based on the hybrid surrogate model RK-HSM can greatly improve the dust-cleaning effect.

**Author Contributions:** Conceptualization, S.S., L.L., Z.Y. and C.Y.; Data curation, S.S., L.L. and Z.Y.; Funding acquisition, L.L. and Z.Y.; Methodology, S.S.; Resources, L.L., Z.Y., C.Y., Y.Z. and Y.C.; Software, S.S. and Y.Z.; Validation, S.S., W.C., C.Y. and Y.C.; Writing—original draft, S.S.; Writing—review and editing, L.L., Z.Y., W.C., C.Y., Y.Z. and Y.C. All authors have read and agreed to the published version of the manuscript.

**Funding:** This work is partially supported by National Natural Science Foundation of China (Grant number 52175461); Intelligent Manufacturing Project of Tianjin (Grant number 20201199).

**Institutional Review Board Statement:** Not applicable.

**Informed Consent Statement:** Not applicable.

**Data Availability Statement:** Not applicable.

**Conflicts of Interest:** The authors declare no conflict of interest.

## References

1. Liu, J.; Mao, N.; Sun, X.; Wang, J.; Chang, D.; Lin, X. Review of History, Present Situation and Development Trend of Bag Dust Removal Technology in China. *China Environ. Prot. Ind.* **2022**, *1*, 47–58.
2. Kang, F.; Cheng, H.; Leng, H.; Zen, S.; Xu, Z.; Li, X.; Lu, J.; Lin, Y.; Chen, H. Performance Optimization of Rectangular Flat Pleated Filter with Slit Nozzle for Dust Cleaning. *Powder Technol.* **2020**, *376*, 320–331. [CrossRef]
3. Chen, L.; Liu, Z.; Sun, Y.; Qian, F.; Han, Y.; Lu, J. Experimental Study on the Dust-Cleaning Performance of New Structure Microporous Membrane Filter Plate. *Atmosphere* **2022**, *13*, 817. [CrossRef]
4. Park, S.; Joe, Y.; Shim, J.; Park, H.; Shin, W. Non-uniform Filtration Velocity of Process Gas Passing Through a Long Bag Filter. *J. Hazard. Mater.* **2019**, *365*, 440–447. [CrossRef] [PubMed]
5. Fan, B.; Li, F.; Wang, H.; Huang, G. Research on Dust-Cleaning Performance of Pulse Bag Filter by Response Surface Methodology. In *Mechanics and Materials Science, Proceedings of the 2016 International Conference on Mechanics and Materials Science MMS 2016, Guangzhou, China, 15–16 October 2016*; World Scientific: Singapore, 2016; pp. 412–424.
6. Sun, Z.; Yao, Y.; Guo, D.; Yang, G.; Yao, S.; Zhang, Y.; Chen, D.; Li, G.; Shang, K.; Jia, L. Research Progress in Aerodynamic Optimization of High-speed Trains. *Chin. J. Theor. Appl. Mech.* **2021**, *53*, 51–74.
7. Xu, Z.; Guo, X. Optimization of Bentonite Parameters for Shield Tunneling Based on Response Surface Method. *Chin. J. Geotech. Eng.* **2021**, *43*, 194–200.
8. Thakre, U.; Mote, R. Uncertainty Quantification and Statistical Modeling of Selective Laser Sintering Process Using Polynomial Chaos Based Response Surface Method. *J. Manuf. Process.* **2022**, *81*, 893–906. [CrossRef]
9. Sun, W.; Cheng, W. Material Properties Updating of Honeycomb Sandwich Plates Using a Global Optimization Technique Based on Response Surface Model. *J. Mater. Eng.* **2019**, *47*, 159–166.
10. Xu, K.; Wang, G.; Zhang, L.; Wang, L.; Yun, F.; Sun, W.; Wang, X.; Chen, X. Multi-objective Optimization of Jet Pump Based on RBF Neural Network Model. *J. Mar. Sci. Eng.* **2021**, *9*, 236. [CrossRef]
11. Wang, W.; Li, J.; Liu, G.; Wei, J.; Zhang, Z.; Cheng, M. Optimization Design of Drum Brake Stability Based on Kriging Surrogate Model. *J. Vib. Shock* **2021**, *40*, 134–138.
12. Li, X.; Cui, H.; Hu, C.; Liu, X.; Huang, S. Optimal Design of Thrust Characteristics of Flat-Type Permanent Magnet Linear Synchronous Motor. *Trans. China Electrotech. Soc.* **2021**, *36*, 916–923.
13. Jie, H.; Wu, Y.; Ding, J. An Adaptive Metamodel-based Global Optimization Algorithm for Black-box Type Problems. *Eng. Optim.* **2015**, *47*, 1459–1480. [CrossRef]
14. Li, Z.; Wang, D.; Shi, T.; Zheng, D. Optimization Design for Permanent Magnet Drive Based on Entropy-weight and Hybrid Surrogate Model. *Electr. Mach. Control* **2016**, *20*, 102–108.
15. Xie, H.; Yang, Y.; Tong, L.; Zeng, W.; Xie, S. Multi-objective Optimization of the Suspension Parameters for High Speed Rail Vehicle Based Ona Hybrid Surrogate Model. *J. Railw. Sci. Eng.* **2016**, *13*, 2056–2063.
16. He, Y.; Song, B.; Cao, Y. Multi-Step Structural Optimization Design of Multi-Bubble Pressure Cabin in the Autonomous Underwater Vehicle with Blended-Wing-Body. *J. Northwestern Polytech. Univ.* **2018**, *36*, 664–670. [CrossRef]
17. Andersen, B.; Nielsen, N.; Walther, J. Numerical and Experimental Study of Pulse-jet Cleaning in Fabric Filters. *Powder Technol.* **2016**, *291*, 284–298. [CrossRef]
18. Liu, X.; Shen, H. Effect of Venturi Structures on the Cleaning Performance of a Pulse Jet Baghouse. *Appl. Sci.* **2019**, *9*, 3687. [CrossRef]
19. Zhang, Q.; Liu, D.; Wang, M.; Shu, Y.; Xu, H.; Chen, H. Characteristics and Evaluation Index of Pulse-jet Dust Cleaning of Filter Cartridge. *Process Saf. Environ. Prot.* **2022**, *157*, 362–374. [CrossRef]
20. Song, X.; Lv, L.; Li, J.; Sun, W.; Zhang, J. An Advanced and Robust Ensemble Surrogate Model: Extended Adaptive Hybrid Functions. *J. Mech. Des.* **2018**, *140*, 041402. [CrossRef]
21. Wang, M.; Huang, Q. A New Hybrid Uncertain Analysis Method for Structural-acoustic Systems with Random and Interval Parameters. *Comput. Struct.* **2016**, *175*, 15–28. [CrossRef]

22. He, Z.; Xiong, X.; Yang, B.; Li, H. Aerodynamic Optimisation of a High-speed Train Head Shape Using an Advanced Hybrid Surrogate-based Nonlinear Model Representation Method. *Optim. Eng.* **2020**, *23*, 59–84. [CrossRef]
23. Yuan, Z.; Kong, L.; Gao, D.; Tong, X.; Feng, Y.; Yang, G.; Yang, Z.; Li, S. Multi-objective Approach to Optimize Cure Process for Thick Composite Based on Multi-field Coupled Model with RBF Surrogate Model. *Compos. Commun.* **2021**, *24*, 100671. [CrossRef]
24. Dou, S.; Li, J.; Kang, F. Health Diagnosis of Concrete Dams Using Hybrid FWA with RBF-based Surrogate Model. *Water Sci. Eng.* **2019**, *12*, 188–195. [CrossRef]
25. Farazandeh, E.; Mirzaei, D. A Rational RBF Interpolation with Conditionally Positive Definite Kernels. *Adv. Comput. Math.* **2021**, *47*, 74. [CrossRef]
26. Chen, W.; Hong, Y.; Lin, J. The Sample Solution Approach for Determination of the Optimal Shape Parameter in the Multiquadric Function of the Kansa Method. *Comput. Math. Appl.* **2018**, *75*, 2942–2954. [CrossRef]
27. Long, T.; Liu, J.; Wang, G.; Liu, L.; Shi, R.; Guo, X. Discuss on Approximate Optimization Strategies Using Design of Computer Experiments and Metamodels for Flight Vehicle Design. *J. Mech. Eng.* **2016**, *52*, 79–105. [CrossRef]
28. Haeri, A.; Fadaee, M. Efficient Reliability Analysis of Laminated Composites Using Advanced Kriging Surrogate Model. *Compos. Struct.* **2016**, *149*, 26–32. [CrossRef]
29. Xing, J.; Luo, Y.; Gao, Z. A Global Optimization Strategy Based on the Kriging Surrogate Model and Parallel Computing. *Struct. Multidiscip. Optim.* **2020**, *62*, 405–417. [CrossRef]
30. Hao, P.; Feng, S.; Liu, H.; Wang, Y.; Wang, B.; Wang, B. A Novel Nested Stochastic Kriging Model for Response Noise Quantification and Reliability Analysis. *Comput. Methods Appl. Mech. Eng.* **2021**, *384*, 113941. [CrossRef]
31. Noack, M.; Yager, K.; Fukuto, M.; Doerk, G.; Li, R.; Sethian, J. A Kriging-based Approach to Autonomous Experimentation with Applications to X-ray Scattering. *Sci. Rep.* **2019**, *9*, 11809. [CrossRef] [PubMed]
32. Yang, Y.; Sun, W.; Su, G. A Novel Support-Vector-Machine-Based Grasshopper Optimization Algorithm for Structural Reliability Analysis. *Buildings* **2022**, *12*, 855. [CrossRef]
33. Schwab, B.; Janzen, S.; Magnan, N.; Thompson, W. Constructing a Summary Index Using the Standardized Inverse-covariance Weighted Average of Indicators. *Stata J.* **2020**, *20*, 952–964. [CrossRef]
34. Li, W.; Lu, L.; Xie, X.; Yang, M. A Novel Extension Algorithm for Optimized Latin Hypercube Sampling. *J. Stat. Comput. Simul.* **2017**, *87*, 2549–2559. [CrossRef]
35. Zhang, L.; Li, T.; Zhang, J.; Piao, R. Optimization on the Crosswind Stability of Trains Using Neural Network Surrogate Model. *Chin. J. Mech. Eng.* **2021**, *34*, 86. [CrossRef]
36. Mehdiyev, N.; Enke, D.; Fettke, P.; Loos, P. Evaluating Forecasting Methods by Considering Different Accuracy Measures. *Procedia Comput. Sci.* **2016**, *95*, 264–271. [CrossRef]
37. Zhou, Y.; Sun, J.; Wang, S.; Hao, Y.; Liu, C. Risk Assessment of Small-signal Instability for Renewable Power System Based on High-dimensional Model Representation Method. *Autom. Electr. Power Syst.* **2022**, *46*, 73–82.
38. Katoch, S.; Chauhan, S.; Kumar, V. A Review on Genetic Algorithm: Past, Present, and Future. *Multimed. Tools Appl.* **2021**, *80*, 8091–8126. [CrossRef]
39. Wang, Z.; Wang, H.; Cheng, M. Fuzzing Testing Sample Set Optimization Scheme Based on Heuristic Genetic Algorithm. *J. Beijing Univ. Aeronaut. Astronaut.* **2022**, *48*, 217–224.

*energies*

MDPI

*Article*

# Contrast Estimation in Vibroacoustic Signals for Diagnosing Early Faults of Short-Circuited Turns in Transformers under Different Load Conditions

**Jose R. Huerta-Rosales [1], David Granados-Lieberman [2], Juan P. Amezquita-Sanchez [1], Arturo Garcia-Perez [3], Maximiliano Bueno-Lopez [4] and Martin Valtierra-Rodriguez [1,*]**

1 ENAP-Research Group, CA-Sistemas Dinámicos y Control, Laboratorio de Sistemas y Equipos Eléctricos (LaSEE), Facultad de Ingeniería, Universidad Autónoma de Querétaro (UAQ), Campus San Juan del Río, Río Moctezuma 249, Col. San Cayetano, San Juan del Río 76807, Querétaro, Mexico

2 ENAP-Research Group, CA-Fuentes Alternas y Calidad de la Energía Eléctrica, Departamento de Ingeniería Electromecánica, Tecnológico Nacional de México, Instituto Tecnológico Superior de Irapuato (ITESI), Carretera Irapuato-Silao km 12.5, Colonia El Copal, Irapuato 36821, Guanajuato, Mexico

3 ENAP-Research Group, División de Ingenierías, Universidad de Guanajuato, Campus Irapuato-Salamanca, Carretera Salamanca-Valle de Santiago km 3.5 + 1.8 km, Comunidad de Palo Blanco, Salamanca 36885, Guanajuato, Mexico

4 Department of Electronics, Instrumentation, and Control, Universidad del Cauca, Popayán 190002, Cauca, Colombia

* Correspondence: martin.valtierra@enap-rg.org

**Citation:** Huerta-Rosales, J.R.; Granados-Lieberman, D.; Amezquita-Sanchez, J.P.; Garcia-Perez, A.; Bueno-Lopez, M.; Valtierra-Rodriguez, M. Contrast Estimation in Vibroacoustic Signals for Diagnosing Early Faults of Short-Circuited Turns in Transformers under Different Load Conditions. *Energies* **2022**, *15*, 8508. https://doi.org/10.3390/en15228508

Academic Editors: Yuling He, David Gerada, Conggan Ma and Haisen Zhao

Received: 7 October 2022
Accepted: 11 November 2022
Published: 14 November 2022



**Abstract:** The transformer is one of the most important electrical machines in electrical systems. Its proper operation is fundamental for the distribution and transmission of electrical energy. During its service life, it is under continuous electrical and mechanical stresses that can produce diverse types of damage. Among them, short-circuited turns (SCTs) in the windings are one of the main causes of the transformer fault; therefore, their detection in an early stage can help to increase the transformer life and reduce the maintenance costs. In this regard, this paper proposes a signal processing-based methodology to detect early SCTs (i.e., damage of low severity) through the analysis of vibroacoustic signals in steady state under different load conditions, i.e., no load, linear load, nonlinear load, and both linear and nonlinear loads, where the transformer is adapted to emulate different conditions, i.e., healthy (0 SCTs) and with damage of low severity (1 and 2 SCTs). In the signal processing stage, the contrast index is analyzed as a fault indicator, where the Unser and Tamura definitions are tested. For the automatic classification of the obtained indices, an artificial neural network is used. It showed better results than the ones provided by a support vector machine. Results demonstrate that the contrast estimation is suitable as a fault indicator for all the load conditions since 89.78% of accuracy is obtained if the Unser definition is used.

**Keywords:** artificial neural networks; contrast estimation; fault diagnosis; short-circuited turns; transformer fault; vibroacoustic signals

## 1. Introduction

The condition monitoring and fault diagnosis of electrical machines have become essential tools for industrial processes since their reliability and safety can be improved [1]. Among the electric machines, the transformer can suffer various internal faults such as winding and core deformations, broken clamping structures, short-circuit turns (SCT), bending, and others [2]. For this machine, the winding is an essential operating component and, at the same time, it is one of the most vulnerable components; in fact, the associated faults to windings are nearly 40% [3], whereas in the most cases the initial damage can lead to the complete fault of the transformer, increasing repairment costs and people risks.

In this regard, the development and application of early fault detection methods are of paramount importance [4,5].

In literature, different techniques to carry out the monitoring and diagnosis of transformer windings by analyzing different types of signals have been proposed [6–9]. However, the vibration signal analysis has proven to be an effective tool for monitoring and diagnosing internal faults in transformers [6], since the vibrational response changes if the winding mechanical properties are modified [7], e.g., when a fault condition appears. Some studies measure acoustic signals instead of vibration signals [8,9]. It is observed that acoustic signals are produced by the vibration of a source, thus, they contain the same information about the source behavior. In this way, the vibration and acoustic signals are interrelated and homologous [8,9].

The machinery diagnosis that employs vibroacoustic (VA) signals typically consists of three essential steps: acquisition and preprocessing of data, feature extraction from the data, and classification based on the extracted features. Successful methods have shown that feature extraction is the most crucial step in machinery monitoring and diagnosis [7,10]. In literature, several techniques extract features from VA signals in distinct domains (e.g., time, frequency, or time-frequency) for assessing the transformer condition [11–22]. A widely employed tool to change from time-domain to frequency-domain is the Fourier Transform (FT). The FT allows extracting frequential information related to the system condition. In [11], they employ the FT in VA signals and estimate the total harmonic distortion (THD) to assess the transformer healthy condition. Meanwhile, in [9], the THD, dominant frequencies, and the ratio of frequencies from the spectrum of acoustic signals are estimated. They establish a warning threshold that allows identifying a direct current bias state. Although the FT presents promising results, it has some limitations. These limitations compromise the detection of changes in frequency over time, mainly due to the nonstationary nature of the signal and its high-level noise, among others [12]. To overcome these limitations, the short-time Fourier Transform (STFT) is introduced. This technique provides a two-dimensional representation of the analyzed signal in time, i.e., a time-frequency representation (TFR) in the time-frequency domain. In [13], the STFT is used to analyze a VA signal, where the changes in frequency are associated to the anomalous behavior in the transformer. Nevertheless, the STFT has the resolution trade-off issue, i.e., it can gain frequency resolution but, at the same time, loss resolution in time, and vice versa, which can compromise the accuracy of the results; in addition, if the signal presents a severe level of noise, the performance decreases due to the spectral leakage problem, compromising the diagnosis as well [14]. Another effective way to construct a TFR is using the Wavelet transform (WT) and its variations. The WT has better time and frequency localization, making it more effective than STFT for some applications [15]. The work presented in [16] uses the WT to analyze the transient state of the transformer from its vibration signals. Also, a WT-based work called Empirical WT is used in [17], in which it is employed to build a TFR and estimate the multiscale entropy as a fault feature. On the other hand, in [18], the Wavelet packet is applied to detect DC bias in transformers, obtaining very effective results. Yet, the success of a WT-based approach depends on the proper selection of both the mother wavelet and the decomposition level [19], which is different in each application.

As mentioned above, the changing of domain (i.e., time or frequency) always requires the application of a transformation technique, implying a computational cost. In this regard, feature extraction in the time domain has been explored. The work presented in [20] diagnoses a transformer under a SCT fault by employing different fractal algorithms. They extract information from the vibration signals to detect diverse SCT fault severities. In [21], several statistical features are extracted from the vibration signals, then, the most representative ones are selected to diagnose the transformer under different SCT fault severities. Also, a mathematical statistics-based method for the analysis of vibration signals is presented in [22], where the probability distribution is used to identify the SCT fault. Although promising results have been obtained, there are still some issues that have not been completely addressed, e.g., damage detection of low severity (i.e., 1 or 2 SCTs) under different load conditions, which is of paramount importance since online condition monitoring systems for predictive maintenance can be developed. It is worth noting that the detection of early faults is a challenging task since the behavior of the transformer slightly changes. The challenge increases if both the high-level noise in the VA signals and the electrical stresses associated to the different load conditions are considered. Therefore, the development and application of new methods are still current needs.

Recently, the contrast index used in image processing for analyzing texture [23,24] has been used for diagnosing broken rotor bars in induction motors through the analysis of electric current signals [25], showing that it can provide information about variations into a signal waveform, which is congruent as it detects variations in color and brightness in an image. In this regard, as the main contribution, this work explores the application of the contrast index to diagnose early SCT faults, where a complete methodology (i.e., signal processing and automatic pattern recognition) is proposed. Also, unlike other works, the proposal is assessed when the transformer is under different load conditions, increasing its applicability in real operating scenarios; besides, it is worth noting that the detection of early SCT faults is not an easy task due to imperceptible changes in the transformer and the high noise into the VA signals. In order to develop and validate the proposed method, the VA signals from a transformer capable to emulate different early SCT severities, i.e., 0, 1, and 2 SCTs, where 0 SCTs represents the healthy condition, are firstly acquired. At this stage, four different load conditions, i.e., (1) no load, (2) linear load, (3) nonlinear load, and (4) both linear and nonlinear loads, are considered. In general, these loads represent real life operating conditions [5]. Once the signals have been acquired, two contrast definitions, i.e., Unser and Tamura, are computed and compared in order to observe which one provides the best sensibility to the fault. Finally, the different values of contrast for the different SCT conditions are processed by an artificial neural network (ANN) for automatic pattern recognition, where the performance of a support vector machine (SVM) is also compared. Results demonstrate the effectiveness of the proposal. The rest of the paper is organized as follows. Section 2 presents the theoretical background for transformer vibrations, contrast index, and ANNs. The proposed methodology is described in Section 3. Section 4 shows the experimental setup and the obtained results. A discussion of works is presented in Section 5. Finally, Section 6 draws the obtained conclusions.

## 2. Theoretical Background

### 2.1. Transformer Vibration

Transformer vibration is from two principal sources: windings and core. Although the vibration components of each source are distinct, the winding component is smaller than the component from the core [9].

The winding vibration is principally caused by the electric force, which is generated by the current that circulates through the windings and its interaction with the magnetic leakage flux [26]. The forces created by winding vibration have two components: axial and radial, where the force is proportional to the square of the current as shown in Equation (1) and its fundamental frequency is twice the fundamental frequency of the current signal [27].

$$F_{winding} \propto I^2 \quad (1)$$

On the other hand, the main cause of core vibration is the magnetostriction phenomenon. This phenomenon changes the shape of a ferromagnetic material when it is under the influence of a magnetic field. Then, in Equation (2), the force magnitude from core vibration is proportional to the square of the voltage [28]. The fundamental frequency due to the magnetostriction is twice the fundamental frequency of the voltage signal, but high frequency harmonics can appear due to the nonlinear magnetostriction behavior. Also, the force direction is perpendicular to the core plane [29].

$$F_{core} \propto V^2 \quad (2)$$

As mentioned above, the vibration signals and acoustic signals are homologous. In this regard, any change in the VA signal is directly related to the transformer performance. Therefore, the characterization of these changes through time-domain indicators (e.g., the contrast index) can help to determine the transformer condition.

### 2.2. *Contrast*

The contrast index is a textural feature that measures the variations present in an image [30]. In a simple way, it is the difference between a pair of black and white dots. The contrast value can be maximum when two pair of dots have opposite values in the gray scale, e.g., 0 and 255, and be minimum when their values are similar. Although this index is intended for images, it can also be applied to 1D signals. Two simple ways to estimate the contrast are presented by Unser [31] and Tamura et al. [32].

#### 2.2.1. Unser Contrast

Unser employs the sum and difference of histograms to estimate the contrast [31]. In order to do so, firstly, the variation between two points separated by $d$ samples in a signal $L(n)$ is computed as follows:

$$D_{n,d} = L(n) - L(n+d) \tag{3}$$

where $n \in \{0, 1, \ldots, N-1\}$. With this value, the histogram of differences is computed as:

$$h_d(j) = card\{n \in N, D_{n,d} = j\} \tag{4}$$

where $j = -N_g + 1, -N_g + 2, \ldots, N_g - 2, N_g - 1$ for at least $N_g$ levels of gray. $card\{\ \}$ refers to the number of elements of a set.

The normalized difference histogram is then given by:

$$P_D = \frac{h_d(j)}{T} \tag{5}$$

where $T$ is the total number of counts. It is computed by:

$$T = \sum_j h(j) \tag{6}$$

Finally, the contrast can be computed as:

$$C = \sum_j j^2 P_D(j) \tag{7}$$

#### 2.2.2. Tamura Contrast

Tamura et al. propose the contrast in function of [32]:

1. Dynamic range of gray levels;
2. Polarization of the distribution of black and white on the histogram of gray levels or the ratio between the black and white areas.

In this regard, the kurtosis provides the polarization factor as:

$$\alpha_4 = \frac{\mu_4}{\sigma^4} \tag{8}$$

where $\mu_4$ is the fourth moment around the mean and $\sigma^2$ is the variance that represents the distribution of the gray levels. Therefore, the Tamura contrast is defined as:

$$C = \frac{\sigma}{(\alpha_4)^n} \tag{9}$$

where $n$ is a positive number, which is experimentally defined as $n = 1/4$.

### 2.3. Artificial Neural Networks

ANNs are a computational structure designed to emulate a neurological structure of a human brain, which are constructed by individual elements with similar behavior to the biologic neuron also called neurons, having the capability to learn and solve problems [33]. One of the most employed ANN architectures is the feed-forward neural network due to its reduced computational load and its simplicity to be used as a classifier [33]. Figure 1a shows the ANN architecture. This architecture includes an input layer, hidden layers, and an output layer. The information moves from the input layer the output layer. In each layer, it can have a single or multiple neurons. In Equation (10), the mathematical function of a neuron is shown, whereas its structure appears in Figure 1b. The neuron consists of a summation of multiplications between the inputs $x_i$ and weights $w_i$, which emulates the synapsis process, and a bias $b$. The result of the summation is evaluated by a function called the activation function $f(\cdot)$, which is a nonlinear function that provides the capability of modeling nonlinear relationships. Finally, a training data stage is carried out to characterize the ANN weights, i.e., a set of inputs and the desired outputs are presented to the ANN. Consequently, comparing the desired outputs and the computed outputs, the error can be estimated. Employing a training rule, the weights can be adjusted to minimize the error in a subsequent iteration. This process is repeated until the error is acceptable.

$$O_1 = f\left(\sum_{i=1}^{l} w_i x_i + b\right) \quad (10)$$

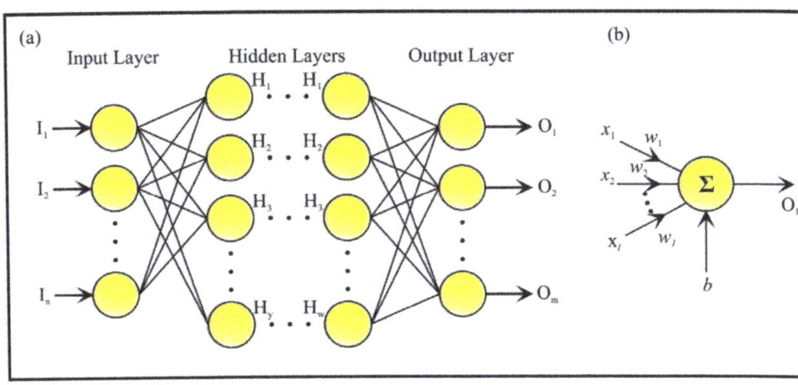


**Figure 1.** ANN architecture: (**a**) ANN with hidden layers and (**b**) a single neuron.

## 3. Proposed Methodology

Figure 2 shows the flowchart for the proposed methodology. In general, as depicted by the red dashed rectangle, it is a three-step process: (1) signal processing which consists of the signal segmentation and the gray-scale normalization, (2) estimation of contrast as fault index, and (3) design and validation of a classification algorithm. To test the proposal, VA signals from a transformer under early SCT fault conditions and different load conditions are analyzed. In order to do so, firstly, VA signals are acquired from a single-phase transformer which can emulate different early SCT fault conditions: 0, 1, and 2 SCTs, where 0 SCTs is the healthy condition. To select the SCT condition, a conductor cable is connected to different taps with a resistor R. This resistor helps to limit the short-circuit current. For carrying out the transformer energization and connecting/disconnecting the different loads, three solid-state relays are used (see the switches in Figure 2). Once the VA signals are acquired, they are segmented, taking 1 s of the signal in each different load case: (1) no load, (2) linear load, (3) linear and nonlinear loads, and (4) nonlinear load.

This process is repeated for all the SCT conditions previously mentioned. Then, as part of the proposed methodology, the VA signals are normalized into a gray-scale, i.e., the VA signal takes values from 0 to 255. For each new signal, two different contrast indicators, i.e., Tamura and Unser, are computed to obtain the best fault indicator. Finally, in the automatic classification step, two different algorithms, i.e., ANN and SVM, are evaluated to detect and classify the SCT transformer condition under different load conditions. Results demonstrates that the Unser contrast and the ANN algorithm present the best results for detection and classification of SCT fault conditions regardless the load type.

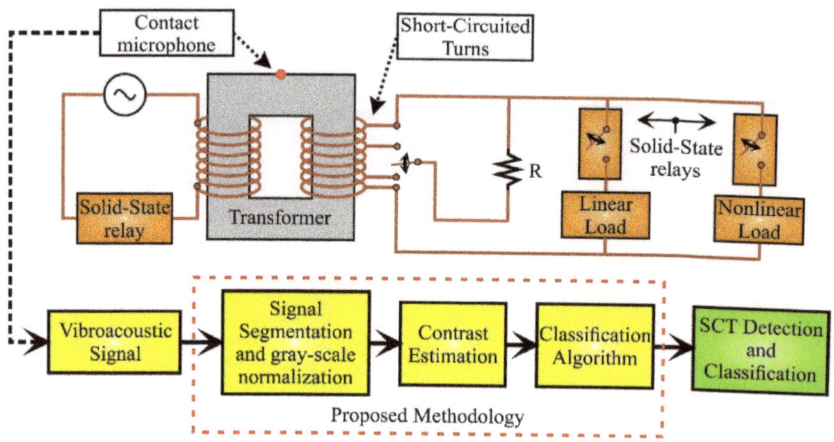


**Figure 2.** Flowchart for the proposed methodology.

## 4. Experimental Setup and Results

This section shows the experimental setup and the obtained results

### *4.1. Experimental Setup*

The experimental setup for the proposed work is shown in Figure 3. For the tests, a single-phase transformer of 1.5 kVA operated to 220 V with a relation 2:1 is employed. The transformer is modified to emulate early SCT fault conditions in the secondary winding. In order to do so, the transformer winding is unwound and, then, various taps are pulled out while rewinding. To protect the winding during the SCT fault, a resistor, R, of 2 ohms is used.

For the transformer energization and the connection/disconnection of the linear and nonlinear loads, three solid-state relays, model SAP4050D, are used. The linear and nonlinear loads represent approximately the 93% of the total load of the transformer, i.e., 900 W and 500 W, respectively. The linear load is a resistive array of 15 ohms and the nonlinear load is the combination of a resistive array of 24 ohms and a rectifier. The rectifier is composed by an array of diodes of 800 V and 8 A with a capacitor filter of 210 μF and 450 V. To measure the VA signals from the transformer and provide a low-cost, easy-to-use, and commercial solution, a contact microphone piezoelectric pickup (TP-6 model from RECKLESS) is used. Its operating principle is based on a piezoelectric diaphragm that measures the vibrations (i.e., acceleration in $m/s^2$) from the contact zone with 6.4 kHz of maximum bandwidth which is good enough to capture the frequency range of interest, i.e., 1 kHz. It is located at the top center of the transformer core; in this location, the core and winding vibrations as well as the generated noise can be received in a similar way from a geometrical point of view. The data are acquired by using a data acquisition system (DAS) based on the National Instruments NI-USB-6211 board which is configured with a sampling frequency of 7812.5 samples/s. For each condition, i.e., 0, 1, and 2 SCTs, and the different combinations of load, 20 tests are carried out. All the data acquisition and their processing are on MATLAB software by using a personal computer (PC), featuring Windows 10, a processor 17-4510U at 2.6 GHz, and 8 GB of RAM.

Figure 4 shows an example of the acquired signals. The acquisition time is of 10 s; thus, the transient and steady states can be captured. Current signals (bottom images) show clearly the changes associated to load condition. There are four load conditions: (1) no load (S0), (2) linear load (S1), (3) both linear and nonlinear loads (S2), and (4) nonlinear load (S4), where the activation/deactivation times are 2, 4, 6, and 8 s, respectively. From these time-windows, sections of 1 s are selected. This time-window contains enough information about the steady state and avoids overlaps with other load scenarios (see the red dashed rectangles in Figure 4). Once the signals are segmented, the data processing (i.e., the contrast index estimation and the classification based on a pattern recognition algorithm) is carried out.

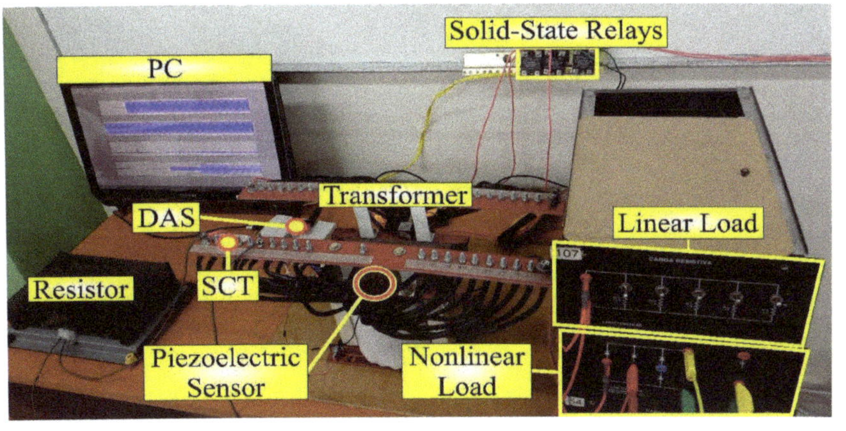


**Figure 3.** Experimental setup.

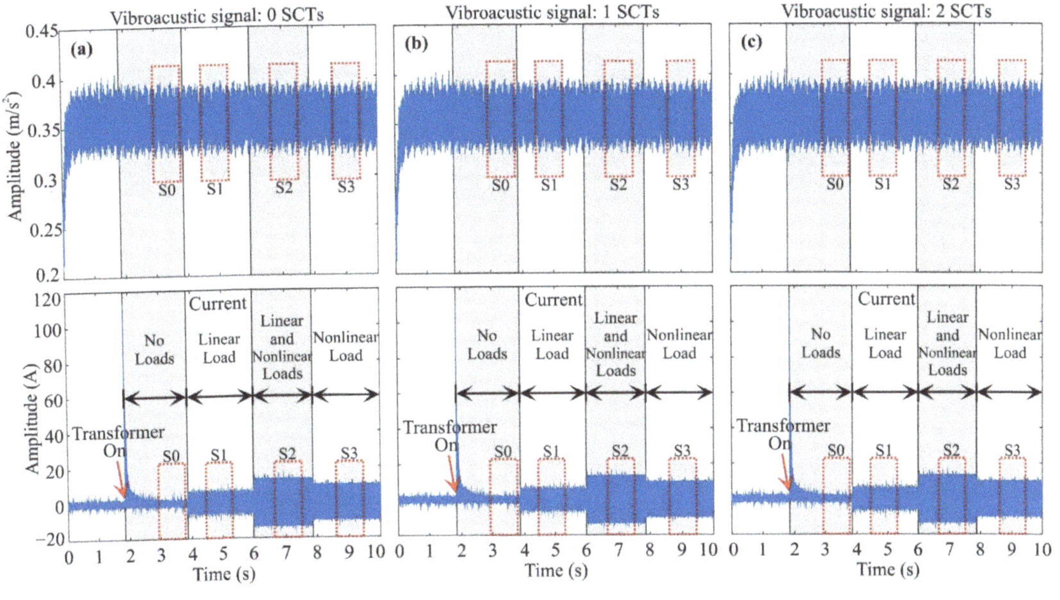


**Figure 4.** Signals acquired from the transformer: VA signals for: (**a**) 0 SCTs, (**b**) 1 SCTs, and (**c**) 2 SCTs.

## 4.2. Results

### 4.2.1. Gray-Scale Normalization

For the contrast estimation, the signals have to be normalized to a grayscale, i.e., to values from 0 to 255. Through this linear conversion, the minimum value of the VA signal is mapped to zero, whereas the maximum value is mapped to 255. Also, rounding down is carried out only to consider integer numbers (see Figure 5e–h). In general, this change in the VA signal values does not modify its behavior but it helps to maintain the same reference level, minimizing and generalizing the impact of different load levels.

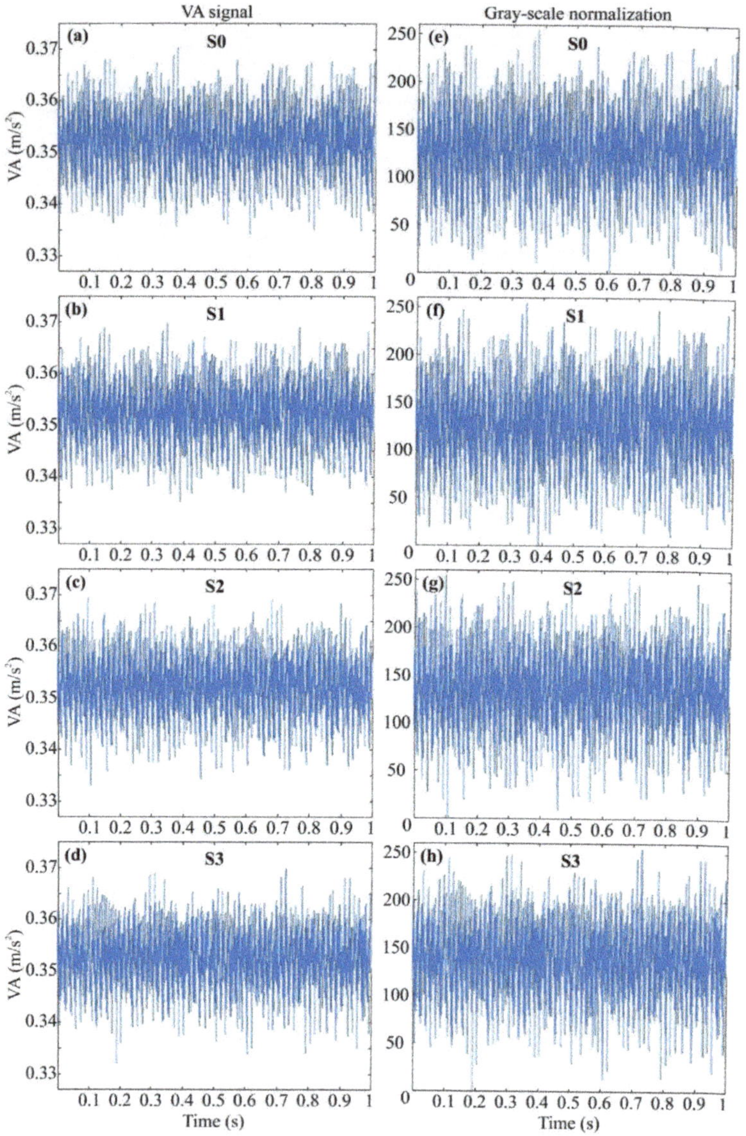


**Figure 5.** Example of a VA acquired signal and its gray-scale normalization: (**a**) section S0, (**b**) section S1, (**c**) section S2, (**d**) section S3, and gray-scale normalization for sections (**e**) S0, (**f**) S1, (**g**) S2, and (**h**) S3.

#### 4.2.2. Unser and Tamura Contrast Estimations

Two contrast definitions, i.e., Unser and Tamura, are used with the purpose of determining which approach is the most adequate to characterize variations associated to an early SCT fault condition in VA signals from a transformer.

The Unser [31] definition uses the normalized differences histograms, i.e., Equation (7), to compute the contrast. In order to do so, a distance of separation between two data points from the VA signal, i.e., $d$, has to be selected. After different trial-and-error tests, a value of $d = 15$ is selected since it presented the best results as will be discussed at the end of this section. Table 1 and Figure 6 show the obtained results. Table 1 presents the mean ($\mu$) and the standard deviation ($\sigma$) for the values of contrast by including the three SCT conditions (i.e., 0, 1, and 2 SCTs), the four load conditions (i.e., S0, S1, S2, and S3), and the 20 tests carried out for each condition. From the results (Table 1 or Figure 6a), it can be observed that the mean of the contrast value does not exhibit considerable variations for the different load combinations, i.e., the load does not influence the contrast index estimation.

In a similar way, the Tamura contrast index is estimated for the SCT fault conditions and their different load conditions, but with the difference that the Tamura contrast is a statistical-based calculation, i.e., Equation (9). As can be observed in both Table 1 and Figure 6, this index also presents a similar behavior for the different load combinations. Although the estimated contrast values for both Unser and Tamura indexes are not on the same scale, both graphs are quite similar, which reaffirms the low influence of the load in the contrast estimation.

**Table 1.** Mean ($\mu$) and standard deviation ($\sigma$) for the contrast values, different load combinations, and different SCT fault conditions.

| | | Unser Contrast, $d = 15$ | | | | Tamura Contrast | | | |
|---|---|---|---|---|---|---|---|---|---|
| | | S0 | S1 | S2 | S3 | S0 | S1 | S2 | S3 |
| 0 SCTs | $\mu$ | 3862.5413 | 3426.7146 | 3597.4812 | 3511.6746 | 38.1231 | 35.4044 | 36.5590 | 36.0443 |
| | $\sigma$ | 683.7739 | 690.3487 | 652.7069 | 848.0176 | 4.6503 | 4.6849 | 4.4967 | 5.4222 |
| 1 SCTs | $\mu$ | 6732.2393 | 6363.5508 | 6543.8623 | 6294.7972 | 53.8102 | 51.9030 | 53.0096 | 51.8991 |
| | $\sigma$ | 1441.4742 | 1220.7590 | 1301.9952 | 1195.6256 | 7.5780 | 6.6916 | 6.9990 | 6.5757 |
| 2 SCTs | $\mu$ | 8699.9884 | 8304.6975 | 8529.3658 | 8147.1655 | 62.6589 | 60.8933 | 62.1086 | 60.5648 |
| | $\sigma$ | 1084.0015 | 1022.1635 | 1049.8129 | 1016.2967 | 4.3483 | 4.1768 | 4.2930 | 4.2054 |

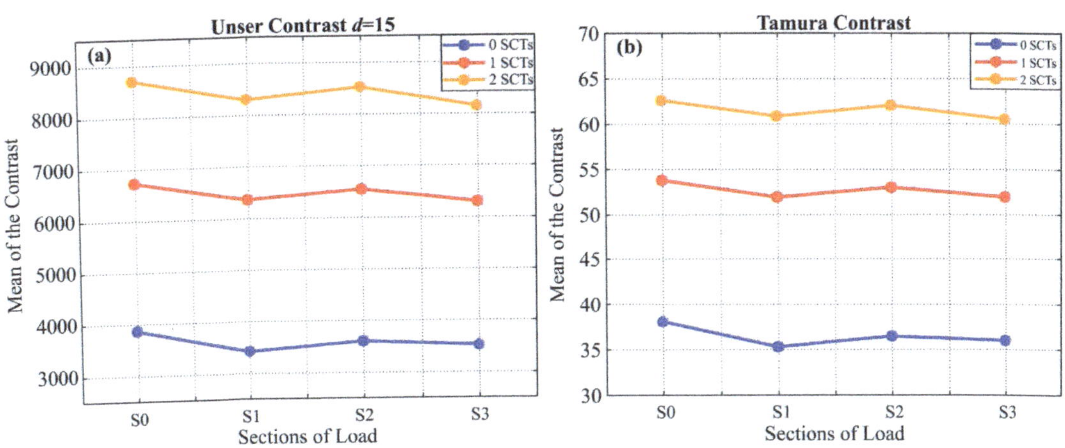


**Figure 6.** Mean ($\mu$) of the contrast values for different loads: (**a**) Unser contrast with $d = 15$ and (**b**) Tamura contrast.

As the values of the Unser contrast does not change with the load, the values for each section (load combinations: S0, S1, S2, and S3) can produce a bigger cluster of SCT conditions.

Figure 7 shows the distributions of the clusters for each SCT condition. From this figure, it is also observed that the contrast indices are sensitive to the SCT severity since their value increases when the SCT fault severity increases. In Table 2, the values of the mean ($\mu$) and the deviation standard ($\sigma$) for these clusters are shown. All these results characterize in a numerical way the behavior of each SCT condition. It is worth noting that some conventional time features (e.g., peak factor, kurtosis, standard deviation, among others) have also been analyzed. However, although some of them present a certain degree of separation, their clusters are overlapped due to their more significant variances which are also affected by the load condition, resulting in lower classification accuracy.

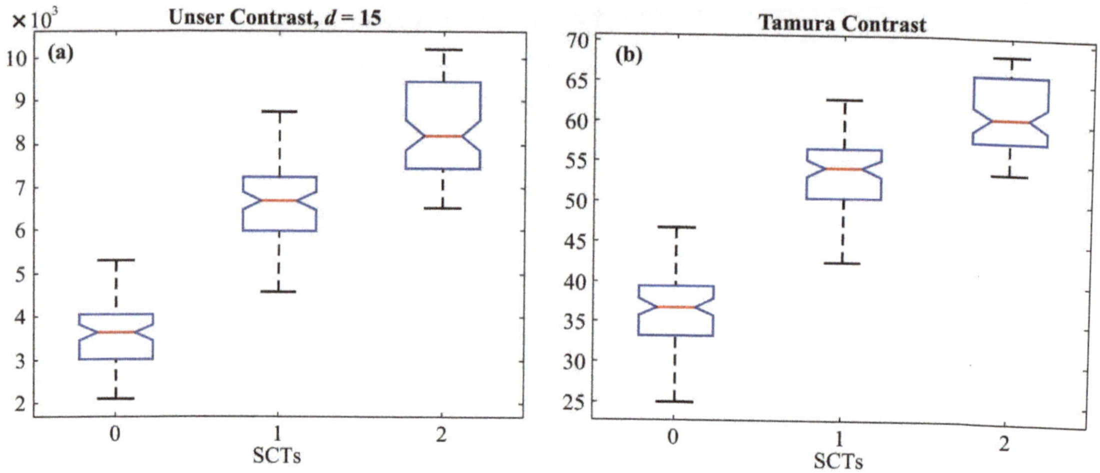


**Figure 7.** Data distributions of the contrast values for the VA signals with different SCT fault conditions: (**a**) Unser contrast and (**b**) Tamura contrast.

**Table 2.** Mean ($\mu$) and standard deviation ($\sigma$) for the different contrast distributions of the different SCT fault conditions.

| | Unser Contrast, $d = 15$ | | | Tamura Contrast | | |
|---|---|---|---|---|---|---|
| | 0 SCTs | 1 SCTs | 2 SCTs | 0 SCTs | 1 SCTs | 2 SCTs |
| $\mu$ | 3599.6029 | 6483.6124 | 8420.3043 | 36.5327 | 52.6555 | 61.5564 |
| $\sigma$ | 740.9540 | 130.6383 | 1064.5340 | 4.9304 | 7.0183 | 4.3420 |

Once the concept of grouping the contrast values for different load conditions (i.e., the contrast index does not change with the load as was shown in Figure 7) is described, the results for different values of $d$ in the Unser contrast definition are shown for clarity purposes (see Figure 8).

In Figure 8, the clusters for different values of $d$ are shown. For $d = 5$, the clusters do not show separation among them. The separation can be observed from $d = 10$, where the clusters maintain a similar separation but in a different scale ($y$-axis). Despite this fact, as mentioned above, the most suitable value for $d$ is 15. This is not merely for the separation observed between the clusters but also for the accuracy obtained when the classification algorithm is applied.

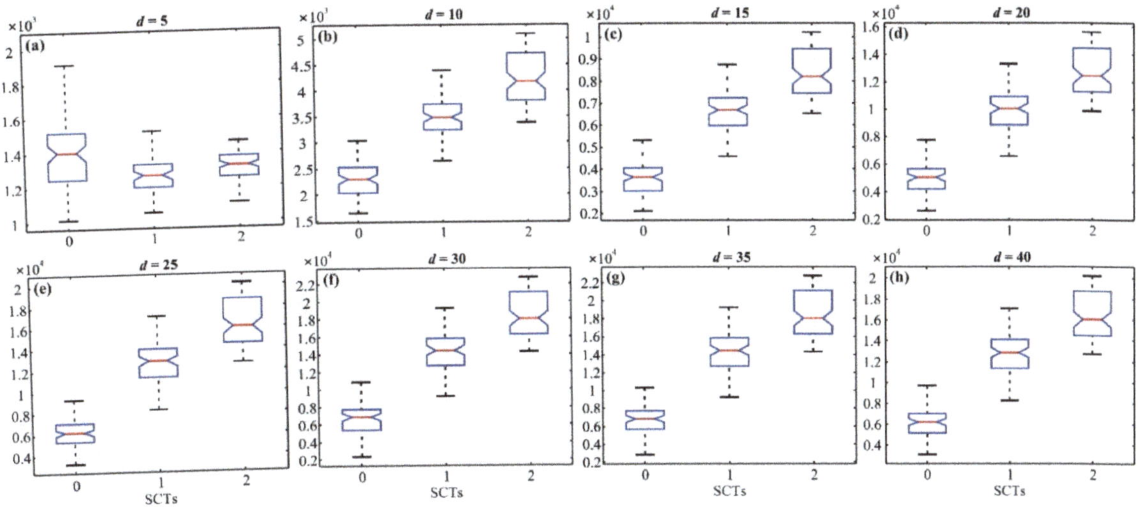


**Figure 8.** Unser contrast clusters for different values of *d*: (**a**) *d* = 5, (**b**) *d* = 10, (**c**) *d* = 15; (**d**) *d* = 20, (**e**) *d* = 25, (**f**) *d* = 30, (**g**) *d* = 35, and (**h**) *d* = 40.

### 4.2.3. Classification Results

As shown in Figure 7, the contrast index is sensitive to the fault; however, there are some overlaps between the different conditions, which can produce uncertainty and, consequently, misclassification. In order to automate the diagnosis process and improve the classification accuracy, a pattern recognition algorithm is proposed. In this regard, two different classifiers, i.e., ANN and SVM, are tested to detect and classify the early SCT fault. These classifiers are applied to both contrast definitions: Unser and Tamura. The application of both classifiers helps to compare their performance, selecting the best method and obtaining the highest classification accuracy. To obtain the parameters of the classifiers that achieve more accurate results, an exhaustive analysis in a trial-and-error process is carried out, i.e., the algorithms with different parameters are applied and the parameters with the best results are selected. In addition, for both classifiers, the training and validation processes are carried out using the k-fold validation, where k is set to 5 due to the number of samples. Figure 9 shows the accuracy percentage for each k-fold iteration and Table 3 shows their mean ($\mu$).

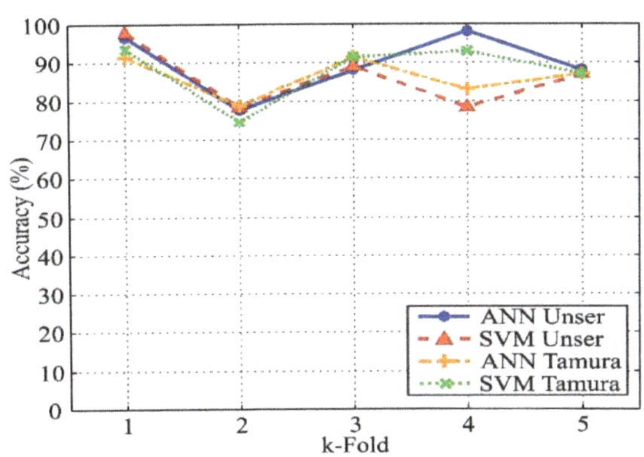


**Figure 9.** Accuracy results for the ANN and SVM methods during the k-fold validation.

**Table 3.** Accuracy percentage mean ($\mu$) for the ANN and SVM classifiers by considering different contrast estimation methods.

| | **Unser with $d = 15$** | **Tamura** |
|---|---|---|
| ANN (%) | 89.78 | 86.38 |
| SVM (%) | 86.1 | 85.96 |

As can be seen in Table 3, the ANN for Unser contrast has the best performance among the classifiers, i.e., 89.78%. The parameters that provide the accuracy previously mentioned in the ANN configuration are: one neuron as input, two hidden layers with 5 and 25 neurons, respectively, and three neurons in the output layer. The activation functions are log-sigmoid for all the layers. The difference with the ANN used for the Tamura contrast is that in its first hidden layer there are ten neurons, i.e., the hidden layers are of 10 and 25 neurons. On the other hand, when a SVM classifier is employed, the parameters with the most significant influence on the classification accuracy are the kernel scale, $\delta$, and the penalty function, $C$, [34]. Consequently, the resulting SVM parameters ($\delta$, $C$) for the Unser contrast are 0.03 and 25, respectively. The parameter $d$ for the Unser contrast definition also changes for the SVM, taking the value of 25. For the Tamura contrast, the SVM parameters ($\delta$, $C$) are 0.4 and 35, respectively.

Table 4 shows the confusion matrix for the k-fold validation when k = 1 (see Figure 9), in which 91.7% of effectiveness is obtained. It is worth noting that 16 tests, i.e., 20%, from the 80 available tests are used for the testing. In these results, the healthy condition is clearly identified; however, the 1 and 2 SCTs conditions present two errors. These values represent for the healthy condition (0 SCTs) a recall and a specificity of 1, which is a perfect score to diagnose the healthy condition. Despite these results, the effectiveness that is considered corresponds to the average obtained during the k-fold validation, i.e., 89.78%.

**Table 4.** Confusion matrix for SCTs classification.

| **SCTs** | **0** | **1** | **2** | |
|---|---|---|---|---|
| 0 | 16 | 0 | 0 | |
| 1 | 0 | 14 | 2 | |
| 2 | 0 | 2 | 14 | |
| Accuracy | 100% | 87.5% | 87.5% | Average = 91.7% |

## 5. Discussion

After designing and validating the proposal by using experimental data, two major advantages are observed: (i) its sensitivity to early SCT faults (i.e., it can react to little changes in the vibration patterns) and (ii) its robustness to the load changes (i.e., the contrast index barely changes when different load conditions appear).

Table 5 shows a comparison between the proposal and other works reported in the literature. Although in [28] an early fault condition is presented, it does not report a pattern recognition algorithm for automatic diagnosis unlike the proposed work that implements a neural network with 89.78% of effectiveness. In [20,35], different severity levels of SCT fault are analyzed; however, the impact of the load in the transformer is not taken into account, which compromises and limits their practical application. On the contrary, the proposal is tested under four load conditions: (i) no load, (ii) linear load, (iii) nonlinear load, and (iv) both linear and nonlinear loads. On the other hand, the works presented in [17,36] also demonstrate that vibration signals can provide information to detect faults in windings; however, the analysis of different fault severities and their automatic detection are not presented. Finally, from the works presented in Table 5, the proposal can be considered as the lowest complex method since the contrast index in the time domain is only required to process the VA signal.

Although promising results have been obtained, some opportunities of research still remain. For instance, in this work, the tests are carried out in a controlled environment;

therefore, real operating conditions such as load intermittences, power quality issues, vibrations generated by other faults, and vibrations from external sources, among others, can negatively affect the performance of the proposal. In this regard, the obtained results can be considered as preliminary.

**Table 5.** Comparison of the proposed work and other methods.

| Work | Method | Signal | Fault Detected in Windings | Early Detection/Severities (SCTs) | Load Conditions | Automatic Classification |
|---|---|---|---|---|---|---|
| Proposed Work | Contrast index | Vibrations | SCTs | Yes 1, 2 | 4 | ANN |
| [17] | Empirical Wavelet transform, HT, and entropies | Vibrations | Winding deformation | No | - | - |
| [20] | Fractal algorithms, ANOVA, Data mining | Vibrations | SCTs | No 5, 10, . . . , 35 | No load | Decision trees, Naïve Bayes, k-nearest neighbor |
| [28] | FFT and total harmonic distortion | Vibrations | SCTs | Yes Initiation | 1 | - |
| [35] | Complete ensemble empirical mode decomposition, Shannon entropy, RMS, and energy index | Current | SCTs | No 5, 10, . . . , 40 | - | - |
| [36] | Short time Fourier transform and RMS | Vibrations | Winding loosening | No | 1 | - |

- Not mentioned.

## 6. Conclusions

This paper presents a methodology for detecting and classifying the transformer condition under different early SCT fault conditions (i.e., 1 and 2 SCTs) from its VA signals. To develop, validate, and test the methodology, a transformer capable of emulating the conditions previously mentioned is used, where different load conditions (no load, linear load, nonlinear load, and both loads) are considered to represent a more realistic scenario and open the possibility to offer an online fault detection method.

In this regard, the main findings are:

- The proposed method can diagnose early SCT fault conditions, i.e., 0, 1, and 2 SCTs. Detection of early SCT faults helps to increase the transformer life, reduce the breakdown maintenance, and avoid possible catastrophic failures.
- One image processing feature used for the texture analysis, i.e., the contrast, allows characterizing the variations in a VA signal to detect early SCTs in a transformer.
- Two contrast definitions, i.e., Unser and Tamura, are tested, where the Unser definition with d = 15 demonstrated to provide the best results.
- The contrast measure is unaffected by the different load combinations connected to the transformer, when early SCTs are present.
- The classification effectiveness from both the Unser contrast definition and the ANN as classifier (i.e., 89.78%) is higher than the one obtained by the SVM (i.e., 86.1%). The

Tamura contrast definition with an ANN obtained 86.38% of effectiveness and with a SVM 85.96% of effectiveness.

Forthcoming work will focus on employing the methodology to diagnose three-phase transformers under different operation conditions, e.g., under unbalance and harmonic content in the power supply, including different types of loads and other faults such as core faults, partial discharges, and winding loosening, among others. In addition, the combination of other signal preprocessing techniques, statistical time features, and pattern recognition algorithms, as well as the fusion of sensors, i.e., the information provided by different types of signals, e.g., current and voltage signals, to improve the fault detection and classification accuracy will be also explored.

**Author Contributions:** Conceptualization, J.R.H.-R. and M.V.-R.; methodology and software, J.R.H.-R. and M.V.-R.; investigation, validation, and formal analysis, all authors; data curation, J.R.H.-R., D.G.-L. and M.B.-L.; writing—original draft preparation, J.R.H.-R. and M.V.-R.; writing—review and editing, all authors; visualization, A.G.-P. and M.V.-R.; supervision, D.G.-L., J.P.A.-S. and M.V.-R.; project administration, D.G.-L., M.B.-L. and M.V.-R.; resources, funding acquisition, A.G.-P., J.P.A.-S., M.B.-L. and M.V.-R. All authors have read and agreed to the published version of the manuscript.

**Funding:** This research received no external funding.

**Data Availability Statement:** The data presented in this study are not publicly available due to privacy issues.

**Acknowledgments:** This work was partially supported by the Mexican Council of Science and Technology (CONACyT) by the scholarship 734987.

**Conflicts of Interest:** The authors declare no conflict of interest.

## References


1. Kudelina, K.; Asad, B.; Vaimann, T.; Rassõlkin, A.; Kallaste, A.; Van Khang, H. Methods of Condition Monitoring and Fault Detection for Electrical Machines. *Energies* **2021**, *14*, 7459. [CrossRef]
2. Huerta-Rosales, J.R.; Granados-Lieberman, D.; Amezquita-Sanchez, J.P.; Camarena-Martinez, D.; Valtierra-Rodriguez, M. Vibration Signal Processing-Based Detection of Short-Circuited Turns in Transformers: A Nonlinear Mode Decomposition Approach. *Mathematics* **2020**, *8*, 575. [CrossRef]
3. Secic, A.; Krpan, M.; Kuzle, I. Vibro-Acoustic Methods in the Condition Assessment of Power Transformers: A Survey. *IEEE Access* **2019**, *7*, 83915–83931. [CrossRef]
4. Zhao, Z.; Tang, C.; Chen, Y.; Zhou, Q.; Yao, C.; Islam, S. Interpretation of Transformer Winding Deformation Fault by the Spectral Clustering of FRA Signature. *Int. J. Electr. Power Energy Syst.* **2021**, *130*, 106933. [CrossRef]
5. Granados-Lieberman, D.; Razo-Hernandez, J.R.; Venegas-Rebollar, V.; Olivares-Galvan, J.C.; Valtierra-Rodriguez, M. Harmonic PMU and Fuzzy Logic for Online Detection of Short-Circuited Turns in Transformers. *Electr. Power Syst. Res.* **2021**, *190*, 106862. [CrossRef]
6. Hu, Y.; Zheng, J.; Huang, H. Experimental Research on Power Transformer Vibration Distribution under Different Winding Defect Conditions. *Electronics* **2019**, *8*, 842. [CrossRef]
7. Zheng, J.; Huang, H.; Pan, J. Detection of Winding Faults Based on a Characterization of the Nonlinear Dynamics of Transformers. *IEEE Trans. Instrum. Meas.* **2018**, *68*, 206–214. [CrossRef]
8. Kim, M.; Lee, S. Power Transformer Voltages Classification with Acoustic Signal in Various Noisy Environments. *Sensors* **2022**, *22*, 1248. [CrossRef]
9. Zhou, Y.; Wang, B. Acoustic Multi-Parameter Early Warning Method for Transformer DC Bias State. *Sensors* **2022**, *22*, 2906. [CrossRef]
10. Contreras-Valdes, A.; Amezquita-Sanchez, J.P.; Granados-Lieberman, D.; Valtierra-Rodriguez, M. Predictive Data Mining Techniques for Fault Diagnosis of Electric Equipment: A Review. *Appl. Sci.* **2020**, *10*, 950. [CrossRef]
11. Bartoletti, C.; Desiderio, M.; Di Carlo, D.; Fazio, G.; Muzi, F.; Sacerdoti, G.; Salvatori, F. Vibro-Acoustic Techniques to Diagnose Power Transformers. *IEEE Trans. Power Deliv.* **2004**, *19*, 221–229. [CrossRef]
12. Romero-Troncoso, R.D.J. Multirate Signal Processing to Improve FFT-Based Analysis for Detecting Faults in Induction Motors. *IEEE Trans. Ind. Inform.* **2017**, *13*, 1291–1300. [CrossRef]
13. Borucki, S.; Cichoń, A.; Majchrzak, H.; Zmarzły, D. Evaluation of the Technical Condition of the Active Part of the High Power Transformer Based on Measurements and Analysis of Vibroacoustic Signals. *Arch. Acoust.* **2017**, *42*, 313–320. [CrossRef]
14. Valtierra-Rodriguez, M.; Rivera-Guillen, J.R.; Basurto-Hurtado, J.A.; Jesus De-Santiago-Perez, J.; Granados-Lieberman, D.; Amezquita-Sanchez, J.P. Convolutional Neural Network and Motor Current Signature Analysis during the Transient State for Detection of Broken Rotor Bars in Induction Motors. *Sensors* **2020**, *20*, 3721. [CrossRef]

15. Goyal, D.; Pabla, B.S. The Vibration Monitoring Methods and Signal Processing Techniques for Structural Health Monitoring: A Review. *Arch. Comput. Methods Eng.* **2016**, *23*, 585–594. [CrossRef]
16. Borucki, S.; Cichon, A. Wavelet Analysis of Vibroacoustic Signals Registered during the Transformer Start-Up. In Proceedings of the IEEE 2012 International Conference on High Voltage Engineering and Application, Shanghai, China, 17–20 September 2012; pp. 579–582.
17. Zhao, M.; Xu, G. Feature Extraction of Power Transformer Vibration Signals Based on Empirical Wavelet Transform and Multiscale Entropy. *IET Sci. Meas. Technol.* **2018**, *12*, 63–71. [CrossRef]
18. Wu, X.; Li, L.; Zhou, N.; Lu, L.; Hu, S.; Cao, H.; He, Z. Diagnosis of DC Bias in Power Transformers Using Vibration Feature Extraction and a Pattern Recognition Method. *Energies* **2018**, *11*, 1775. [CrossRef]
19. Amezquita-Sanchez, J.P.; Adeli, H. A New Music-Empirical Wavelet Transform Methodology for Time-Frequency Analysis of Noisy Nonlinear and Non-Stationary Signals. *Digit. Signal Process. A Rev. J.* **2015**, *45*, 55–68. [CrossRef]
20. Valtierra-Rodriguez, M. Fractal Dimension and Data Mining for Detection of Short-Circuited Turns in Transformers from Vibration Signals. *Meas. Sci. Technol.* **2020**, *31*, 025902. [CrossRef]
21. Huerta-Rosales, J.R.; Granados-Lieberman, D.; Garcia-Perez, A.; Camarena-Martinez, D.; Amezquita-Sanchez, J.P.; Valtierra-Rodriguez, M. Short-Circuited Turn Fault Diagnosis in Transformers by Using Vibration Signals, Statistical Time Features, and Support Vector Machines on Fpga. *Sensors* **2021**, *21*, 3958. [CrossRef]
22. Zhang, Z.; Wu, Y.; Zhang, R.; Jiang, P.; Liu, G.; Ahmed, S.; Dong, Z. Novel Transformer Fault Identification Optimization Method Based on Mathematical Statistics. *Mathematics* **2019**, *7*, 288. [CrossRef]
23. Chen, J.; Yu, W.; Tian, J.; Chen, L.; Zhou, Z. Image Contrast Enhancement Using an Artificial Bee Colony Algorithm. *Swarm Evol. Comput.* **2018**, *38*, 287–294. [CrossRef]
24. Cai, J.; Gu, S.; Zhang, L. Learning a Deep Single Image Contrast Enhancer from Multi-Exposure Images. *IEEE Trans. Image Process.* **2018**, *27*, 2049–2062. [CrossRef]
25. Ferrucho-Alvarez, E.R.; Martinez-Herrera, A.L.; Cabal-Yepez, E.; Rodriguez-Donate, C.; Lopez-Ramirez, M.; Mata-Chavez, R.I. Broken Rotor Bar Detection in Induction Motors through Contrast Estimation. *Sensors* **2021**, *21*, 7446. [CrossRef]
26. Wu, Y.; Zhang, Z.; Xiao, R.; Jiang, P.; Dong, Z.; Deng, J. Operation State Identification Method for Converter Transformers Based on Vibration Detection Technology and Deep Belief Network Optimization Algorithm. *Actuators* **2021**, *10*, 56. [CrossRef]
27. Zhang, F.; Ji, S.; Shi, Y.; Zhan, C.; Zhu, L. Investigation on Vibration Source and Transmission Characteristics in Power Transformers. *Appl. Acoust.* **2019**, *151*, 99–112. [CrossRef]
28. Bagheri, M.; Nezhivenko, S.; Naderi, M.S.; Zollanvari, A. A New Vibration Analysis Approach for Transformer Fault Prognosis over Cloud Environment. *Int. J. Electr. Power Energy Syst.* **2018**, *100*, 104–116. [CrossRef]
29. García, B.; Burgos, J.C.; Alonso, Á.M. Transformer Tank Vibration Modeling as a Method of Detecting Winding Deformations—Part I: Theoretical Foundation. *IEEE Trans. Power Deliv.* **2006**, *21*, 157–163. [CrossRef]
30. Haralick, R.M.; Shanmugam, K.; Dinstein, I. Textural Features for Image Classification. *IEEE Trans. Syst. Man. Cybern.* **1973**, *SMC-3*, 610–621. [CrossRef]
31. Unser, M. Sum and Difference Histograms for Texture Classification. *IEEE Trans. Pattern Anal. Mach. Intell.* **1986**, *PAMI-8*, 118–125. [CrossRef]
32. Tamura, H.; Mori, S.; Yamawaki, T. Textural Features Corresponding to Visual Perception. *IEEE Trans. Syst. Man. Cybern.* **1978**, *8*, 460–473. [CrossRef]
33. Amezquita-Sanchez, J.P.; Valtierra-Rodriguez, M.; Camarena-Martinez, D.; Granados-Lieberman, D.; Romero-Troncoso, R.J.; Dominguez-Gonzalez, A. Fractal Dimension-Based Approach for Detection of Multiple Combined Faults on Induction Motors. *J. Vib. Control* **2016**, *22*, 3638–3648. [CrossRef]
34. Yang, Z.; Zhou, Q.; Wu, X.; Zhao, Z. A Novel Measuring Method of Interfacial Tension of Transformer Oil Combined PSO Optimized SVM and Multi Frequency Ultrasonic Technology. *IEEE Access* **2019**, *7*, 182624–182631. [CrossRef]
35. Mejia-Barron, A.; Valtierra-Rodriguez, M.; Granados-Lieberman, D.; Olivares-Galvan, J.C.; Escarela-Perez, R. The Application of EMD-Based Methods for Diagnosis of Winding Faults in a Transformer Using Transient and Steady State Currents. *Meas. J. Int. Meas. Confed.* **2018**, *117*, 371–379. [CrossRef]
36. Borucki, S. Diagnosis of Technical Condition of Power Transformers Based on the Analysis of Vibroacoustic Signals Measured in Transient Operating Conditions. *IEEE Trans. Power Deliv.* **2012**, *27*, 670–676. [CrossRef]

energies

MDPI

*Article*

# Analysis of the Characteristics of Stator Circulating Current Inside Parallel Branches in DFIGs Considering Static and Dynamic Air-Gap Eccentricity

**Yu-Ling He [1], Xiang-Ao Liu [1], Ming-Xing Xu [1,*], Wen Zhang [1], Wen-Jie Zheng [1], De-Rui Dai [1], Gui-Ji Tang [1], Shu-Ting Wan [1] and David Gerada [2]**

1 Hebei Key Laboratory of Electric Machinery Health Maintenance and Failure Prevention, North China Electric Power University, Baoding 071003, China
2 Power Electronics, Machines and Control Group, University of Nottingham, Nottingham NG7 2RD, UK
* Correspondence: 120202102055@ncepu.edu.cn

**Abstract:** In this article, the stator winding circulating current inside parallel branches (CCPB) of a doubly fed induction generator (DFIG) is comprehensively investigated. Different from other studies, this study not only focuses on the CCPB in radial static air-gap eccentricity (RSAGE) and radial dynamic air-gap eccentricity (RDAGE) but also takes the radial hybrid air-gap eccentricity (RHAGE) cases into account. Firstly, the detailed expressions of CCPB in normal and radial air-gap eccentricity (RAGE) are obtained. Then, the finite element analysis (FEA) and experimental studies are performed on a four-pole DFIG with a rated speed of 1470 rpm in order to verify the theoretical analysis. It is shown that the RAGE increases the amplitude of the CCPB and brings new frequency components to the CCPB. For RSAGE, the CCPB brings new frequency components, which are $f_1$ (50) and $f_\mu$ (540/640). For RDAGE, the newly generated frequency components are $f_1 \pm f_r$ (25/75), $f_u \pm f_r$ (515/565/615/665, and $k = \pm 1$). For RHAGE, the newly added frequency components in RSAGE and RDAGE are present at the same time. In addition, the more the RAGE degree is, the larger the amplitude of characteristic frequency components will be. The results obtained in this paper can be used as a supplementary criterion for diagnosing DFIG eccentric faults.

**Keywords:** doubly fed induction generator (DFIG); radial static air-gap eccentricity (RSAGE); radial dynamic air-gap eccentricity (RDAGE); radial hybrid air-gap eccentricity (RHAGE); circulating current inside parallel branches (CCPB)

**Citation:** He, Y.-L.; Liu, X.-A.; Xu, M.-X.; Zhang, W.; Zheng, W.-J.; Dai, D.-R.; Tang, G.-J.; Wan, S.-T.; Gerada, D. Analysis of the Characteristics of Stator Circulating Current Inside Parallel Branches in DFIGs Considering Static and Dynamic Air-Gap Eccentricity. *Energies* **2022**, *15*, 6152. https://doi.org/10.3390/en15176152

Academic Editor: Armando Pires

Received: 18 July 2022
Accepted: 23 August 2022
Published: 24 August 2022



## 1. Introduction

In contrast to traditional fossil energy sources such as oil and coal, wind energy is a new energy source, which is renewable and clean [1]. Against the backdrop of the double-carbon goal, the cumulative installed capacity of China's wind generators has maintained a steady growth trend. At present, the onshore wind generators are mainly double-fed induction generators (DFIGs). However, due to their complex structure, high assembly requirements, and severe operating environment, eccentricity failure often occurs in DFIGs. Therefore, it is essential to study the operating characteristics of DFIG under eccentric faults [2,3].

Radial air-gap eccentricity (RAGE) is a common machinery failure, which is produced by various factors [4,5]. When eccentric failures occur, additional induction current components are generated in the generator stator winding due to variations in the air-gap length, resulting in a decrease in the quality of the output electrical energy of the generator. In addition, RAGE will cause generator vibration to intensify, resulting in serious production liability accidents, such as rotor shaft bending, a shortened generator life, and even burnout of the generator [6,7]. Based on the above reasons, it is necessary to study RAGE faults, which will benefit the early diagnosis and treatment of faults.

During the actual generator operation, the eccentricity of DFIGs can be divided into three categories, which are radial static air-gap eccentricity (RSAGE), radial dynamic air-gap eccentricity (RDAGE), and radial hybrid air-gap eccentricity (RHAGE), respectively. Particularly, RHAGE is the composite of RSAGE and RDAGE [8].

Currently, many scholars have researched the characteristics of the eccentricity of the generator. Y. Da used the search coil to detect the magnetic field characteristics of the electrical machinery after RSAGE [9]. S. Attestog studied the magnetic field characteristics after an RDAGE failure [10]. D.G. Dorrell detected eccentric faults in wound rotor induction motors and suppressed unbalanced magnetic pull by using pole-specific search coils and auxiliary windings [11]. Based on the work of the predecessors, several studies [6,12,13] used non-embedded search coils to detect various types of air-gap eccentric faults. In addition, the use of vibration characteristics to detect the eccentric failure of the generator is also favored by researchers. Wan Shu-ting studied the vibration characteristics of stator and rotor under the eccentricity fault of the turbine generator [14]. D. Zarko studied the unbalanced magnetic force of the rotor under the eccentricity of the generator and measured the axis trajectory of the rotor [15]. Y.-L. He studied the vibration characteristics of the rotor under 3D eccentricity [16]. The winding vibration characteristics of the generator under eccentric failure were also addressed [17].

Another widely used approach is to detect the changes in voltage/current amplitude and frequency to determine whether an eccentric failure occurs. R.N. Andriamalala detected eccentricity faults by detecting fault signals in the stator voltage [18]. J. Faiz used the frequency spectrum detection of line currents as an indicator for eccentricity fault diagnosis [19]. C. Bruzzese used a split-phase current to detect eccentricity faults in synchronous machines [20] and DFIGs [21]. Xiang Gong proposed a pulse detection algorithm to detect eccentricity faults by identifying excitations from the spectrum of simultaneously sampled stator current signals [22]. A.A. Salah used the changes in the magnitude of stator current components to detect eccentricity faults [23]. E. Hamatwi detected short-circuit and eccentricity faults in DFIGs by collecting and analyzing real-time stator current signals [24].

The stator winding circulating current inside parallel branches (CCPB) of generators under faults is also addressed by researchers. As early as 1999, A. Foggia measured the CCPB of synchronous generators under eccentricity and short-circuit faults [25]. P. Rodriguez pointed out the advantages of using the CCPB of a synchronous motor stator as an early indicator of motor faults (RSAGE and RDAGE) [26]. Wan Shuting studied stator CCPB characteristics under turbo-generator eccentricity [27]. M.M. Mafruddin studied CCPB characteristics under RSAGE failure of synchronous generators [28]. Xu studied the influence of the degree and location of short circuits between the turns of the generator rotor on the CCPB [29].

The above references provide a valuable research idea for this paper. Table 1 is used to show the previous research work more clearly. From Table 1, it can be seen that few studies have investigated the CCPB characteristics of DFIGs under RHAGE failure. In fact, the occurrence of eccentricity faults can have a significant impact on the stator CCPB. Studying CCPB changes in DFIG can help determine the type and extent of eccentricity.

**Table 1.** Overview of previous work in the literature.

| Reference | Fault Type | Detection Object | Research Method |
|---|---|---|---|
| [21] | eccentricity | DFIG | split-phase current |
| [24] | eccentricity and short circuit | DFIG | stator current |
| [25] | eccentricity and short circuit | synchronous generator | CCPB |
| [26] | RSAGE and RDAGE | synchronous motor | CCPB |
| [27] | RHAGE | turbo-generator | CCPB |
| [28] | RSAGE | synchronous generator | CCPB |
| [29] | short circuit | synchronous generator | CCPB |
| This work | RHAGE | DFIG | CCPB |

In this paper, we present a comprehensive analysis of the characteristic of CCPB in RSAGE, RDAGE, and RHAGE faults. The structure of this paper is as follows: In Section 2, we mainly describe the theoretical derivation process, while in Section 3, the finite element model calculations and experimental verification are carried out. Finally, Section 4 is the summary of this paper.

## 2. Theoretical Analysis

*Theoretical Model*

Magnetic flux density (MFD) can be obtained by multiplying the magnetomotive force (MMF) by the permeance per unit area (PPUA) as follows:

$$b(\alpha_m, t) = f(\alpha_m, t)\Lambda(\alpha_m, t) \tag{1}$$

where $b$ is MFD, $f$ is MMF, and $\Lambda$ is PPUA.

The air-gap MMF in the spatial angle $\alpha_m$ can be written as

$$f(\alpha_m, t) = f_p(\alpha_m, t) + \sum_{v} f_v(\alpha_m, t) + \sum_{\mu} f_\mu(\alpha_m, t) \tag{2}$$

where $f_p$ $(\alpha_m,\ t)$, $f_v$ $(\alpha_m,\ t)$, and $f_\mu$ $(\alpha_m,\ t)$ are the main wave composite MMF, the stator winding harmonic MMF, and the rotor winding harmonic MMF, respectively. In addition, $p$ is the number of pole pairs of the main wave composite MMF. $v$ and $\mu$ are the number of pole pairs of the stator and rotor winding tooth harmonic MMF, respectively.

The specific expression of each part of the MMF is

$$\begin{cases} f_p(\alpha_m, t) = F_0 \cos(p\alpha_m - \omega_1 t - \varphi_p) \\ f_v(\alpha_m, t) = F_v \cos(v\alpha_m - \omega_1 t - \varphi_v) \\ f_\mu(\alpha_m, t) = F_\mu \cos(\mu\alpha_m - \omega_\mu t - \varphi_\mu) \\ v = \pm k_1 Z_1 + p, k_1 = 1, 2 \cdots \\ \mu = \pm k_2 Z_2 + p, k_2 = 1, 2 \cdots \end{cases} \tag{3}$$

where $F_0$, $F_v$, and $F_\mu$ are the amplitude of the main wave composite MMF, the stator winding harmonic MMF, and the rotor winding harmonic MMF, respectively. $\varphi_p$ is the initial phase angle of the main wave composite MMF. $\varphi_v$ and $\varphi_\mu$ are the initial phase angles of the stator $v$ and rotor $\mu$ subharmonic MMFs. $\omega_1$ is the angular frequency of the main wave synthesized MMF, and $\omega_\mu$ is the angular frequency of the rotor $\mu$ order harmonic MMF relative to the stator. $Z_1$ is the number of stator slots, and $Z_2$ is the number of rotor slots.

The $\omega_\mu$ can be expressed as

$$\omega_\mu = \omega_1[1 + k_2 Z_2(1 - s)/p] \tag{4}$$

where $s$ is the slip ratio of the DFIG.

The air-gap length affects the magnetic PPUA, which in turn affects the air-gap MFD. Firstly, a geometric model of the generator stator and rotor motion during eccentricity is established, and the expression of air-gap length is obtained, as indicated in Figure 1. Given that the stator and rotor cross-sectional edges are approximately ideal circles, and the shape and position of the stator and rotor do not alter in the axial direction, there is no axial eccentricity. According to the rotor movement characteristics, the eccentricity faults are divided into RSAGE, RDAGE, and RHAGE, as shown in Figure 1. In this article, RSAGE is the situation where $O'$ ($O''$) deviates from $O$, RDAGE is the case where $O'$ deviates from $O$ ($O''$), and the numerals behind are offset distances (the unit is millimeter). RHAGE is the coexistence of both RSAGE and RDAGE (RHAGE0.1 = RSAGE0.1 + RDAGE0.1). Details are as follows:

(1) $O$, $O'$, and $O''$ coincide when there is no eccentricity;
(2) In RSAGE, $O'$ coincides with $O''$ but not with $O$;
(3) In RDAGE, $O$ and $O''$ coincide but not with $O'$;
(4) In RHAGE, $O$, $O'$, and $O''$ do not coincide.

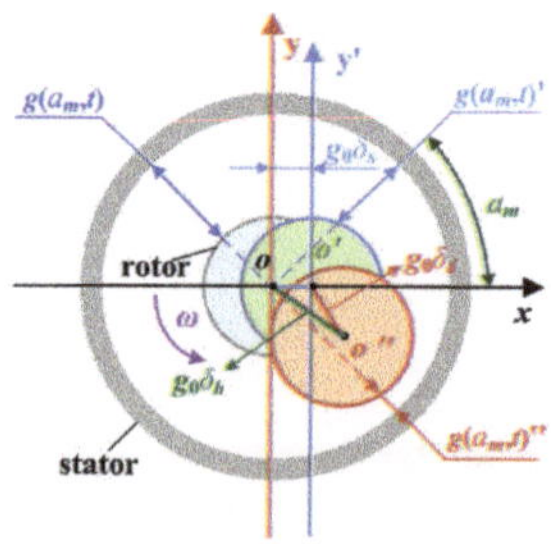


**Figure 1.** Air-gap under the generator is RAGE.

$O$, $O'$, and $O''$ in different cases are shown in Figure 2.

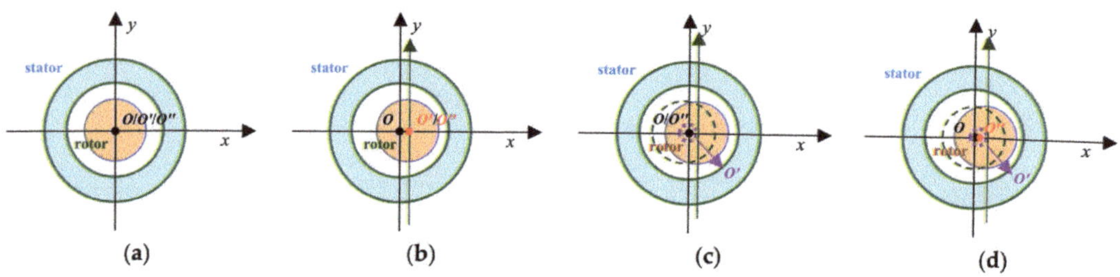


**Figure 2.** Air-gap of generator: (**a**) normal; (**b**) RSAGE; (**c**) RDAGE; (**d**) RHAGE.

PPUA should depend on the radial air-gap length, which is affected by RAGE. According to Figure 1, the radial air-gap length can be written as

$$g(a_m,t)=\begin{cases} g_0 \cdots\cdots\cdots\cdots \text{normal} \\ g_0(1-\delta_s\cos\alpha_m)\cdots\cdots\cdots \text{RSAGE} \\ g_0[1-\delta_d\cos(\omega_r t-\alpha_m)]\cdots\cdots \text{RDAGE} \\ g_0[1-\delta_s\cos\alpha_m-\delta_d\cos(\omega_r t-\alpha_m)]\cdots \text{RHAGE} \end{cases} \tag{5}$$

where $g_0$ is the air-gap length in normal conditions, and $\alpha_m$ is the circumferential angle of the air gap. $\delta_s$ and $\delta_d$ are the values of static eccentricity and dynamic eccentricity, respectively. $\omega_r$ is the rotational frequency of the rotor under RDAGE.

Then, based on Equation (5), PPUA can be obtained as

$$\Lambda(a_m,t)=\frac{\mu_0}{g(a_m,t)}\approx\begin{cases} \Lambda_0+\sum\limits_{k_1}\lambda_{k_1}+\sum\limits_{k_2}\lambda_{k_2}+\sum\limits_{k_1}\sum\limits_{k_2}\lambda_{k_1}\lambda_{k_2}\cdots\cdots\cdots \text{normal} \\ (\Lambda_0+\sum\limits_{k_1}\lambda_{k_1}+\sum\limits_{k_2}\lambda_{k_2}+\sum\limits_{k_1}\sum\limits_{k_2}\lambda_{k_1}\lambda_{k_2})(1+\delta_s\cos\alpha_m+\delta_s^2\cos^2\alpha_m)\cdots\cdots \text{RSAGE} \\ (\Lambda_0+\sum\limits_{k_1}\lambda_{k_1}+\sum\limits_{k_2}\lambda_{k_2}+\sum\limits_{k_1}\sum\limits_{k_2}\lambda_{k_1}\lambda_{k_2})[1+\delta_d\cos(\omega_r t-\alpha_m)]\cdots\cdots \text{RDAGE} \\ (\Lambda_0+\sum\limits_{k_1}\lambda_{k_1}+\sum\limits_{k_2}\lambda_{k_2}+\sum\limits_{k_1}\sum\limits_{k_2}\lambda_{k_1}\lambda_{k_2})[1+\delta_s\cos\alpha_m+\delta_d\cos(\omega_r t-\alpha_m)]\cdots \text{RHAGE} \end{cases} \tag{6}$$

where $\Lambda_0$ is the constant part of the air-gap permeance. $\lambda_{k1}$ is the harmonic permeance caused when the stator is slotted, and the rotor surface is smooth. $\lambda_{k2}$ is the harmonic permeance caused when the rotor is slotted, and the stator surface is smooth. $\lambda_{k1}$ $\lambda_{k2}$ is the harmonic permeance caused by a simultaneous slotting interaction of the stator and the rotor.

MFD can be obtained by feeding Equations (3) and (6) into Equation (1) as follows:

$$
\begin{array}{l}
B(\alpha m,t)=f(\alpha m,t)\Lambda(\alpha m,t) \\
=\left\{\begin{array}{l}
F_0\Lambda_0\cos\left(p\alpha m-\omega_1 t-\varphi p\right)+\sum\limits_{v}F_v\Lambda_0\cos(v\alpha m-\omega_1 t-\varphi v)+\sum\limits_{\mu}F_\mu\Lambda_0\cos\left(\mu\alpha m-\omega_\mu t-\varphi_\mu\right)+\sum\limits_{k_1}\frac{F_0\Lambda_0\lambda_{k_1}}{2}\cos(v\alpha m-\omega_1 t-\varphi v) \\
+\sum\limits_{k_2}\frac{F_0\Lambda_0\lambda_{k_2}}{2}\cos\left(\mu\alpha m-\omega_\mu t-\varphi_\mu\right)\cdots\cdots\text{normal} \\
F_0\Lambda_0\cos\left(p\alpha m-\omega_1 t-\varphi p\right)+\sum\limits_{v}F_v\Lambda_0\cos(v\alpha m-\omega_1 t-\varphi v)+\sum\limits_{\mu}F_\mu\Lambda_0\cos\left(\mu\alpha m-\omega_\mu t-\varphi_\mu\right)+\sum\limits_{k_1}\frac{F_0\Lambda_0\lambda_{k_1}}{2}\cos(v\alpha m-\omega_1 t-\varphi v) \\
+\sum\limits_{k_2}\frac{F_0\Lambda_0\lambda_{k_2}}{2}\cos\left(\mu\alpha m-\omega_\mu t-\varphi_\mu\right)+\frac{F_0\Lambda_0\delta_s}{2}\cos\left[(p\pm1)\alpha m-\omega_1 t-\varphi p\right]+\sum\limits_{v}\frac{F_v\Lambda_0\delta_s}{2}\cos[(v\pm1)\alpha m-\omega_1 t-\varphi v] \\
+\sum\limits_{\mu}\frac{F_\mu\Lambda_0\delta_s}{2}\cos\left[(\mu\pm1)\alpha m-\omega_\mu t-\varphi_\mu\right]+\sum\limits_{k_1}\frac{F_0\Lambda_0\lambda_{k_1}\delta_s}{2}\cos[(v\pm1)\alpha m-\omega_1 t-\varphi v]+\sum\limits_{k_2}\frac{F_0\Lambda_0\lambda_{k_2}\delta_s}{2}\cos\left[(\mu\pm1)\alpha m-\omega_\mu t-\varphi_\mu\right]\cdots\cdots\text{RSAGE} \\
F_0\Lambda_0\cos\left(p\alpha m-\omega_1 t-\varphi p\right)+\sum\limits_{v}F_v\Lambda_0\cos(v\alpha m-\omega_1 t-\varphi v)+\sum\limits_{\mu}F_\mu\Lambda_0\cos\left(\mu\alpha m-\omega_\mu t-\varphi_\mu\right)+\sum\limits_{k_1}\frac{F_0\Lambda_0\lambda_{k_1}}{2}\cos(v\alpha m-\omega_1 t-\varphi v) \\
+\sum\limits_{k_2}\frac{F_0\Lambda_0\lambda_{k_2}}{2}\cos\left(\mu\alpha m-\omega_\mu t-\varphi_\mu\right)+\frac{F_0\Lambda_0\delta_d}{2}\cos\left[(p\pm1)\alpha m-(\omega_1\pm\omega_r)t-\varphi p\right]+\sum\limits_{v}\frac{F_v\Lambda_0\delta_d}{2}\cos[(v\pm1)\alpha m-(\omega_1\pm\omega_r)t-\varphi v] \\
+\sum\limits_{\mu}\frac{F_\mu\Lambda_0\delta_d}{2}\cos\left[(\mu\pm1)\alpha m-(\omega_\mu\pm\omega_r)t-\varphi_\mu\right]+\sum\limits_{k_1}\frac{F_0\Lambda_0\lambda_{k_1}\delta_d}{2}\cos[(v\pm1)\alpha m-(\omega_1\pm\omega_r)t-\varphi v]+\sum\limits_{k_2}\frac{F_0\Lambda_0\lambda_{k_2}\delta_d}{2}\cos\left[(\mu\pm1)\alpha m-(\omega_\mu\pm\omega_r)t-\varphi_\mu\right]\cdots\text{RDAGE} \\
F_0\Lambda_0\cos\left(p\alpha m-\omega_1 t-\varphi p\right)+\sum\limits_{v}F_v\Lambda_0\cos(v\alpha m-\omega_1 t-\varphi v)+\sum\limits_{\mu}F_\mu\Lambda_0\cos\left(\mu\alpha m-\omega_\mu t-\varphi_\mu\right)+\sum\limits_{k_1}\frac{F_0\Lambda_0\lambda_{k_1}}{2}\cos(v\alpha m-\omega_1 t-\varphi v) \\
+\sum\limits_{k_2}\frac{F_0\Lambda_0\lambda_{k_2}}{2}\cos\left(\mu\alpha m-\omega_\mu t-\varphi_\mu\right)+\frac{F_0\Lambda_0\delta_s}{2}\cos\left[(p\pm1)\alpha m-\omega_1 t-\varphi p\right]+\sum\limits_{v}\frac{F_v\Lambda_0\delta_s}{2}\cos[(v\pm1)\alpha m-\omega_1 t-\varphi v]+ \\
\sum\limits_{\mu}\frac{F_\mu\Lambda_0\delta_s}{2}\cos\left[(\mu\pm1)\alpha m-\omega_\mu t-\varphi_\mu\right]+\sum\limits_{k_1}\frac{F_0\Lambda_0\lambda_{k_1}\delta_s}{2}\cos[(v\pm1)\alpha m-\omega_1 t-\varphi v]+\sum\limits_{k_2}\frac{F_0\Lambda_0\lambda_{k_2}\delta_s}{2}\cos\left[(\mu\pm1)\alpha m-\omega_\mu t-\varphi_\mu\right] \\
+\frac{F_0\Lambda_0\delta_d}{2}\cos\left[(p\pm1)\alpha m-(\omega_1\pm\omega_r)t-\varphi p\right]+\sum\limits_{v}\frac{F_v\Lambda_0\delta_d}{2}\cos[(v\pm1)\alpha m-(\omega_1\pm\omega_r)t-\varphi v]+\sum\limits_{\mu}\frac{F_\mu\Lambda_0\delta_d}{2}\cos\left[(\mu\pm1)\alpha m-(\omega_\mu\pm\omega_r)t-\varphi_\mu\right] \\
+\sum\limits_{k_1}\frac{F_0\Lambda_0\lambda_{k_1}\delta_d}{2}\cos[(v\pm1)\alpha m-(\omega_1\pm\omega_r)t-\varphi v]+\sum\limits_{k_2}\frac{F_0\Lambda_0\lambda_{k_2}\delta_d}{2}\cos\left[(\mu\pm1)\alpha m-(\omega_\mu\pm\omega_r)t-\varphi_\mu\right]\cdots\cdots\text{RHAGE}
\end{array}\right.
\end{array}
\tag{7}
$$

According to Equation (7), RSAGE introduces extra MFD harmonic components, whose spatial coefficients are $p \pm 1$, $v \pm 1$, and $\mu \pm 1$, and the corresponding frequencies are $\omega_1$, $\omega_1$, and $\omega_\mu$. The stator winding adopts a double Y-shaped connection, and each phase has two parallel branches, as shown in Figure 3.

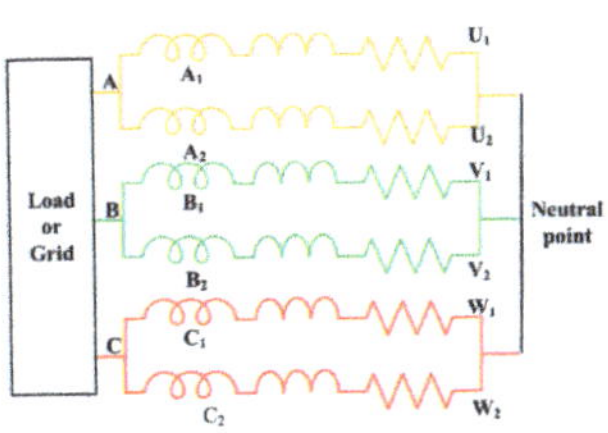


**Figure 3.** Schematic diagram of a double Y-shaped connection of the stator winding.

In Figure 3, $U_1U_2$ is the two branches of the A-phase winding. $V_1V_2$ is the two branches of the B-phase winding. $W_1W_2$ is the two branches of the C-phase winding.

Using the knowledge of electrical machinery, the high order and small amplitude harmonics can be ignored, and the instantaneous value of the induced electromotive force of a single parallel branch of the generator stator winding is determined as

$$
\begin{array}{l}
E(\alpha_m,t)=qw_ck_{w1}b(\alpha_m,t)lv=qw_ck_{w1}b(\alpha_m,t)l(2\tau f)=2qw_ck_{w1}\tau lf\Lambda_0 \\
\times\left\{\begin{array}{l}
\left\{F_0\cos(p\alpha_m-\omega_1 t-\varphi_p)+\sum\limits_{v}F_v\cos(v\alpha_m-\omega_1 t-\varphi_v)+\sum\limits_{\mu}F_\mu\cos(\mu\alpha_m-\omega_\mu t-\varphi_\mu)\right\}\cdots\cdots\text{normal} \\
\left\{F_0\cos(p\alpha_m-\omega_1 t-\varphi_p)+\sum\limits_{v}F_v\cos(v\alpha_m-\omega_1 t-\varphi_v)+\sum\limits_{\mu}F_\mu\cos(\mu\alpha_m-\omega_\mu t-\varphi_\mu)+\frac{F_0\delta_s}{2}\cos[(p\pm1)\alpha_m-\omega_1 t-\varphi_p]+\sum\limits_{v}\frac{F_v\delta_s}{2}\cos[(v\pm1)\alpha_m-\omega_1 t-\varphi_v]+\sum\limits_{\mu}\frac{F_\mu\delta_s}{2}\cos[(\mu\pm1)\alpha_m-\omega_\mu t-\varphi_\mu]\right\}\cdots\cdots\text{RSAGE} \\
\left\{F_0\cos(p\alpha_m-\omega_1 t-\varphi_p)+\sum\limits_{v}F_v\cos(v\alpha_m-\omega_1 t-\varphi_v)+\sum\limits_{\mu}F_\mu\cos(\mu\alpha_m-\omega_\mu t-\varphi_\mu)+\frac{F_0\delta_d}{2}\cos[(p\pm1)\alpha_m-(\omega_1\pm\omega_r)t-\varphi_p]+\sum\limits_{v}\frac{F_v\delta_d}{2}\cos[(v\pm1)\alpha_m-(\omega_1\pm\omega_r)t-\varphi_v]+\sum\limits_{\mu}\frac{F_\mu\delta_d}{2}\cos[(\mu\pm1)\alpha_m-(\omega_\mu\pm\omega_r)t-\varphi_\mu]\right\}\cdots\text{RDAGE} \\
\left\{\begin{array}{l}
F_0\cos(p\alpha_m-\omega_1 t-\varphi_p)+\sum\limits_{v}F_v\cos(v\alpha_m-\omega_1 t-\varphi_v)+\sum\limits_{\mu}F_\mu\cos(\mu\alpha_m-\omega_\mu t-\varphi_\mu)+\frac{F_0\delta_s}{2}\cos[(p\pm1)\alpha_m-\omega_1 t-\varphi_p]+\sum\limits_{v}\frac{F_v\delta_s}{2}\cos[(v\pm1)\alpha_m-\omega_1 t-\varphi_v]+\sum\limits_{\mu}\frac{F_\mu\delta_s}{2}\cos[(\mu\pm1)\alpha_m-\omega_\mu t-\varphi_\mu]+ \\
\frac{F_0\delta_d}{2}\cos[(p\pm1)\alpha_m-(\omega_1\pm\omega_r)t-\varphi_p]+\sum\limits_{v}\frac{F_v\delta_d}{2}\cos[(v\pm1)\alpha_m-(\omega_1\pm\omega_r)t-\varphi_v]+\sum\limits_{\mu}\frac{F_\mu\delta_d}{2}\cos[(\mu\pm1)\alpha_m-(\omega_\mu\pm\omega_r)t-\varphi_\mu]
\end{array}\right\}\cdots\cdots\text{RHAGE}
\end{array}\right.
\end{array}
\tag{8}
$$

where $f$ is the mechanical rotation frequency of the rotor. $l$ is the air-gap length. $q$ is the number of slots per pole per phase. $\tau$ is the pole pitch. $w_c$ is the number of turns of a single coil. $k_{w1}$ is the fundamental winding factor, the expression of which is

$$k_{w1} = k_{y1} \times k_{q1} = \sin\left(90^{\circ} \times y/\tau\right) \times \sin(q\alpha_1/2)/(q\sin(\alpha_1/2)) \tag{9}$$

In Equation (9), $k_{y1}$ is the fundamental wave pitch factor, $k_{q1}$ is the fundamental wave distribution factor, $\alpha_1$ is the slot angle, and $y$ is the stator winding pitch.

The corresponding sides of the two parallel branches of the generator have a certain law in the spatial distribution. The equivalent circuit of the parallel branch of the A-phase stator winding can be drawn as shown in Figure 4.

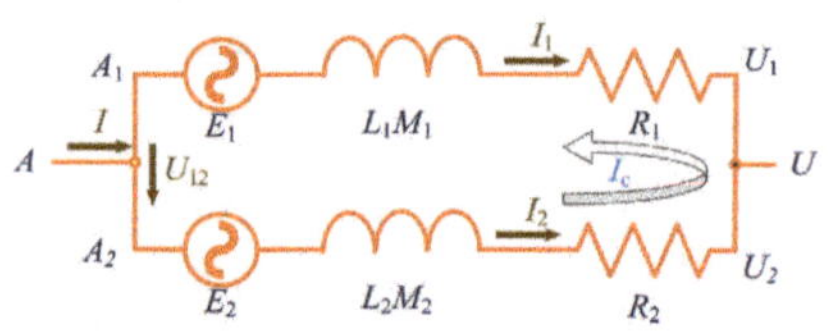


**Figure 4.** Circulating current loop between parallel branches of stator winding.

Where R1, R2, L1, and L2 are the resistance and self-inductance of two parallel branches of the A-phase, respectively. M1 and M2 are the mutual inductances of each branch and other branches, respectively. I1 and I2 are the currents corresponding to the two branches, Ic is the circulating current.

The induced electromotive force of the two branches can be expressed as

$$\left\{\begin{array}{ll}
\left[\begin{array}{l}
E_{a1}(\alpha_m,t)=2qw_ck_{w1}\tau lf\Lambda_0=\left\{F_0\cos\left(p\alpha_m-\omega_1t-\varphi_p\right)+\sum\limits_{v}F_v\cos(v\alpha_m-\omega_1t-\varphi_v)+\sum\limits_{\mu}F_\mu\cos\left(\mu\alpha_m-\omega_\mu t-\varphi_\mu\right)\right\}\\
E_{a2}(\alpha_m,t)=2qw_ck_{w1}\tau lf\Lambda_0=\left\{F_0\cos\left(p(\alpha_m+\pi)-\omega_1t-\varphi_p\right)+\sum\limits_{v}F_v\cos(v(\alpha_m+\pi)-\omega_1t-\varphi_v)+\sum\limits_{\mu}F_\mu\cos\left(\mu(\alpha_m+\pi)-\omega_\mu t-\varphi_\mu\right)\right\}
\end{array}\right] & \cdots\text{normal}\\
\left[\begin{array}{l}
E_{a1}(\alpha_m,t)=2qw_ck_{w1}\tau lf\Lambda_0=\left\{\begin{array}{c}F_0\cos\left(p\alpha_m-\omega_1t-\varphi_p\right)+\sum\limits_{v}F_v\cos(v\alpha_m-\omega_1t-\varphi_v)+\sum\limits_{\mu}F_\mu\cos\left(\mu\alpha_m-\omega_\mu t-\varphi_\mu\right)\\
+\frac{F_0\delta_s}{2}\cos\left[(p\pm1)\alpha_m-\omega_1t-\varphi_p\right]+\sum\limits_{v}\frac{F_v\delta_s}{2}\cos[(v\pm1)\alpha_m-\omega_1t-\varphi_v]+\sum\limits_{\mu}\frac{F_\mu\delta_s}{2}\cos\left[(\mu\pm1)\alpha_m-\omega_\mu t-\varphi_\mu\right]\end{array}\right\}\\
E_{a2}(\alpha_m,t)=2qw_ck_{w1}\tau lf\Lambda_0=\left\{\begin{array}{c}F_0\cos\left(p(\alpha_m+\pi)-\omega_1t-\varphi_p\right)+\sum\limits_{v}F_v\cos(v(\alpha_m+\pi)-\omega_1t-\varphi_v)+\sum\limits_{\mu}F_\mu\cos\left(\mu(\alpha_m+\pi)-\omega_\mu t-\varphi_\mu\right)\\
+\frac{F_0\delta_s}{2}\cos\left[(p\pm1)(\alpha_m+\pi)-\omega_1t-\varphi_p\right]+\sum\limits_{v}\frac{F_v\delta_s}{2}\cos[(v\pm1)(\alpha_m+\pi)-\omega_1t-\varphi_v]+\sum\limits_{\mu}\frac{F_\mu\delta_s}{2}\cos\left[(\mu\pm1)(\alpha_m+\pi)-\omega_\mu t-\varphi_\mu\right]\end{array}\right\}
\end{array}\right] & \cdots\text{RSAGE}\\
\left[\begin{array}{l}
E_{a1}(\alpha_m,t)=2qw_ck_{w1}\tau lf\Lambda_0=\left\{\begin{array}{c}F_0\cos\left(p\alpha_m-\omega_1t-\varphi_p\right)+\sum\limits_{v}F_v\cos(v\alpha_m-\omega_1t-\varphi_v)+\sum\limits_{\mu}F_\mu\cos\left(\mu\alpha_m-\omega_\mu t-\varphi_\mu\right)+\frac{F_0\delta_d}{2}\cos\left[(p\pm1)\alpha_m-(\omega_1\pm\omega_r)t-\varphi_p\right]\\
+\sum\limits_{v}\frac{F_v\delta_d}{2}\cos[(v\pm1)\alpha_m-(\omega_1\pm\omega_r)t-\varphi_v]+\sum\limits_{\mu}\frac{F_\mu\delta_d}{2}\cos\left[(\mu\pm1)\alpha_m-(\omega_\mu\pm\omega_r)t-\varphi_\mu\right]\end{array}\right\}\\
E_{a2}(\alpha_m,t)=2qw_ck_{w1}\tau lf\Lambda_0=\left\{\begin{array}{c}F_0\cos\left(p(\alpha_m+\pi)-\omega_1t-\varphi_p\right)+\sum\limits_{v}F_v\cos(v(\alpha_m+\pi)-\omega_1t-\varphi_v)+\sum\limits_{\mu}F_\mu\cos\left(\mu(\alpha_m+\pi)-\omega_\mu t-\varphi_\mu\right)+\frac{F_0\delta_d}{2}\cos\left[(p\pm1)(\alpha_m+\pi)-(\omega_1\pm\omega_r)t-\varphi_p\right]\\
+\sum\limits_{v}\frac{F_v\delta_d}{2}\cos[(v\pm1)(\alpha_m+\pi)-(\omega_1\pm\omega_r)t-\varphi_v]+\sum\limits_{\mu}\frac{F_\mu\delta_d}{2}\cos\left[(\mu\pm1)(\alpha_m+\pi)-(\omega_\mu\pm\omega_r)t-\varphi_\mu\right]\end{array}\right\}
\end{array}\right] & \cdots\text{RDAGE}\\
\left[\begin{array}{l}
E_{a1}(\alpha_m,t)=2qw_ck_{w1}\tau lf\Lambda_0=\left\{\begin{array}{c}F_0\cos\left(p\alpha_m-\omega_1t-\varphi_p\right)+\sum\limits_{v}F_v\cos(v\alpha_m-\omega_1t-\varphi_v)+\sum\limits_{\mu}F_\mu\cos\left(\mu\alpha_m-\omega_\mu t-\varphi_\mu\right)+\frac{F_0\delta_s}{2}\cos\left[(p\pm1)\alpha_m-\omega_1t-\varphi_p\right]\\
+\sum\limits_{v}\frac{F_v\delta_s}{2}\cos[(v\pm1)\alpha_m-\omega_1t-\varphi_v]+\sum\limits_{\mu}\frac{F_\mu\delta_s}{2}\cos\left[(\mu\pm1)\alpha_m-\omega_\mu t-\varphi_\mu\right]+\frac{F_0\delta_d}{2}\cos\left[(p\pm1)\alpha_m-(\omega_1\pm\omega_r)t-\varphi_p\right]\\
+\sum\limits_{v}\frac{F_v\delta_d}{2}\cos[(v\pm1)\alpha_m-(\omega_1\pm\omega_r)t-\varphi_v]+\sum\limits_{\mu}\frac{F_\mu\delta_d}{2}\cos\left[(\mu\pm1)\alpha_m-(\omega_\mu\pm\omega_r)t-\varphi_\mu\right]\end{array}\right\}\\
E_{a2}(\alpha_m,t)=2qw_ck_{w1}\tau lf\Lambda_0=\left\{\begin{array}{c}F_0\cos\left(p(\alpha_m+\pi)-\omega_1t-\varphi_p\right)+\sum\limits_{v}F_v\cos(v(\alpha_m+\pi)-\omega_1t-\varphi_v)+\sum\limits_{\mu}F_\mu\cos\left(\mu(\alpha_m+\pi)-\omega_\mu t-\varphi_\mu\right)+\frac{F_0\delta_s}{2}\cos\left[(p\pm1)(\alpha_m+\pi)-\omega_1t-\varphi_p\right]\\
+\sum\limits_{v}\frac{F_v\delta_s}{2}\cos[(v\pm1)(\alpha_m+\pi)-\omega_1t-\varphi_v]+\sum\limits_{\mu}\frac{F_\mu\delta_s}{2}\cos\left[(\mu\pm1)(\alpha_m+\pi)-\omega_\mu t-\varphi_\mu\right]+\frac{F_0\delta_d}{2}\cos\left[(p\pm1)(\alpha_m+\pi)-(\omega_1\pm\omega_r)t-\varphi_p\right]\\
+\sum\limits_{v}\frac{F_v\delta_d}{2}\cos[(v\pm1)(\alpha_m+\pi)-(\omega_1\pm\omega_r)t-\varphi_v]+\sum\limits_{\mu}\frac{F_\mu\delta_d}{2}\cos\left[(\mu\pm1)(\alpha_m+\pi)-(\omega_\mu\pm\omega_r)t-\varphi_\mu\right]\end{array}\right\}
\end{array}\right] & \cdots\text{RHAGE}
\end{array}\right. \tag{10}$$

where $w_{a1} = w_{a2} = w_c$, $R_1 = R_2$, $L_1 = L_2$, $I_1 = I_2$, and $E_1 = E_2$. Therefore, the CCPB of the stator can be expressed as

$$\begin{aligned} U_{a12}(\alpha_m,t) &= E_{a1}(\alpha_m,t) + j\omega L_{a1}I_{a1} + R_{a1}I_{a1} + j\omega\sum_i M_{a1i}I_i \\ &\quad - j\omega\sum_k M_{a2k}I_k - R_{a2}I_{a2} - j\omega L_{a2}I_{a2} - E_{a2}(\alpha_m,t) \end{aligned} \tag{11}$$

Feeding Equation (10) into Equation (11), we can obtain the potential difference between the two parallel branches of the generator stator winding before and after SAGE, which can be expressed as

$$\begin{array}{l}
U_{a12}(\alpha_m,t)=2qw_ck_{w1}\tau lf\Lambda_0\\
\times\left\{\begin{array}{ll}
0 & \cdots\text{normal}\\
F_0\delta_s\cos\left[(p\pm1)\alpha_m-\omega_1t-\varphi_p\right]+\sum\limits_{v}F_v\delta_s\cos[(v\pm1)\alpha_m-\omega_1t-\varphi_v]+\sum\limits_{\mu}F_\mu\delta_s\cos\left[(\mu\pm1)\alpha_m-\omega_\mu t-\varphi_\mu\right] & \cdots\text{RSAGE}\\
\begin{array}{l}F_0\delta_d\cos\left[(p\pm1)\alpha_m-(\omega_1\pm\omega_r)t-\varphi_p\right]\\
+\sum\limits_{v}F_v\delta_d\cos[(v\pm1)\alpha_m-(\omega_1\pm\omega_r)t-\varphi_v]+\sum\limits_{\mu}F_\mu\delta_d\cos\left[(\mu\pm1)\alpha_m-\left(\omega_\mu\pm\omega_r\right)t-\varphi_\mu\right]\end{array} & \cdots\text{RDAGE}\\
\begin{array}{l}F_0\delta_s\cos\left[(p\pm1)\alpha_m-\omega_1t-\varphi_p\right]\\
+\sum\limits_{v}F_v\delta_s\cos[(v\pm1)\alpha_m-\omega_1t-\varphi_v]+\sum\limits_{\mu}F_\mu\delta_s\cos\left[(\mu\pm1)\alpha_m-\omega_\mu t-\varphi_\mu\right]+F_0\delta_d\cos\left[(p\pm1)\alpha_m-(\omega_1\pm\omega_r)t-\varphi_p\right]\\
+\sum\limits_{v}F_v\delta_d\cos[(v\pm1)\alpha_m-(\omega_1\pm\omega_r)t-\varphi_v]+\sum\limits_{\mu}F_\mu\delta_d\cos\left[(\mu\pm1)\alpha_m-\left(\omega_\mu\pm\omega_r\right)t-\varphi_\mu\right]\end{array} & \cdots\text{RHAGE}
\end{array}\right.
\end{array} \tag{12}$$

For the sake of analyzing the effect on CCPB characteristics before and after the SAGE fault, the components introduced in different air-gap eccentricity faults were classified according to the same frequency, as shown in Table 2.

**Table 2.** Circulating current characteristics of stator parallel branches before and after eccentricity.

| Cases | Amplitude | Number of Pole Pairs | Freq. | Impact Factor |
|---|---|---|---|---|
| Normal | - | - | - | - |
| RSAGE | $2qw_ck_{w1}\tau lf\Lambda_0\delta_sF_0$ | $p\pm1$ | $\omega_1$ | $q, w_c, k_{w1}, \tau, l, f, F_0, \Lambda_0, \delta_s$ |
| | $2qw_ck_{w1}\tau lf\Lambda_0\delta_sF_v$ | $v\pm1$ | $\omega_1$ | $q, w_c, k_{w1}, \tau, l, f, F_v, \Lambda_0, \delta_s$ |
| | $2qw_ck_{w1}\tau lf\Lambda_0\delta_sF_\mu$ | $\mu\pm1$ | $\omega_\mu$ | $q, w_c, k_{w1}, \tau, l, f, F_\mu, \Lambda_0, \delta_s$ |
| RDAGE | $2qw_ck_{w1}\tau lf\Lambda_0\delta_dF_0$ | $p\pm1$ | $\omega_1\pm\omega_r$ | $q, w_c, k_{w1}, \tau, l, f, F_0, \Lambda_0, \delta_d$ |
| | $2qw_ck_{w1}\tau lf\Lambda_0\delta_dF_v$ | $v\pm1$ | $\omega_1\pm\omega_r$ | $q, w_c, k_{w1}, \tau, l, f, F_v, \Lambda_0, \delta_d$ |
| | $2qw_ck_{w1}\tau lf\Lambda_0\delta_dF_\mu$ | $\mu\pm1$ | $\omega_\mu\pm\omega_r$ | $q, w_c, k_{w1}, \tau, l, f, F_\mu, \Lambda_0, \delta_d$ |
| RHAGE | $2qw_ck_{w1}\tau lf\Lambda_0\delta sF_0$ | $p\pm1$ | $\omega_1$ | $q, w_c, k_{w1}, \tau, l, f, F_0, \Lambda_0, \delta_s$ |
| | $2qw_ck_{w1}\tau lf\Lambda_0\delta_dF_0$ | | $\omega_1\pm\omega_r$ | $q, w_c, k_{w1}, \tau, l, f, F_0, \Lambda_0, \delta_d$ |
| | $2qw_ck_{w1}\tau lf\Lambda_0\delta sF_v$ | $v\pm1$ | $\omega_1$ | $q, w_c, k_{w1}, \tau, l, f, F_v, \Lambda_0, \delta_s$ |
| | $2qw_ck_{w1}\tau lf\Lambda_0\delta s\ F_v$ | | $\omega_1\pm\omega_r$ | $q, w_c, k_{w1}, \tau, l, f, F_v, \Lambda_0, \delta_d$ |
| | $2qw_ck_{w1}\tau lf\Lambda_0\delta_dF_\mu$ | $\mu\pm1$ | $\omega_\mu$ | $q, w_c, k_{w1}, \tau, l, f, F_\mu, \Lambda_0, \delta_s$ |
| | $2qw_ck_{w1}\tau lf\Lambda_0\delta_dF_\mu$ | | $\omega_\mu\pm\omega_r$ | $q, w_c, k_{w1}, \tau, l, f, F_\mu, \Lambda_0, \delta_d$ |

According to Table 2 and Equation (12), there is no CCPB of the stator winding under normal operation. The CCPB of the stator winding appears after the occurrence of RAGE. It is mainly composed of three parts: The first part is caused by an additional magnetic field with a pole pair number of $p\pm1$ and a corresponding frequency of $\omega_1$ that will be introduced by the static eccentricity fault. The second part is caused by an additional magnetic field with a pole pair number of $v\pm1$ and a corresponding frequency of $\omega_1$ caused by the static eccentricity fault. The third part is caused by the additional magnetic field with a pole pair number of $\mu\pm1$ and a corresponding frequency of $\omega_\mu$ caused by the static eccentricity fault. RDAGE also introduces new frequency components compared with normal conditions, which are $f_1\pm f_r$ and $f_u\pm f_r$, respectively. The $f_1\pm f_r$ frequency component includes magnetic pole log numbers of $p\pm1$ and $v\pm1$, and the $f_u\pm f_r$ frequency component includes a magnetic polar log of $\mu\pm1$. The RHAGE frequency ingredient is an overlay of RSAGE and RDAGE.

In order to clarify the influence of the variables in MFD and CCPB expressions, the changes in the frequency components and amplitude of the MFD and CCPB before and after the different types of eccentricity are listed in Table 3.

**Table 3.** Frequency components and trends in normal and RAGE cases (theory).

| Condition | MFD | CCPB | Trend |
|---|---|---|---|
| normal | $f_1, f_\mu$ | $f_1, f_\mu$ | - |
| RSAGE0.1<br>RSAGE0.2<br>RSAGE0.3 | $f_1, f_\mu$ | $f_1, f_\mu$ | increase |
| RDAGE0.1<br>RDAGE0.2<br>RDAGE0.3 | $f_1, f_\mu, f_1\pm f_r, f_\mu\pm f_r$ | $f_1, f_\mu, f_1\pm f_r, f_\mu\pm f_r$ | increase |
| RHAGE0.1<br>RHAGE0.2<br>RHAGE0.3 | $f_1, f_\mu, f_1\pm f_r, f_\mu\pm f_r$ | $f_1, f_\mu, f_1\pm f_r, f_\mu\pm f_r$ | increase |

According to Table 3, RSAGE only changes the amplitude of the MFD and CCPB without changing their frequency components. Conversely, RDAGE and RHAGE faults change the frequency components and amplitude of the MFD and CCPB at the same time. In addition, with an increase in the eccentric volume, the amplitude of the frequency component of the MFD (the side of the decrease in air gap) and CCPB also increases.

## 3. FEA and Experiment Validation

### 3.1. FEA and Experiment Setup

In this study, FEA and experiments were performed on a double-fed asynchronous wind turbine with two pairs of poles. The main parameters of the generator are shown in Table 4. The simulation model established by using ANSYS Electronic is shown in Figure 5, and the experimental units are shown in Figure 6.

**Table 4.** Parameters of DFIG prototype generator.

| Parameters | Value | Parameters | Value |
|---|---|---|---|
| Rated capacity | 5.5 kW | Rated rotating speed | $n_r$ = 1500 rpm |
| Stator core length | $l$ = 155 mm | Stator external diameter | 210 mm |
| Parallel branches | $\alpha$ = 2 | Rotor external diameter | 134 mm |
| air-gap length | 1 mm | Power factor | cos $\varphi$ = 0.8 |
| Rated voltage | 380 V | Stator slots | $Z_1$ = 36 |
| Pole pairs | $p$ = 2 | Rotor slots | $Z_2$ = 24 |

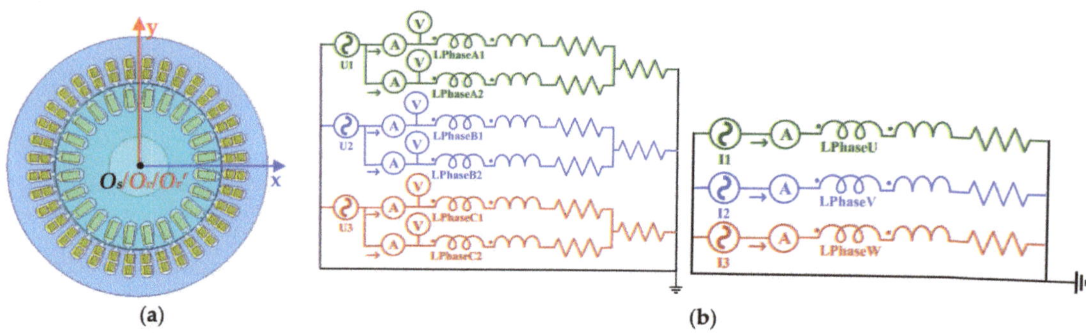


**Figure 5.** FEA model: (**a**) DFIG two-dimensional model; (**b**) external circuit model.

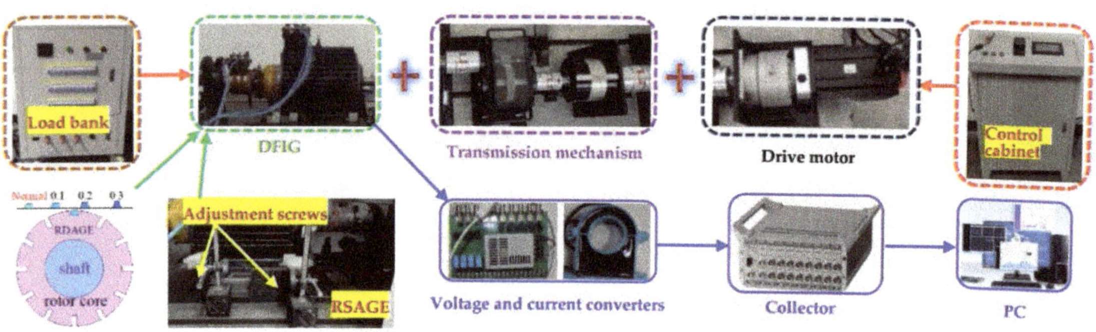


**Figure 6.** Experimental device of DFIG system.

According to the working principle of the ANSYS Electronics ACT, the following assumptions can be made: The origin of the coordinate axis (the geometric center of the stator) is defined as $O_s$, and the center of rotation of the rotor is defined as $O_r$. The rotor rotates at an angular velocity $\omega$ around $O_r$. In RDAGE, the rotation of the rotor also simultaneously occurs around a point with the angular velocity $\omega$, so the center of the rotational trajectory of $O_r$ is defined as $O_r'$. The RSAGE, RDAGE, and RHAGE can be obtained by combining $O_s$, $O_r$, and $O_r'$. In addition, the eccentricity fault level was set to 0.1, 0.2, and 0.3 in each case. In this paper, the positive direction of the X-axis was chosen as the eccentricity direction.

Since the air-gap eccentricity fault is a mechanical fault, it has no effect on the external circuit. Therefore, all the external circuits used in this paper are shown in Figure 5b.

The actual speed of the simulation model of this paper is 1470 rpm, that is, the slip rate $s = (1500 - 1470)/1500 = 2\%$. Therefore, the value of $f_u = 540/640$ ($k_2 = \pm 1$). This article only lists the frequency components for which the absolute value of $k_2$ is less than 2.

The overall structure of the generator is shown in Figure 6. The degree of the RSAGE of the generator was achieved by the radial displacement of the stator. The radial displacement of the stator was controlled by adjusting four screws on the front and back, and the specific offset distance was measured by dial indicators, as indicated in Figure 6. In addition, the degree of RDAGE was adjusted by exchanging the wedges embedded in the grooves of the rotor core, as seen in Figure 6. Specifically, normal wedges flushed with the circumferential surface of the rotor, and 0.1 mm, 0.2 mm, and 0.3 mm wedges above the circumferential surface were prepared in advance; the degree of the RDAGE eccentricity increased as the height of the replacement wedges increased. The external equipment of the generator (control cabinet and load bank) is shown in Figure 6. The experiments and FEA calculations had the same parameter settings and were conducted four times in the following order:

(1) Common normal condition;
(2) RSAGE of 0.1, 0.2, and 0.3 mm;
(3) RDAGE of 0.1, 0.2, and 0.3 mm;
(4) RHAGE of 0.1, 0.2, and 0.3 mm.

### 3.2. FEA Results' Discussion

The air-gap MFD results in different cases are shown in Figure 7 and Table 5. The analysis of the simulation time domain results revealed that as the extent of the degree of eccentricity increased, the amplitude of the local magnetic field density of the air gap increased.

**Table 5.** Simulation results of MFD.

| Type | Freq. | Normal ($\times 10^{-3}$) | RSAGE ($\times 10^{-3}$) | | | RDAGE ($\times 10^{-3}$) | | | RHAGE ($\times 10^{-3}$) | | |
|---|---|---|---|---|---|---|---|---|---|---|---|
| | | | 0.1 | 0.2 | 0.3 | 0.1 | 0.2 | 0.3 | 0.1 | 0.2 | 0.3 |
| $f_1$ | 50 | 709.10 | 751.80 | 799.4 | 856.45 | 697.20 | 716.45 | 717.15 | 754.60 | 809.90 | 879.55 |
| $f_u$ | 540 | 91.80 | 104.23 | 118.12 | 137.55 | 92.64 | 94.39 | 94.60 | 106.05 | 121.59 | 149.06 |
| | 640 | 23.25 | 25.452 | 27.21 | 30.14 | 23.19 | 24.20 | 25.19 | 25.71 | 28.50 | 32.99 |
| $f_1 \pm f_r$ | 25 | - | - | - | - | 13.88 | 32.41 | 48.75 | 19.13 | 46.65 | 85.12 |
| | 75 | - | - | - | - | 15.86 | 40.39 | 59.57 | 22.39 | 50.54 | 86.45 |
| $f_u \pm f_r$ | 515 | - | - | - | - | 8.05 | 13.53 | 18.18 | 10.47 | 20.27 | 36.99 |
| | 564 | - | - | - | - | 11.39 | 20.06 | 27.93 | 13.87 | 28.02 | 52.46 |
| | 615 | - | - | - | - | 2.32 | 2.71 | 3.78 | 2.71 | 4.13 | 6.24 |
| | 664 | - | - | - | - | 3.59 | 6.11 | 8.34 | 4.13 | 7.96 | 15.22 |
| Trend | | - | - | | increase | | | increase | | increase | |

The occurrence of RAGE changed the frequency component of the MFD. The RSAGE condition did not change the composition of the frequency components and only affected the size of the amplitude, as shown in Figure 7a,b. However, RDAGE resulted in $f_1 \pm f_r$ (25/75) $f_u \pm f_r$ (515/564, $k_2 = \pm 1$) for the MFD, as shown in Figure 7c,d. The result of RHAGE was the superposition of RSAGE and RDAGE. Therefore, the frequency components of RHAGE were consistent with RDAGE, as shown in Figure 7e,f and Table 5. The amplitude of these frequency components was enlarged with an increase in eccentricity.

According to the setting of the simulation model speed, one cycle of the DFIG was approximately 40 ms, and the CCPB in this article used a stable waveform of 60–100 ms, as shown in Figure 8 and Table 6. According to Figure 8a–e, the amplitude of the CCPB was close to zero under normal conditions. However, the amplitude of the CCPB increased with the development of RAGE, as indicated in Figure 8 and Table 6.

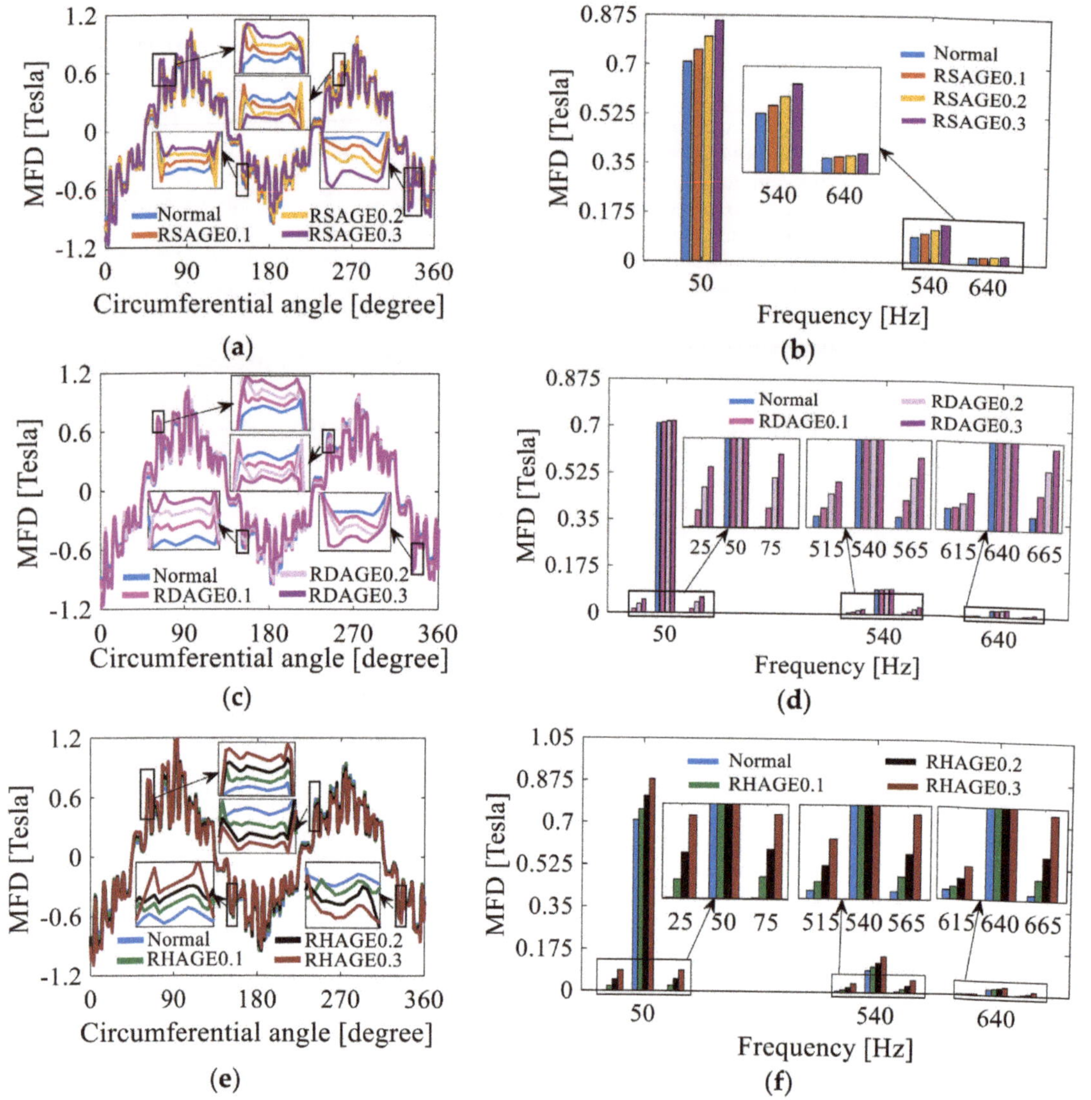


**Figure 7.** The MFD in before and after RAGE conditions: (**a**) the MFD time domain in RSAGE; (**b**) the frequency domain in RSAGE; (**c**) the MFD time domain in RDAGE; (**d**) the frequency domain in RDAGE; (**e**) the MFD time domain in RHAGE; (**f**) the frequency domain in RHAGE.

From Figure 8b,d,f, it can be derived that the CCPB frequency components $f_1$ (50) and $f_u$ (540/640) were induced when RSAGE occurred. Similarly, RDAGE yielded $f_1 \pm f_r$ (25/75) and $f_u \pm f_r$ (515/565/615/665, $k_2 = \pm 1$) for the CCPB frequency components. The frequency components of the CCPB during RHAGR were the superposition of RSAGR and RDAGR. Generally, with the increase in RAGE, the amplitude of each component also increased. The frequency component calculated by using FEA was consistent with the theoretical analysis.

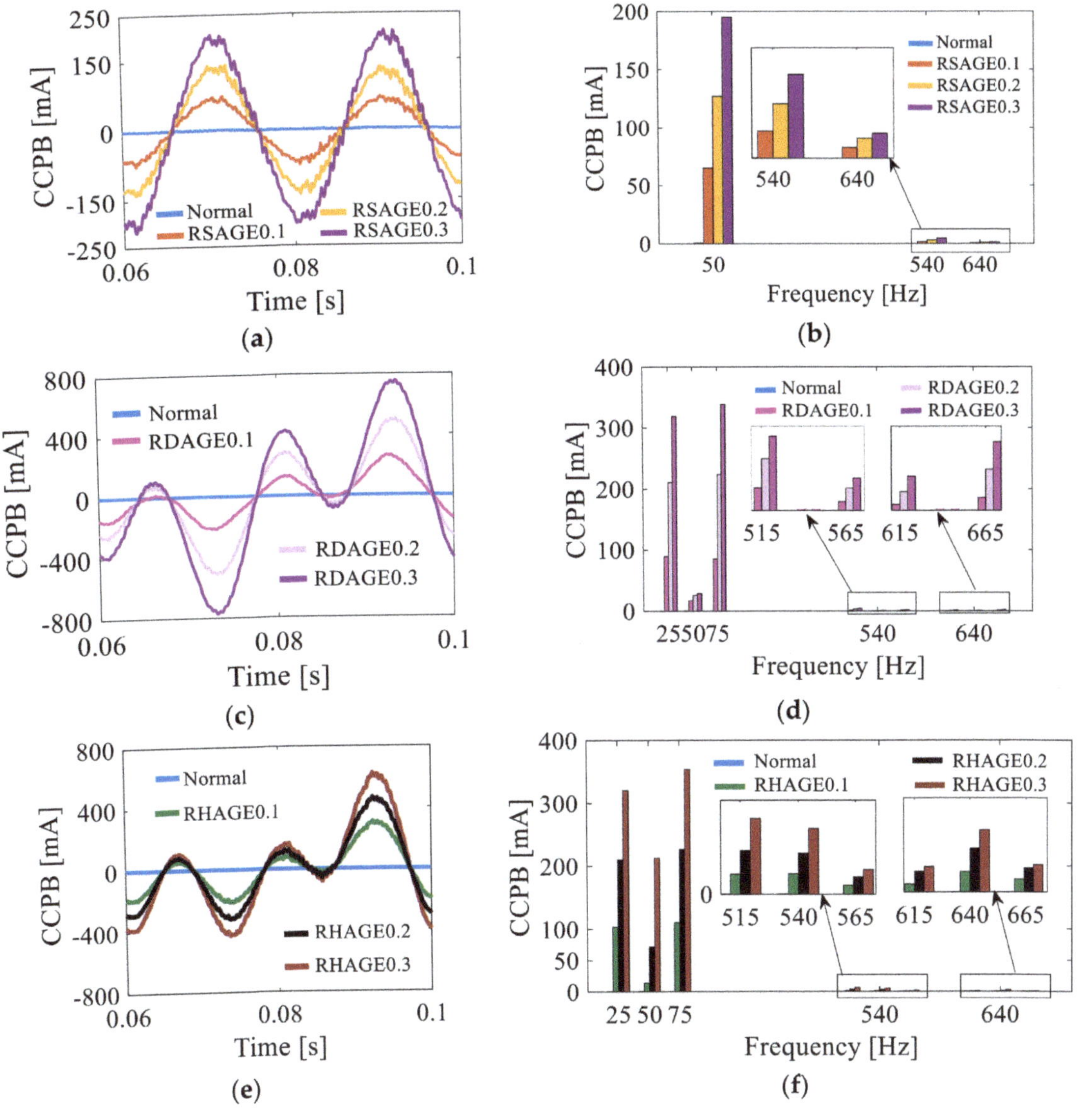


**Figure 8.** The stator CCPB in normal and RAGE conditions: (**a**) the time domain of stator CCPB in RSAGE; (**b**) the frequency domain in RSAGE; (**c**) the time domain of stator CCPB in RDAGE; (**d**) the frequency domain in RDAGE; (**e**) the time domain of stator CCPB in RHAGE; (**f**) the frequency domain in RHAGE.

### *3.3. Experiment Results' Discussion*

The experimental data were obtained on the DFIG experimental unit, as indicated in Figure 6. Regardless of the branches of the A-phase, B-phase, or C-phase, it can be seen that the degree of eccentricity caused an increase in the amplitude of the CCPB, as shown in Figure 9a–c. The change in frequency components also changed the curves' shape. The frequency components of each case are shown in Figure 9. More details can be seen in Table 7.

**Table 6.** Simulation results of CCPB.

| Type | Freq. | Normal | RSAGE ($\times 10^{-3}$) | | | RDAGE ($\times 10^{-3}$) | | | RHAGE ($\times 10^{-3}$) | | |
|---|---|---|---|---|---|---|---|---|---|---|---|
| | | | 0.1 | 0.2 | 0.3 | 0.1 | 0.2 | 0.3 | 0.1 | 0.2 | 0.3 |
| $f_1$ | 50 | - | 65.15 | 127.01 | 195.01 | - | - | - | 14.16 | 71.34 | 212.8 |
| $f_u$ | 540 | - | 1.582 | 3.134 | 4.789 | - | - | - | 1.754 | 3.511 | 5.616 |
| | 640 | - | 0.063 | 1.148 | 1.453 | - | - | - | 0.933 | 1.86 | 2.591 |
| $f_1 \pm f_r$ | 25 | - | - | - | - | 89.38 | 210.9 | 319 | 103.5 | 210.3 | 320.5 |
| | 75 | - | - | - | - | 85.45 | 223.9 | 338.7 | 110.5 | 226.3 | 353.8 |
| $f_u \pm f_r$ | 515 | - | - | - | - | 1.337 | 3.011 | 4.302 | 1.717 | 3.758 | 6.481 |
| | 565 | - | - | - | - | 0.561 | 1.329 | 1.91 | 0.763 | 1.479 | 2.113 |
| | 615 | - | - | - | - | 0.331 | 0.736 | 1.244 | 0.444 | 0.948 | 1.132 |
| | 665 | - | - | - | - | 0.553 | 1.474 | 2.366 | 0.632 | 1.088 | 1.216 |
| Trend | - | - | | increase | | | increase | | | increase | |

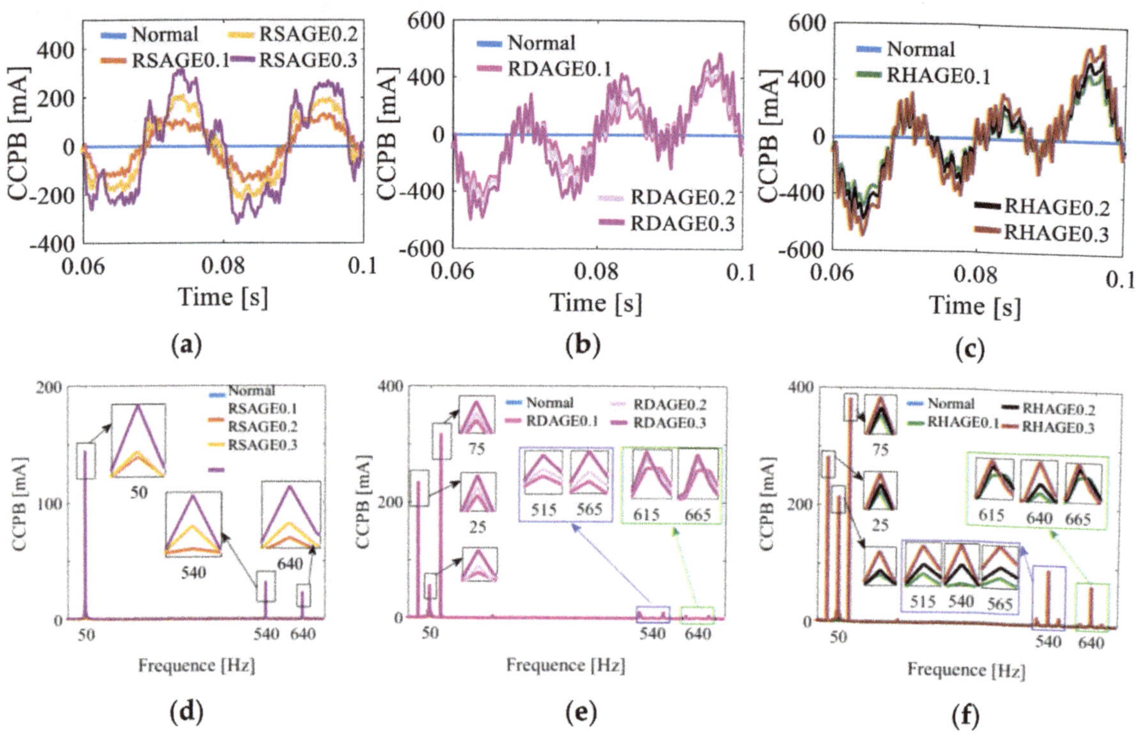


**Figure 9.** The stator CCPB in normal and RAGE conditions: (**a**) the time domain of stator CCPB in RSAGE; (**b**) the time domain of stator CCPB in RDAGE; (**c**) the time domain of stator CCPB in RHAGE; (**d**) the frequency domain in RSAGE; (**e**) the frequency domain in RDAGE; (**f**) the frequency domain in RHAGE.

According to Figure 9d–f, after RSAGR, the components with frequencies $f_1$ (50) and $f_u$ (540/640) appeared in the CCPB. However, RDAGE yielded $f_1 \pm f_r$ (25/75) and $f_u \pm f_r$ (515/565/615/665, $k_2 = \pm 1$) for the CCPB. Additionally, RHAGR was the superposition of RSAGR and RDAGR, as shown in Table 7. At this point, the theoretical derivation, FEA conclusions, and experimental results were basically consistent, thus, the model was well-verified.

**Table 7.** Experiment results of CCPB.

| Type | Freq. | Normal | RSAGE ($\times 10^{-3}$) | | | RDAGE ($\times 10^{-3}$) | | | RHAGE ($\times 10^{-3}$) | | |
|---|---|---|---|---|---|---|---|---|---|---|---|
| | | | 0.1 | 0.2 | 0.3 | 0.1 | 0.2 | 0.3 | 0.1 | 0.2 | 0.3 |
| $f_1$ | 50 | - | 229.2 | 281.4 | 725 | - | - | - | 180.6 | 242 | 536.5 |
| $f_u$ | 540 | - | 11.92 | 77.23 | 163 | - | - | - | 23.13 | 127.6 | 238 |
| | 640 | - | 30.87 | 56.53 | 118.7 | - | - | - | 59.88 | 93.4 | 173.4 |
| $f_1 \pm f_r$ | 25 | - | - | - | - | 387.6 | 467.3 | 586.9 | 465.9 | 561.8 | 706.9 |
| | 75 | - | - | - | - | 530.9 | 636.2 | 794.2 | 637.7 | 764.3 | 954.2 |
| $f_u \pm f_r$ | 515 | - | - | - | - | 10.54 | 16.77 | 29.42 | 12.7 | 20.27 | 35.64 |
| | 565 | - | - | - | - | 10.54 | 16.77 | 29.42 | 12.88 | 20.26 | 35.34 |
| | 615 | - | - | - | - | 10.94 | 13.73 | 15.64 | 13.22 | 16.55 | 18.82 |
| | 665 | - | - | - | - | 10.94 | 13.73 | 15.64 | 13.13 | 16.47 | 18.76 |
| Trend | - | - | | increase | | | increase | | | increase | |

## 4. Conclusions

In this paper, we studied the CCPB with RSAGE, RDAGE, and RHAGE faults in a DFIG, derived the corresponding theoretical formula of the MFD and CCPB, and established a finite element model for simulation. Later, experimental verification was carried out on the fault simulation. The theoretical derivation, finite element simulation results, and experimental results were consistent with each other. The conclusions are as follows:

(1) For the MFD, the increase in eccentricity resulted in a gradual increase in the air-gap magnetic field density. Different kinds of air-gap eccentricity changed the composition of the MFD. For RSAGE, the frequency component was the same as normal. For RDAGE and RHAGE, the frequency component of $f_1 \pm f_r, f_u \pm f_r$ was newly present relative to the normal condition. The increase in eccentricity would cause an increase in the values of the frequency components.

(2) For the CCPB, the change in eccentric species caused a change in the shape of the time domain curve, which was caused by changes in the frequency components. At the same time, the increase in eccentricity caused a significant increase in the CCPB. Similarly, different kinds of air-gap eccentric also changed the composition of CCPB. For RSAGE, the frequency component had new frequency components of $f_1$ and $f_\mu$ compared with normal conditions. For RDAGE, the frequency components of $f_1 \pm f_r$ and $f_u \pm f_r$ were newly present relative to the normal condition. For RHAGE, the frequency components at RSAGE and RDAGE appeared at the same time. The increase in eccentricity would result in an increase in the amplitude of the frequency components.

**Author Contributions:** Conceptualization, Y.-L.H., X.-A.L. and M.-X.X.; methodology, Y.-L.H., W.-J.Z. and M.-X.X.; software, Y.-L.H. and X.-A.L.; validation, D.-R.D. and W.Z.; formal analysis, Y.-L.H.; investigation, Y.-L.H.; resources, Y.-L.H.; data curation, Y.-L.H. and S.-T.W.; writing—original draft preparation, Y.-L.H., X.-A.L., W.-J.Z., D.-R.D., M.-X.X., W.Z. and D.G.; writing—review and editing, Y.-L.H., X.-A.L. and D.G.; visualization, D.-R.D., M.-X.X. and W.Z.; supervision, Y.-L.H. and D.G.; project administration, Y.-L.H. and W.Z.; funding acquisition, Y.-L.H. and G.-J.T. All authors have read and agreed to the published version of the manuscript.

**Funding:** This research was funded by the National Natural Science Foundation of China (No.52177042), in part by the Chinese Fundamental Research Funds for the Central Universities (2018YQ03), in part by the Hebei Provincial Top Youth Talent Support Program ((2018)-27), and in part by the Natural Science Foundation of Hebei Province of China (E2020502031).

**Institutional Review Board Statement:** Not applicable.

**Informed Consent Statement:** Not applicable.

**Data Availability Statement:** Not applicable.

**Acknowledgments:** This work was supported by the Hebei Key Laboratory of Electric Machinery Health Maintenance and Failure Prevention, North China Electric Power University, Baoding, China.

**Conflicts of Interest:** The authors declare no conflict of interest.

## Abbreviations

The following abbreviations are used in this manuscript:

| | |
|---|---|
| DFIG | Doubly fed induction generator |
| RAGE | Radial air-gap eccentricity |
| RSAGE | Radial static air-gap eccentricity |
| RDAGE | Radial dynamic air-gap eccentricity |
| RHAGE | Radial hybrid air-gap eccentricity |
| CCPB | Circulating current inside parallel branches |
| MFD | Magnetic flux density |
| MMF | Magnetomotive force |
| PPUA | Permeance per unit area |
| FEA | Finite element analysis |

## References


1. Energy-Wind Energy. New Wind Energy Research from Rzeszow University of Technology Outlined (Taxonomic Analysis of the Diversity in the Level of Wind Energy Development in European Union Countries). *Energy Weekly News* **2020**, 876.
2. Reja, R.K.; Amin, R.; Tasneem, Z.; Ali, M.F.; Islam, M.R.; Saha, D.K.; Badal, F.R.; Ahamed, M.H.; Moyeen, S.L.; Das, S.K. A review of the evaluation of urban wind resources: Challenges and per-spectives. *Energy Build.* **2022**, *257*, 111781. [CrossRef]
3. Dinh, Q.V.; Doan, Q.V.; Ngo-Duc, T.; Duc, N.D. Offshore wind resource in the context of global climate change over a tropical area. *Appl. Energy* **2022**, *308*, 118369. [CrossRef]
4. Xiahou, K.; Lin, X.; Liu, Y.; Wu, Q.H. Robust Rotor-Current Sensor less Control of Doubly Fed Induction Genera-tors. *IEEE Trans. Energy Convers.* **2018**, *33*, 897–899. [CrossRef]
5. Han, J.; Liu, Z.; Liang, N. Nonlinear Adaptive Robust Control Strategy of Doubly Fed Induction Generator Based on Virtual Synchronous Generator. *IEEE Access* **2020**, *8*, 159887–159896. [CrossRef]
6. He, Y.L.; Zhang, Z.J.; Tao, W.Q.; Wang, X.L.; Gerada, D.; Gerada, C.; Gao, P. A New External Search Coil Based Method to Detect Detailed Static Air-Gap Eccentricity Position in Nonsalient Pole Synchronous Generators. *IEEE Trans. Ind. Electron.* **2021**, *68*, 7535–7544. [CrossRef]
7. Bruzzese, C.; Joksimovic, G. Harmonic Signatures of Static Eccentricities in the Stator Voltages and in the Rotor Current of No-Load Salient-Pole Synchronous Generators. *IEEE Trans. Ind. Electron.* **2011**, *58*, 1606–1624. [CrossRef]
8. He, Y.L.; Xu, M.X.; Xiong, J.; Sun, Y.X.; Wang, X.L.; Gerada, D.; Vakil, G. Effect of 3D Unidirectional and Hybrid SAGE on Electromagnetic Torque Fluctuation Characteristics in Synchronous Generator. *IEEE Access* **2019**, *7*, 100813–100823. [CrossRef]
9. Da, Y.; Shi, X.; Krishnamurthy, M. A New Approach to Fault Diagnostics for Permanent Magnet Synchronous Machines Using Electromagnetic Signature Analysis. *IEEE Trans. Power Electron.* **2013**, *28*, 4104–4112. [CrossRef]
10. Attestog, S.; Khang, H.V.; Robbersmyr, K.G. Detecting Eccentricity and Demagnetization Fault of Permanent Magnet Synchronous Generators in Transient State. In Proceedings of the 2019 22nd International Conference on Electrical Machines and Systems (ICEMS), Harbin, China, 11–14 August 2019; pp. 1–5.
11. Dorrell, D.G.; Salah, A.; Guo, Y. The Detection and Suppression of Unbalanced Magnetic Pull in Wound Rotor Induction Motors Using Pole-Specific Search Coils and Auxiliary Windings. *IEEE Trans. Ind. Appl.* **2017**, *53*, 2066–2076. [CrossRef]
12. Ehya, H.; Nysveen, A.; Nilssen, R.; Liu, Y. Static and dynamic eccentricity fault diagnosis of large salient pole synchronous generators by means of external magnetic field. *IET Electr. Power Appl.* **2021**, *15*, 890–902. [CrossRef]
13. Li, J.; Wang, X.-B.; Luo, Z.; Chen, H.; Yang, Z.-X. Diagnosis of the Mixed Eccentricity Fault in Interior Permanent Magnet Synchronous Motors based on the Open-Circuit External Field. In Proceedings of the 2022 5th International Conference on Energy, Electrical and Power Engineering (CEEPE), Chongqing, China, 22–24 April 2022; pp. 484–489.
14. Wan, S.; He, Y. Investigation on stator and rotor vibration characteristics of turbo-generator under air gap eccentricity fault. *Trans. Can. Soc. Mech. Eng.* **2011**, *35*, 161–176. [CrossRef]
15. Zarko, D.; Ban, D.; Vazdar, I.; Jarica, V. Calculation of Unbalanced Magnetic Pull in a Salient-Pole Synchronous Generator Using Finite-Element Method and Measured Shaft Orbit. *IEEE Trans. Ind. Electron.* **2012**, *59*, 2536–2549. [CrossRef]
16. He, Y.L.; Sun, Y.X.; Xu, M.X.; Wang, X.L.; Wu, Y.C.; Vakil, G.; Gerada, D.; Gerada, C. Rotor UMP characteristics and vibration properties in synchronous generator due to 3D static air-gap eccentricity faults. *IET Electr. Power Appl.* **2020**, *14*, 961–971. [CrossRef]
17. He, Y.-L.; Wang, Y.; Jiang, H.-C.; Gao, P.; Yuan, X.-H.; Gerada, D.; Liu, X.-Y. A Novel Universal Model Considering SAGE for MFD-Based Faulty Property Analysis Under RISC in Synchronous Generators. *IEEE Trans. Ind. Electron.* **2022**, *69*, 7415–7427. [CrossRef]
18. Andriamalala, R.N.; Razik, H.; Baghli, L.; Sargos, F. Eccentricity Fault Diagnosis of a Dual-Stator Winding Induction Machine Drive Considering the Slotting Effects. *IEEE Trans. Ind. Electron.* **2008**, *55*, 4238–4251. [CrossRef]
19. Faiz, J.; Ebrahimi, B.M.; Akin, B.; Toliyat, H.A. Comprehensive Eccentricity Fault Diagnosis in Induction Motors Using Finite Element Method. *IEEE Trans. Magn.* **2009**, *45*, 1764–1767. [CrossRef]

20. Bruzzese, C. Diagnosis of Eccentric Rotor in Synchronous Machines by Analysis of Split-Phase Currents—Part II: Experimental Analysis. *IEEE Trans. Ind. Electron.* **2014**, *61*, 4206–4216. [CrossRef]
21. Bruzzese, C.; Trentini, F.; Santini, E.; Joksimović, G. Sequence Circuit-Based Modeling of a Doubly Fed Induction Wind Generator for Eccentricity Diagnosis by Split-Phase Current Signature Analysis. In Proceedings of the 2018 5th International Symposium on Environment-Friendly Energies and Applications (EFEA), Rome, Italy, 24–26 September 2018; pp. 1–8.
22. Gong, X.; Qiao, W. Current-Based Mechanical Fault Detection for Direct-Drive Wind Turbines via Synchronous Sampling and Impulse Detection. *IEEE Trans. Ind. Electron.* **2015**, *62*, 1693–1702. [CrossRef]
23. Salah, A.A.; Dorrell, D.G. Operating Induction Machine in DFIG Mode Including Rotor Asymmetry. In Proceedings of the 2019 Southern African Universities Power Engineering Conference/Robotics and Mechatronics/Pattern Recognition Association of South Africa (SAUPEC/RobMech/PRASA), Bloemfontein, South Africa, 28–30 January 2019; pp. 469–474.
24. Hamatwi, E.; Barendse, P.; Khan, A. Development of a Test Rig for Fault Studies on a scaled-down DFIG. In Proceedings of the 2021 IEEE Energy Conversion Congress and Exposition (ECCE), Vancouver, BC, Canada, 10–14 October 2021; pp. 3805–3812.
25. Foggia, A.; Torlay, J.-E.; Corenwinder, C.; Audoli, A.; Herigault, J. Circulating current analysis in the parallel-connected windings of synchronous generators under abnormal operating conditions. In Proceedings of the IEEE International Electric Machines and Drives Conference' IEMDC'99' Proceedings (Cat. No.99EX272), Seattle, WA, USA, 9–12 May 1999; pp. 634–636.
26. Rodriguez, P.; Rzeszucinski, P.; Sulowicz, M.; Disselnkoetter, R.; Ahrend, U.; Pinto, C.T.; Ottewill, J.R.; Wildermuth, S. Stator circulating currents as media of fault detection in synchronous motors. In Proceedings of the 2013 9th IEEE International Symposium on Diagnostics for Electric Machines, Power Electronics and Drives (SDEMPED), Valencia, Spain, 27–30 August 2013; pp. 207–214.
27. Shuting, W.; Yuling, H. Analysis on stator circulating current characteristics of turbo-generator under eccentric faults. In Proceedings of the 2009 IEEE 6th International Power Electronics and Motion Control Conference, Wuhan, China, 17–20 May 2009; pp. 2062–2067.
28. Mafruddin, M.M.; Suwarno, S.; AbuSiada, A. Finite Element Simulation of a 126 MW Salient Pole Synchronous Generator with Rotor Eccentricity. In Proceedings of the 2019 2nd International Conference on High Voltage Engineering and Power Systems (ICH-VEPS), Denpasar, Indonesia, 1–4 October 2019; pp. 1–96.
29. Xu, M.; He, Y.; Dai, D.; Liu, X.; Zheng, W.; Zhang, W. Effect of Rotor Interturn Short circuit degree and position on Stator Circulating Current inside Parallel Branches in Generators. In Proceedings of the 2021 IEEE 4th Student Conference on Electric Machines and Systems (SCEMS), Huzhou, China, 1–3 December 2021; pp. 1–7. [CrossRef]

energies

MDPI

Article

# Dynamic Simulation of Starting and Emergency Conditions of a Hydraulic Unit Based on a Francis Turbine

Andrey Achitaev [1], Pavel Ilyushin [2,3], Konstantin Suslov [3,4,*] and Sergey Kobyletski [5]

1 Department of Hydropower, Hydroelectric Power Plants, Electricity Systems and Electricity Networks, Siberian Federal University, 660041 Krasnoyarsk, Russia
2 Energy Research Institute of the Russian Academy of Sciences, 117186 Moscow, Russia
3 Department of Hydropower and Renewable Energy, National Research University "Moscow Power Engineering Institute", 111250 Moscow, Russia
4 Department of Power Supply and Electrical Engineering, Irkutsk National Research Technical University, 664074 Irkutsk, Russia
5 Information and Technological Systems Communication Service, RusHydro, "Sayano-Shushenskaya" HPP, 655600 Sayanogorsk, Russia
* Correspondence: dr.souslov@yandex.ru

**Abstract:** The Francis hydro-turbine is a typical nonlinear system with coupled hydraulic, mechanical, and electrical subsystems. It is difficult to understand the reasons for its operational failures, since the main cause of failures is due to the complex interaction of the three subsystems. This paper presents an improved dynamic model of the Francis hydro-turbine. This study involves the development of a nonlinear dynamic model of a hydraulic unit, given start-up and emergency processes, and the consideration of the effect of water hammer during transients. To accomplish the objectives set, existing models used to model hydroelectric units are analyzed and a mathematical model is proposed, which takes into account the dynamics during abrupt changes in the conditions. Based on these mathematical models, a computer model was developed, and numerical simulation was carried out with an assessment of the results obtained. The mathematical model built was verified on an experimental model. As a result, a model of a hydraulic unit was produced, which factors in the main hydraulic processes in the hydro-turbine.

**Keywords:** hydro-turbine modeling; dynamic modeling; transient processes; hydropower; hydropower plants; mathematical modeling; water flow inertia

**Citation:** Achitaev, A.; Ilyushin, P.; Suslov, K.; Kobyletski, S. Dynamic Simulation of Starting and Emergency Conditions of a Hydraulic Unit Based on a Francis Turbine. *Energies* **2022**, *15*, 8044. https://doi.org/10.3390/en15218044

Academic Editors: João Carlos de Campos Henriques and Yuling He

Received: 16 September 2022
Accepted: 27 October 2022
Published: 29 October 2022



## 1. Introduction

Currently, stochastic renewable (wind and solar) energy sources are rapidly developing in the world, and hydropower plays a key role in ensuring the security and stability of the energy system with a high share of renewable energy sources. Global hydropower continues to increase with relatively higher rates of total installed capacity in the last 10 years alone. This expansion means that it is becoming increasingly more important to investigate various hydropower plants (HPPs). A significant number of studies examine the dynamics of the Francis turbine. In particular, the work by E.Vagnonia, et al. [1] focuses on the study of the dynamics of the Francis turbine under the action of a voltage regulator. The paper presents the results of the studies on the behavior of the generator operating as a synchronous compensator, given the pressure fluctuations in the turbine. The study presented in [2] is devoted to the minimization of fluctuations in the rotation speed at the idle start of the turbine and the optimization of starting the unit with the turbine control algorithm based on the theory of fuzzy logic. This approach provides the flexibility of the regulator settings and reduces vibration during unit start-up. Article [3] provides an analysis of the algorithms and their impact on the turbine operating conditions and the effects of disturbances from the load side. Article [4] presents the construction of dynamic models of the turbine and their transfer functions, which affect the quality of the transient process in the case of various auxiliary systems during the hydraulic unit operation, but do not take into account emergency processes in the power system. There are studies focusing on the modeling of power fluctuation that occurs during the start-up of the unit [5]. However, the authors of this work do not provide a condition for the emergency shutdown of the unit. The authors of [6] analyze the dynamics of the HPP pump-turbine in terms of an improvement of the model of the Francis turbine. Article [7] addresses quantitative analysis and identification of the mechanism of influence of hydraulic damping of low-frequency oscillations in power systems. In [8], the authors present the results of the analysis of complex power systems, which have a large number of various energy sources, and the dynamic simulation of their electromechanical transients. Article [9] presents various transfer functions for turbine models used in pumped-storage plants, given the process constraints. The authors of [10] present some simulation models of a hydro-turbine governing system with different water diversion system topologies, which make it possible to calculate the damping of low-frequency oscillations. The focus of [11] is on the analysis of various nonlinear dynamic characteristics of regulation systems, which ensure the stability of the control system. Article [12] performs an analysis of the stability of the hydro-turbine controller, which is the most important stage in the design of the prime mover operating in the power system, since it is essential for the safety of both the power plant and the power system. Given the nonlinearity of the hydro-turbine, a comprehensive analysis of the stability of the hydro-turbine control system was carried out for the cases of frequency control and power control. However, the algorithms of emergency scenarios are not provided. In [13], authors consider a nonlinear modal method, which was introduced to analyze the dynamic modal interactions between the subsystems of the HPP hydraulic unit, and compare the results obtained by different methods to check the feasibility of the method. There are studies of the multi-frequency dynamic characteristics of a hydropower plant under the coupling effect of the power grid and the turbine regulating system with a surge tank [14]. In article [15], the study examines the hydraulic oscillations between two pump-turbines after the water flow rate is reduced to zero for the parallel operation of the two units. The findings indicate that two pump-turbine units have the same oscillation cycle, but their transient processes are significantly different. In [16], the influence of the conduit geometry on the dynamic parameters of the unit is analyzed. The finite element method is used to evaluate the performance of the turbine. Article [17] analyzes the influence of backlash nonlinearity on the stability of island power systems, where the Nyquist stability criterion and the numerical modeling method are used to control hydraulic turbines. Another focus of the study is on the mechanism of the influence of four important hydromechanical time constants on the stability of a nonlinear system of a hydraulic unit. Article [18] considers simplified mathematical models to help in monitoring the vibration of hydro-generators and illustrates this proposal by modeling a hydro-generator with a capacity of 700 MW. In [19], fluctuations between two hydroelectric power plants in a separate power system of China are investigated, where different excitation systems are considered. Article [20] considers quaternary storage hydropower technology as one of the new advanced technologies, which combines a variable speed pumping unit and a conventional hydropower turbine in a four-component configuration, which allows it to have a greater competitive ability to provide fast power support in a future system with a high degree of penetration of renewable energy sources. The above studies do not take into account the transient processes associated with emergency processes during load shedding. There is a study investigating transient processes during fast unloading of the turbine [21]. The findings suggest that the output real power characteristic is highly dependent on the guide vane opening speed in both field and simulation tests, however, the studies consider load shedding without analyzing different load ranges. The development of such a model would make it possible to create a system of automatic diagnostics in real time. The study [22] focuses on the development of turbine models for wider ranges of hydraulic unit start-up. The contribution of this article is that an accurate real-time equivalent circuit model of HPP with error compensation is proposed to resolve the conflict between real-time online simulation and accuracy under various operating conditions. In the future, this can help build a system for the automatic distribution of operating units, taking into account the model implemented by the station [23,24]. Article [25] is devoted to modeling the turbine of pumped-storage plants operating in a wide range of power. The finite element method is applied to simulate the pulsation. Numerical transient simulation can support the operation of hydropower plants and help avoid hazardous operating conditions, thereby indirectly extending plant run time and improving economic effect.

Currently, there is a whole host of works constructing mathematical models to study a wide variety of processes from electromechanical oscillations to the influence of hydrodynamic processes on the behavior of the unit. According to the analysis of the literature, many previous studies use the finite element method and the construction of transfer functions for modeling individual main components of hydroelectric power plants, but the developed models are simplified and cannot simulate some complex transients during the emergency unloading of the turbine. The design of a modern hydroelectric power plant or theoretical research in the field of hydropower are unthinkable without the widespread use of mathematical modeling at all stages of development [26–29]. Therefore, this paper develops a nonlinear dynamic model of a hydraulic unit, given start-up and emergency processes, and considers the effect of water hammer during transients. To analyze transient processes, a model is implemented in the SimInTech simulation environment, which allows dynamic simulation for a huge number of various technical systems and automatic control devices.

## 2. Construction of a Mathematical Model

Figure 1 shows a block diagram of a hydroelectric power plant (HPP) [3]. Running water from the upstream enters the spiral chamber after passing through the guide vanes, thus contributing to the rotation of the turbine connected to the generator through the shaft. Hydro-generator speed is maintained by changing the opening angle of the guide vanes, which are regulated by an oil servomotor. The system consists of four main parts: piping system, water turbine, servosystem, and generator. In recent years, linear and nonlinear dynamic models of the main parts have been proposed [3,30], but modeling with these models is carried out in ideal situations or is too simplified.

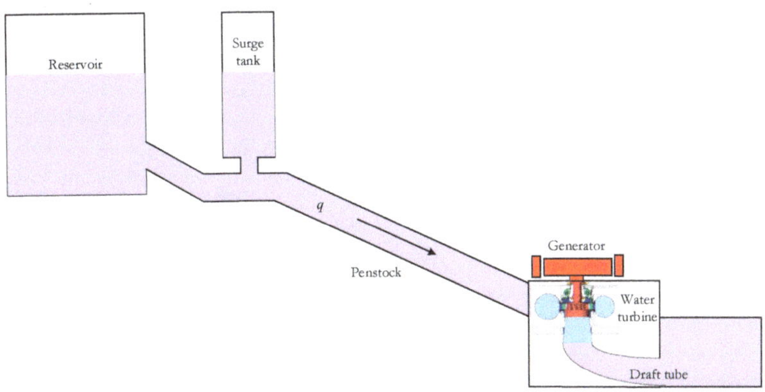


**Figure 1.** A general layout scheme of a hydropower plant. An HPP conduit model (Conduit penstock model).

According to the theory of hydrodynamics, the flow equation is as follows:

$$\begin{bmatrix} H_p(s) \\ Q_p(s) \end{bmatrix} = \begin{bmatrix} \cosh(r\Delta x) & -z_c\sinh(r\Delta x) \\ -\sinh(r\Delta x)/z_c & \cosh(r\Delta x) \end{bmatrix} \begin{bmatrix} H_q(s) \\ Q_q(s) \end{bmatrix} \quad (1)$$

where the subscripts $q$ and $p$ are the designations of the conduit sections for the upper and lower pools of the HPP, $\Delta x = L$ is the conduit length, and $r$ and $z_c$ are HPP the conduit equation components.

## 3. The Mathematical Model Development

### *3.1. Model of a Water Turbine with Elastic Water Hammer*

The turbine equation for a small perturbation around the target point is:

$$\begin{cases} m_t = e_x x_t + e_y y + e_h h \\ q = e_{qx} x_t + e_{qy} y + e_{qh} h \end{cases} \quad (2)$$

Partial derivatives of the turbine torque and flow with respect to the turbine speed have the form [30]:

$$\begin{cases} e_h = \partial m_t / \partial h \\ e_x = \partial m_t / \partial x \\ e_y = \partial m_t / \partial y \\ e_{qh} = \partial q / \partial h \\ e_{qx} = \partial q / \partial x \\ e_{qy} = \partial q / \partial y \end{cases} \quad (3)$$

Transfer function of the turbine and conduit [3] are:

$$G_t(s) = -\frac{e_y}{e_{gh}} \cdot \frac{e_m s^3 - \frac{3}{h_w T_r} s^2 + \frac{24 e_m}{T_r^2} s - \frac{24}{h_w T_r^3}}{s^3 + \frac{3}{e_{qh} h_w T_r} s^2 + \frac{24}{T_r^2 s} + \frac{24}{e_{qh} h_w T_r^3}} \quad (4)$$

where $e_m = e_{qy} e_h / e_y - e_{qh}$ are Francis turbine performance parameters.

### *3.2. Generator Model*

This paper considers the operation of a hydroelectric power plant for an isolated load. The mathematical model can be represented as follows:

$$\begin{cases} \dot{\delta} = x_t \\ \dot{x}_t = (m_t - m_e - D \cdot x_t) / T_a \\ \dot{E}' = \left( E_f - E'_q - (z_d - z'_d) \cdot i_d \right) / T_d \end{cases} \quad (5)$$

where:

$$\begin{cases} i_d = (E'_q - V_s \cos \delta) / z'_d \\ m_e = \frac{V_s}{z'_d} E'_q \sin \delta + \frac{V_s^2}{2} \left( \frac{1}{z_q} - \frac{1}{z'_d} \right) \sin(2\delta) \end{cases} \quad (6)$$

δ is the rotor angle, me is the output electric moment, $D$ is the damper factor, *Ta* is the generator mechanical time constant, $E_f$ is the field voltage, $E'_q$ is the generator voltage, $z_d$ is the positive sequence reactance, $z'_d$ is the positive sequence transient impedance, *Td* is the field winding time constant, $i_d$ is the positive sequence current, vs. is the network voltage, and $Z_q$ is the generator reactance.

## 4. Francis Turbine Control Dynamics

The main expression, which takes into account the control dynamics of the unit, can be represented by an integrating factor that describes the inertia of the water flow in the turbine conduit [27]:

$$\frac{dq}{dt} = \frac{1}{T_R} (h_0 - h) \quad (7)$$

where $q$ is the turbine flow, $h_0$ is the full head of the hydraulic turbine, $h$ is the turbine inlet head, p.u.; $T_B$ is the water hammer time constant (s) calculated from:

$$T_R = \frac{L}{A \cdot g} \tag{8}$$

where $L$ is conduit length, m; $A$ is the conduit area, $m^2$; $g = 9.81\ m^2/s$ is the gravitational acceleration, or:

$$W_q(s) = \frac{1}{T_R s}(h_0 - h) \tag{9}$$

To study the behavior of a hydraulic unit, a simplified mathematical model (Figure 2) is usually used [27]. Its derivation is provided in [31,32]. All formulas are provided for parameters reduced to per units.

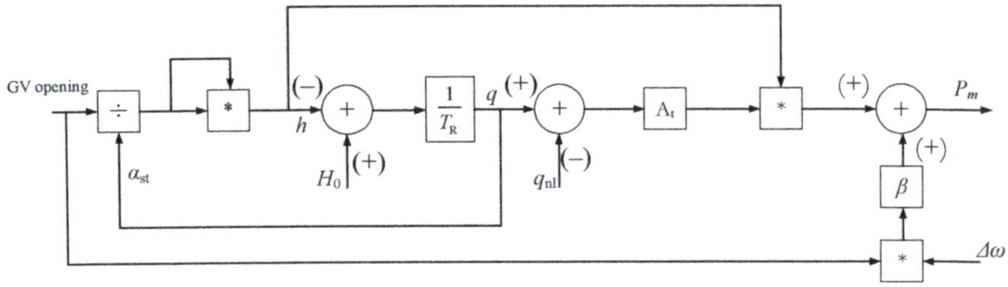


**Figure 2.** A simplified model of a hydraulic turbine.

To calculate the flow through the unit, a simplified equation is used for the liquid outflow through a surface area S, which depends on the opening of the guide vane. This study assumes a linear dependence of change in the flow rate on opening. In actuality, the dependence is not linear and depends on the geometry of the guide vane; however, such an assumption will not affect the quality of the transient process.

$$q = a_0\sqrt{h} \tag{10}$$

where $a_0$ is the guide vane opening, p.u.

Figure 3 shows the model according to the block diagram shown in Figures 1 and 2. To verify the correctness of the resulting model, numerical modeling at the initial stage was carried out with the following conditions: $\Delta w = 0.05$, $q_{nl} = 0.05$, $H_0 = 1$, $T_R = 0.1$, $\beta = 1$.

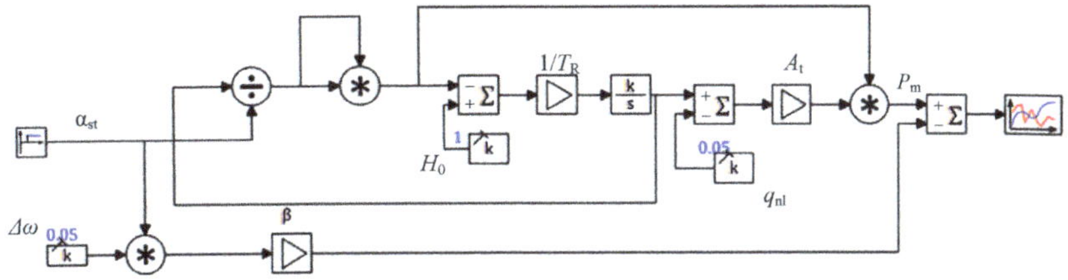


**Figure 3.** The proposed model of the hydraulic turbine.

To test the model, the response of the system to the step input that occurs at the time of 10 s was investigated. A graph of the turbine transient process, with a sharp step change in the opening of the guide vane (GV) ($\alpha_{st}$) from 0.5 to 0.9, is provided in Figure 4. The graph shows a dip at the time of GV opening (t = 10 s), which is explained by a sharp change in the water flow rate in the penstock conduit, causing water hammer. Thus, at the initial moment, most of the flow energy is applied to increase the pressure, and the turbine power drops.

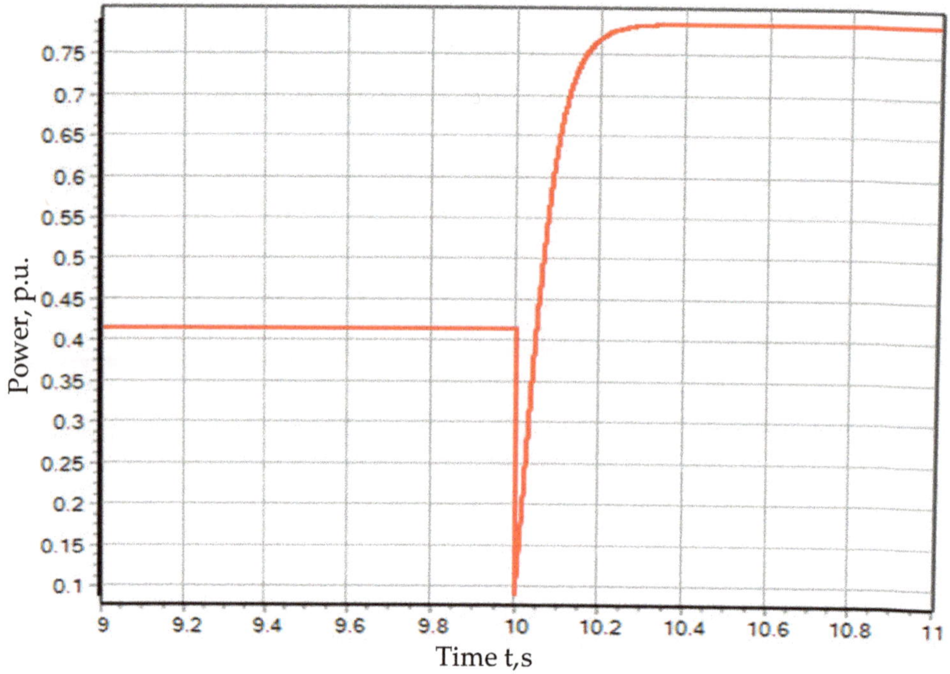


**Figure 4.** Change in power for the guide vane opening from 0.5 to 0.9 p.u.

In practice, such a sharp dip does not occur, due to the inertia of the control system. The movement of the guide vanes takes a few seconds but does not happen instantly. The servomotor can be simulated aperiodically by an expression with time constant $T_s$:

$$W_S(s) = \frac{1}{T_S s + 1} \tag{11}$$

The diagram of a hydro-turbine with an aperiodic inertia element, which takes into account the servomotor, is shown in Figure 5. The result is shown in Figure 6. The power dip caused by the change in setpoint (at the time t = 10 s) is much less compared with the previous result. At the same time, the time of the transient process increased. Such a response of the turbine to a stepped input is more adequate and closer to the behavior of a real-world turbine, in contrast to the transient process shown in Figure 4. All this proves the need to take into account the inertia of the control element, in addition to the inertia of the impeller itself.

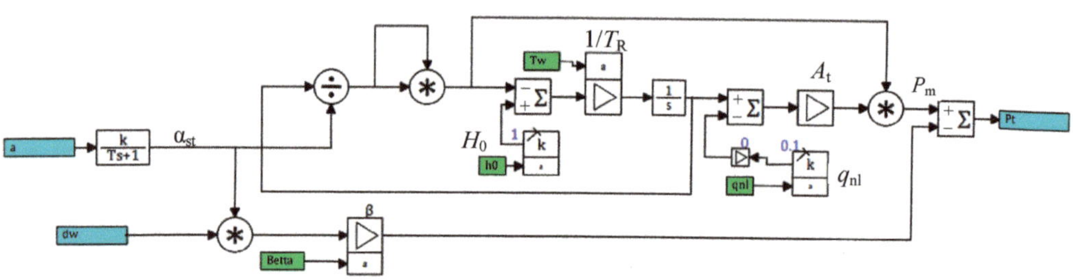


**Figure 5.** A hydraulic turbine diagram with the servomotor.

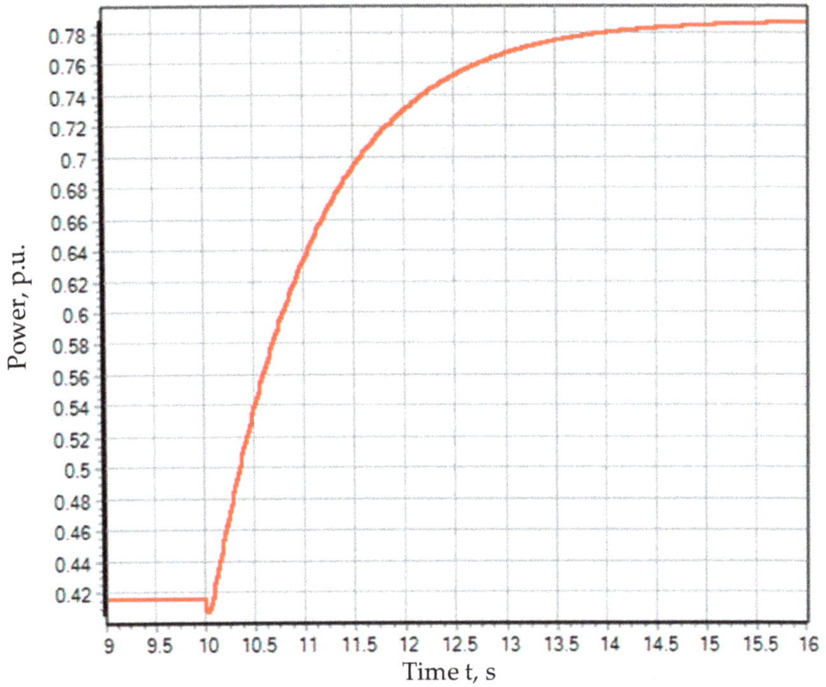


**Figure 6.** Transient process with a smooth opening of the guide vane.

## 5. Comparison of Results with Experimental Data

The modeling results allow the conclusion that the behavior of the model is adequate; however, to make sure of this, it is necessary to compare the results obtained by calculation with real data. To this end, the data obtained in the experiment [33] for the Francis radial–axial turbine are used.

### *5.1. Simulated Plant*

The object of simulation in the work is an experimental stand with a radial–axial turbine, provided by the Norwegian Hydropower Centre for open-source access [34] The stand consists of two tanks simulating the upstream and downstream; the penstock conduit is represented by a pipe closed by a valve. The model turbine diameter is 0.35 m. A magnetic system flow meter is used to measure the water flow. The nominal head of the plant is 12 m. The nominal flow rate at the optimum point is 0.2 $m^3/s$. The nominal power of the hydraulic unit is 24 kW. To measure the water flow parameters, several pressure sensors are placed on the stand: at the turbine inlet (between the guide vanes and the impeller blades), on the turbine blades, and at the turbine outlet (in the suction pipe cone). The sampling frequency of the sensors is 2 KHz. The diagram of this plant is shown in Figure 7.

This experiment involved the investigation of the processes during the turbine acceleration from the minimum opening to the optimal one and the turbine deceleration from the optimal opening angle to the minimum one.

The initial data for the experiment are shown in Table 1.

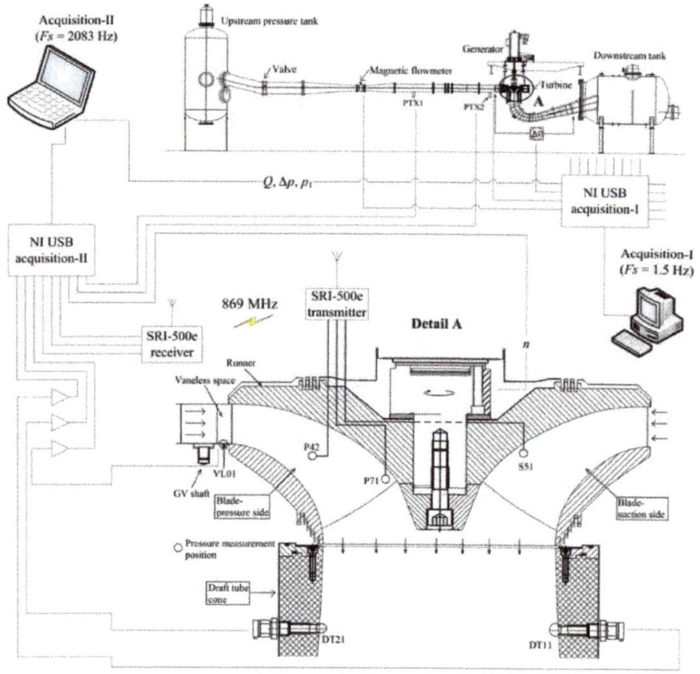


**Figure 7.** Experimental stand of a hydro-turbine plant [34].

**Table 1.** Limiting parameters of the experiment.

| Parameter | Minimum Load | Optimal Load |
|---|---|---|
| Guide vane angle (°) | 0.8 | 9.84 |
| Net head (m) | 12.14 | 11.94 |
| Discharge ($m^3/s$) | 0.022 | 0.200 |
| Torque to the generator (Nm) | 11.16 | 616.13 |
| Friction torque (Nm) | 4.66 | 4.52 |
| Efficiency (%) | 20.94 | 98.39 |
| Runner angular speed (rpm) | 332.8 | 332.59 |
| Casing inlet pressure-abs (kPa) | 221.03 | 215.57 |
| Draft tube outlet pressure-abs (kPa) | 101.85 | 111.13 |
| Water density ($kg/m^3$) | 999.59 | 999.8 |
| Kinematic viscosity ($m^2/s$) | $9.57 \times 10^{-7}$ | $9.57 \times 10^{-7}$ |
| Gravity ($m/c^2$) | 9.82 | 9.82 |

### *5.2. Braking Process*

The study examines the process of braking, in which the opening angle of the guide vane changes from 9.84 to 0.8 degrees. Changing the opening angle of the guide vane is shown in Figure 8.

The change in the flow through the turbine, in time, causes a head surge. This can be seen from the pressure measured at the turbine inlet in Figure 9. When the guide vane reaches the minimum opening angle, the change in the flow stops and the inlet pressure begins to recover. The difference between the steady-state pressure before and after the opening of the guide vanes is explained by a decrease in the speed losses in the conduit, which are virtually absent at the turbine idling.

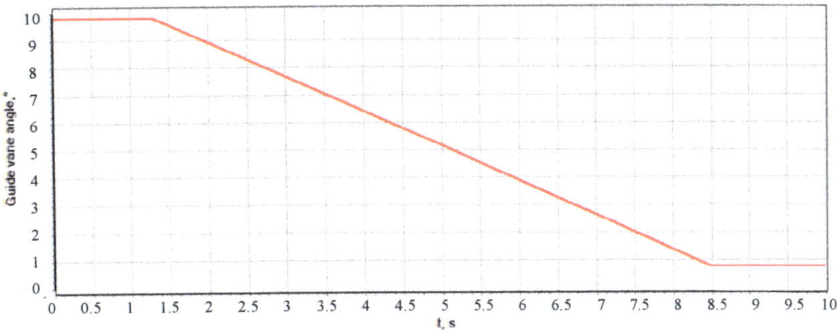


**Figure 8.** The opening angle of the GV versus time (α).

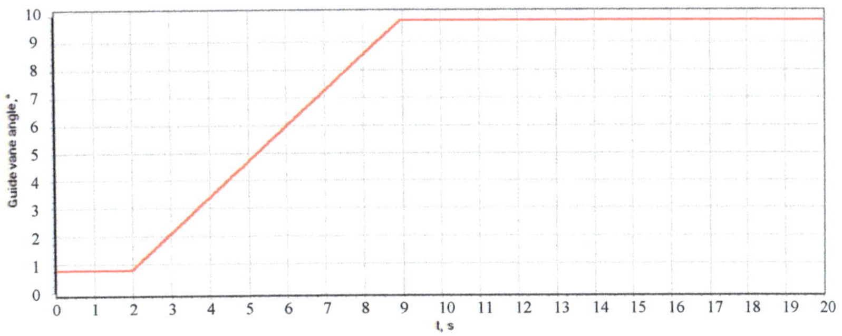


**Figure 9.** The GV opening angle versus time (α).

### 5.3. Acceleration Process

For the case with turbine acceleration, the opening angle changes from 0.8 degrees to 9.84, as shown in Figure 9. The graph of the inlet pressure is shown in Figure 10.

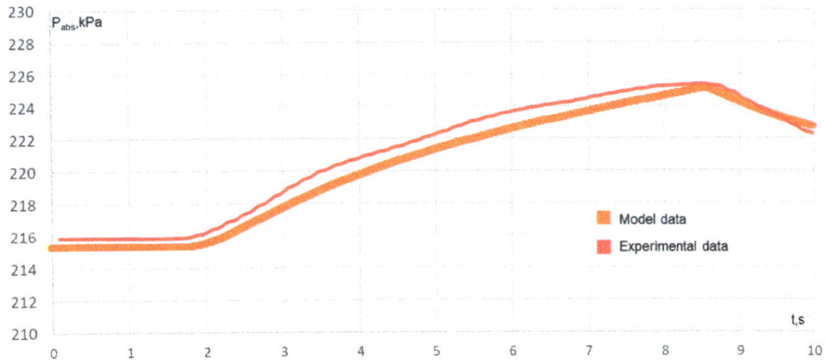


**Figure 10.** Transient process in the case of closing the guide vane and the pressure value before the turbine.

To set the opening of the guide vane, the model employs a piecewise linear approximation. The base value of the guide vane opening is taken to be 12.5°. The maximum and minimum opening values are 9.84 and 0.8, respectively. The opening value is fed to the input of the hydro-turbine model; the results are shown in Figures 10 and 11. Closing the guide vane within 6 s leads to an increase in pressure due to a decreasing flow. When the turn of the guide vanes ends, the pressure value reaches its original value. The basic head value is 12.14 m, and the working head is 11.80 m.

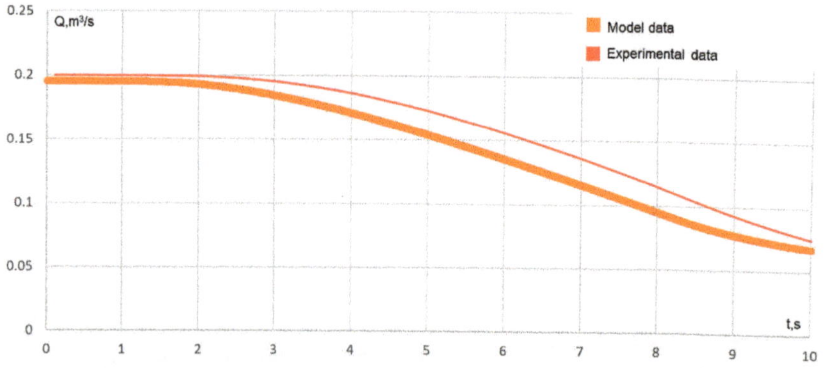


**Figure 11.** Transient process in the case of closing the guide vane and the value of the flow through the turbine.

The results of the load increase modeling are shown in Figures 12 and 13. When the guide vane is opened, the flow energy is used to increase the flow rate and the pressure drops. In contrast to the model characteristic, the steady-state value of the head is different for measurements before and after the guide vane opening. The pressure at the turbine inlet decreases with an increase in the flow, which is caused by water speed losses in the conduit.

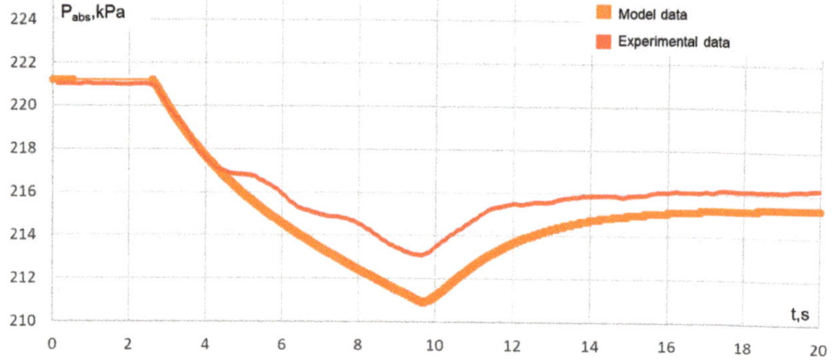


**Figure 12.** Transient process in the case of opening the guide vane before opening the optimum point and the value of the pressure before the turbine.

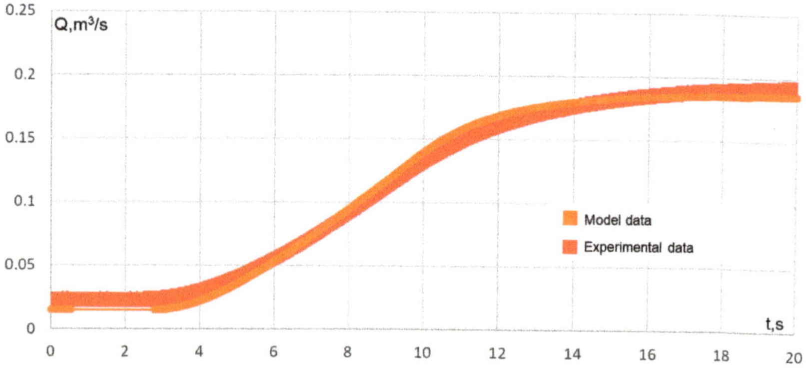


**Figure 13.** Transient process for the case of opening the guide vane before opening the optimum point and the value of the flow through the turbine.

Based on the modeling and comparison of the modeling results with empirical data, we can conclude that the behavior of the model is not only adequate, but also corresponds to the behavior of a real-world turbine.

After the successful verification of the mathematical model, we will model a hydro-turbine as part of a hydroelectric unit. The modeling object will be the above-described radial–axial turbine impeller operating on a shaft with a hydro-generator for an isolated load.

The "bidirectional bus" block (which transmits the values of the power on the shaft, the speed setting, and the actual speed) is used to relate the turbine model with the generator model. The internal structure of the "hydro-turbine" block is shown in Figure 14.

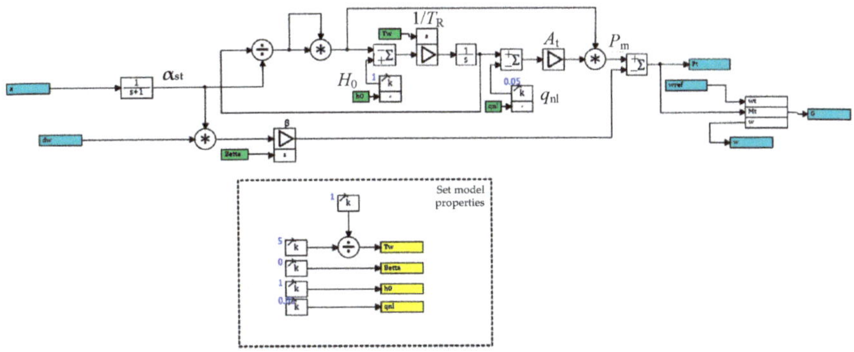


**Figure 14.** A diagram of the hydro-turbine block.

The above-described mathematical model is used to build an electrical circuit (Figure 15). The circuit simulates the energy block "hydraulic unit—transformer" operating for the load.

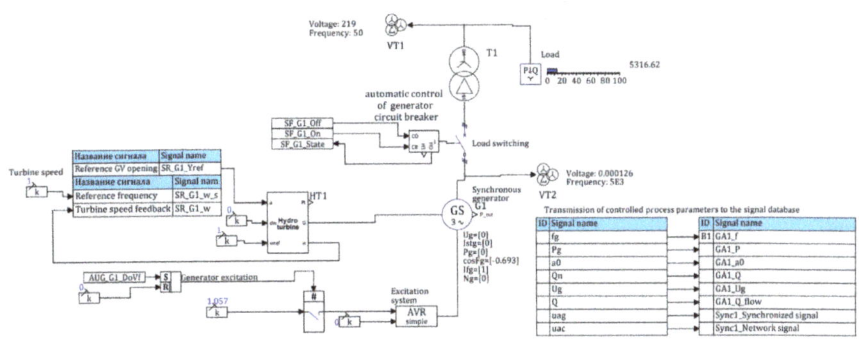


**Figure 15.** Model of the block "hydraulic unit—transformer" operating for the isolated load.

To set the current operating conditions of the hydraulic unit and provide feedback, automatic control circuits are used, which, based on an array of input signals, generate output signals according to a given algorithm.

The following conditions are considered:

1. Shutdown of the hydraulic unit
2. No-load conditions of the hydraulic unit
3. No-load conditions of the generator
4. Generator normal operating conditions

Switching to another operating condition is performed on an appropriate command. The described algorithm is shown in Figure 16.

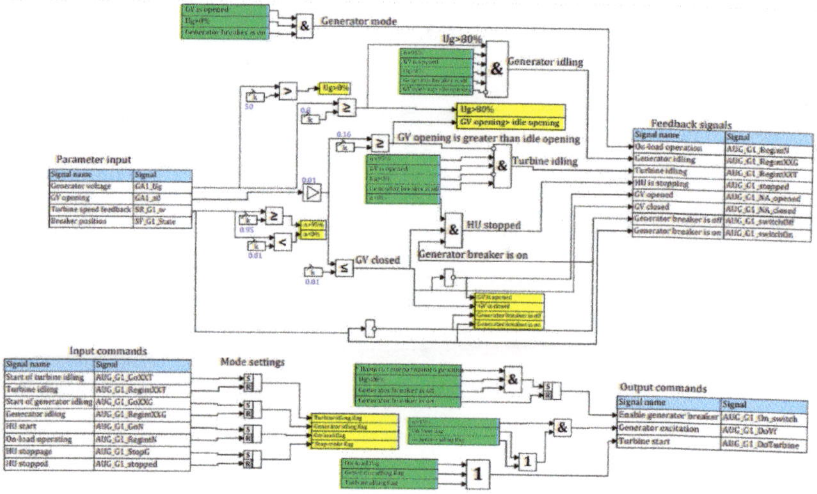


**Figure 16.** An algorithm of automatic control of the hydraulic unit operating conditions.

To control the power of the hydraulic turbine and maintain a given speed, it is necessary to use a controller. Therefore, we use a feedback PID controller. The diagram of such a controller is described in [34].

PID controller maintains the speed of the hydraulic turbine (Figure 17), and, in addition to speed control, implements mechanisms for limiting the maximum and minimum opening of the guide vane. The minimum and maximum opening setpoints depend on the operating conditions of the turbine. For example, when the speed rises from 0 to 100%, the opening limit is 40%, which does not allow the controller to force the opening of the guide vane because of a large speed deviation from the setpoint. Such action of the controller would lead to a significant overshoot and an increase in the time necessary to reach a steady-state value [35–37].

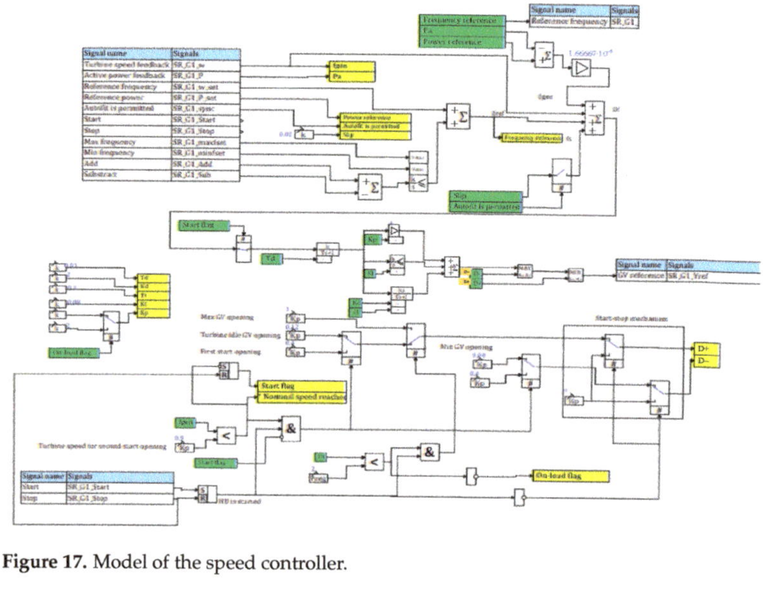

**Figure 17.** Model of the speed controller.

## 6. Analysis of Results

### *6.1. Turbine and Generator No-Load Operating Conditions*

Based on the comparative studies, the mathematical models were verified for various operating conditions of the hydraulic turbine. Relying on the comparative data obtained, we built a model of the hydraulic unit operating for an isolated load. The calculation for a generator with known parameters can be carried out and the previously modeled hydro-turbine can be scaled using known similarity formulas. The similarity equations for hydraulic machines are used to switch from a model turbine to a turbine of a different power [36]:

$$Q_P = Q_1' D_1^2 \sqrt{H} \tag{12}$$

$$n_1' = nD_1 / \sqrt{H} \tag{13}$$

Thus, a hydraulic unit with an active power of 42 MW, a nominal voltage of 10.5 kV, and a rotation speed of 100 rpm was chosen as an object to be simulated.

First of all, we check the no-load operation of the unit. When the "generator no-load operation" command is given, the unit sequentially switches to the turbine no-load operation mode, and then to the generator no-load operation mode, supplying the field current to the rotor winding. Figure 18 shows the parameters of this transient process. At the first time instant, the guide vane (GV) opens for start-up (40%), then, when approaching the nominal speed, it is closed to the opening of no-load conditions. At the same moment, the excitation of the synchronous machine is turned on and the stator voltage rises to the nominal value (10.5 kV). Rotation speed deviation is less than 2%.

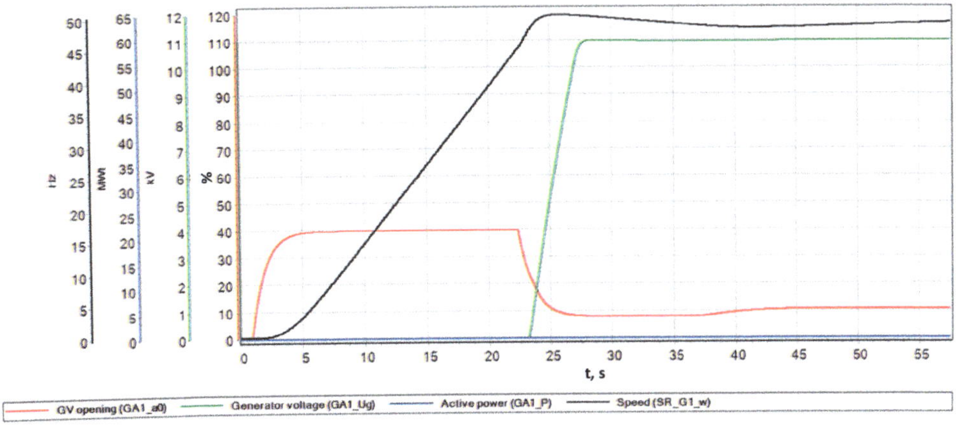


**Figure 18.** Parameters of the transient process during the turbine acceleration to idle speed.

### *6.2. On-Load Operation and Load Increase*

Another operating condition to explore is an increase in the load. Figure 19 shows the transient process during the acceleration of the hydraulic unit to the nominal speed under no load, which is followed by the connection of a 12 MW load [38–40].

When the turbine unit is started and generator reaches the idle speed, the generator breaker is turned on. In this case, a slowly damping oscillatory process occurs with a swing of 12.5 Hz (25%).

In real-world control systems, the controller coefficients are often changed for different operating conditions. Figure 20 presents the frequency response characteristic for the current PID controller (Table 2) and for the desired controller operating under load.

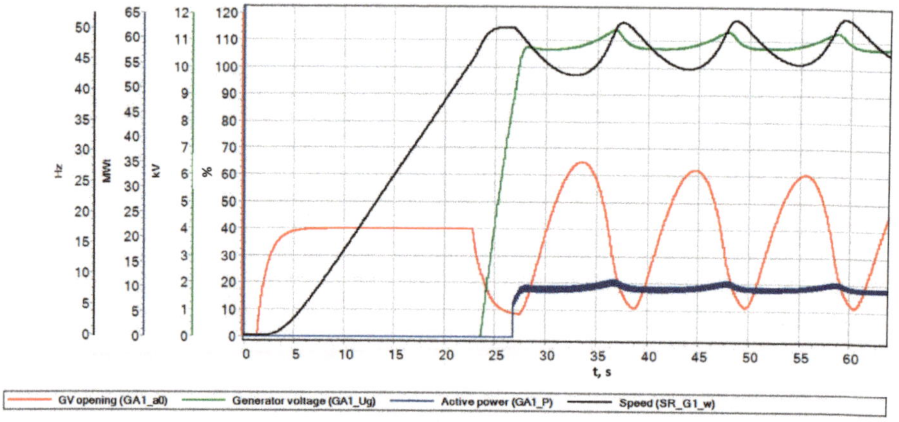


**Figure 19.** Parameters of the transient process during the load surge.

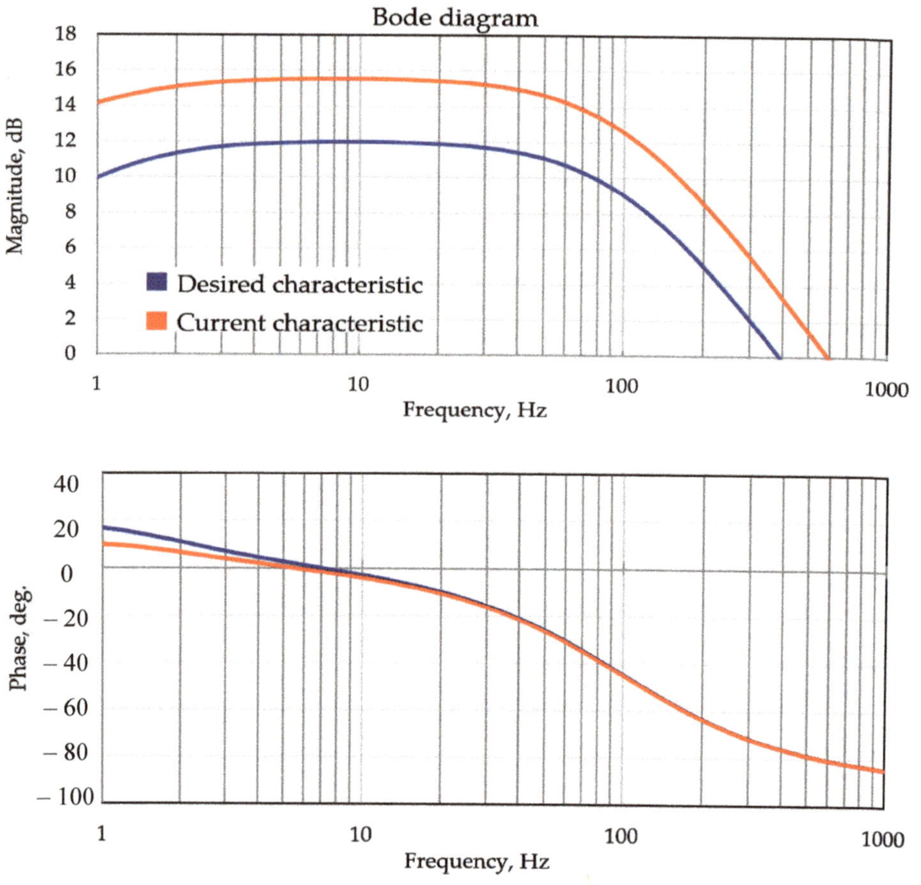


**Figure 20.** Logarithmic amplitude and phase–frequency response curves for the current and desired controllers.

**Table 2.** Final controller settings.

| Parameter | Value |
|---|---|
| Proportional link $P$ | 4-2 |
| Integral link $I$ | 0.08 |
| Differential link $D$ | 2 |
| Time $T_f$ | 0.1 s. |

Analysis of the diagrams indicates that in order to obtain the desired response, it is enough to reduce the proportional coefficient of the PID controller [41–43]. Therefore, we made a proportional change in the coefficient of the controller from 4 to 2. The result of this change is shown in Figure 21. After load surge, the frequency dropped to 43 Hz and recovered to the nominal value within 50 s.

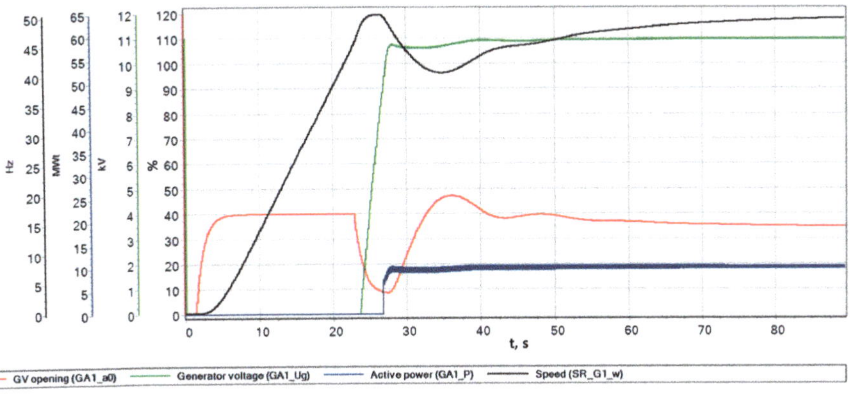


**Figure 21.** Parameters of the transient process.

### 6.3. Emergency Load Shedding

Compared to load surge, load shedding is much more dangerous. To simulate load shedding, we increased the load to the nominal value and when the generator reached the nominal conditions, the load was shed. As soon as the transient process ended and the frequency was nominal, the generator circuit breaker was turned off. The time of tripping corresponds to the simulation time, i.e., 90 s (Figure 22).

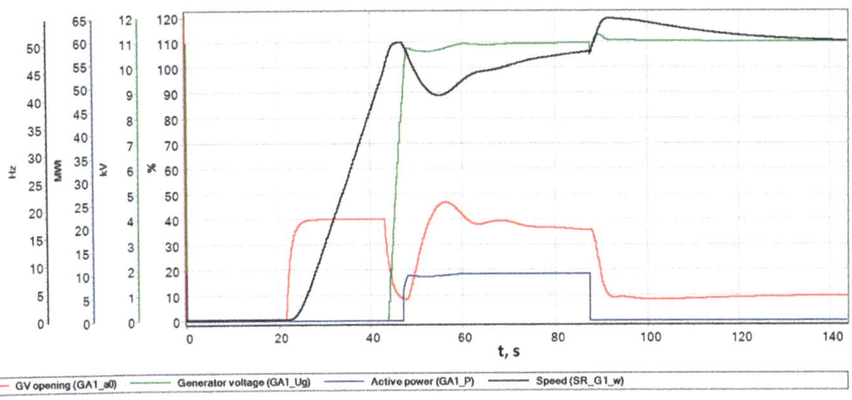


**Figure 22.** Parameters of the transient process during emergency load shedding.

When the load dropped to 12 MW, the unit accelerated to a frequency of 57 Hz (114%); at the same moment, the controller operated and closed the guide vane to the minimum opening. It took approximately a minute for the frequency to recover. The controller responded instantly, at the moment of load shedding, by "landing" the guide vane on the lower opening limit, thereby preventing further acceleration of the rotor of the unit.

## 7. Conclusions

The paper presents a model developed to simulate the electromechanical transient processes of a hydraulic unit operating for an isolated load. With the constructed model, numerical modeling was carried out and the characteristics of transient processes were obtained for various operating conditions of the unit (no-load, load surge, and load shedding). Similar results can be used to study and analyze the operation of existing systems.

The modeling results can be used not only to examine but also to tune the controllers and debug automatic systems used at hydropower plants. Moreover, such real-time simulations can be used to build simulators for the plant personnel to study all the nuances of the process control systems of a hydroelectric unit and its auxiliary equipment.

**Author Contributions:** Conceptualization, A.A., S.K. and K.S.; methodology, S.K. and A.A.; software, A.A.; validation, P.I., A.A., S.K. and K.S.; formal analysis, A.A. and S.K.; investigation, P.I., A.A., S.K. and K.S.; resources, K.S.; data curation, S.K.; writing—original draft preparation, A.A.; writing—review and editing, P.I., A.A., S.K. and K.S.; visualization, A.A.; supervision, K.S.; project administration, P.I. and K.S.; funding acquisition, P.I. and K.S. All authors have read and agreed to the published version of the manuscript.

**Funding:** The research was carried out within the state assignment of Ministry of Science and Higher Education of the Russian Federation (project code: FZZS-2020-0039).

**Data Availability Statement:** Data sharing is not applicable. No new data were created or analyzed in this study. Data sharing is not applicable to this article.

**Conflicts of Interest:** The authors declare no conflict of interest.

## References

1. Vagnonia, E.; Valentinb, D.; Avellana, F. Dynamic Behaviour of a Francis Turbine during Voltage Regulation in the Electrical Power System. Available online: https://www.semanticscholar.org/paper/Dynamic-behaviour-of-a-Francis-turbine-during-in-Vagnonia-Valentinb/b13f7e64e0c79538d11552a84bd9e7fbe3f71951 (accessed on 12 May 2022).
2. Xu, Y.; Zheng, Y.; Du, Y.; Yang, W.; Peng, X.; Li, C. Adaptive Condition Predictive-Fuzzy PID Optimal Control of Start-up Process for Pumped Storage Unit at Low Head Area. *Energy Convers. Manag.* **2018**, *177*, 592–604. [CrossRef]
3. Chen, Z.; Yuan, X.; Yuan, Y.; Lei, X.; Zhang, B. Parameter Estimation of Fuzzy Sliding Mode Controller for Hydraulic Turbine Regulating System Based on HICA Algorithm. *Renew. Energy* **2019**, *133*, 551–565. [CrossRef]
4. Huang, Y.; Yang, W.; Liao, Y.; Zhao, Z.; Ma, W.; Yang, J.; Yang, J. Improved Transfer Function Method for Flexible Simulation of Hydraulic-Mechanical-Electrical Transient Processes of Hydro-Power Plants. *Renew. Energy* **2022**, *196*, 390–404. [CrossRef]
5. Brezovec, M.; Kuzle, I.; Krpan, M.; Holjevac, N. Improved Dynamic Model of a Bulb Turbine-Generator for Analysing Oscillations Caused by Mechanical Torque Disturbance on a Runner Blade. *Int. J. Electr. Power Energy Syst.* **2020**, *119*, 105929. [CrossRef]
6. Zhang, H.; Chen, D.; Wu, C.; Wang, X.; Lee, J.-M.; Jung, K.-H. Dynamic Modeling and Dynamical Analysis of Pump-Turbines in S-Shaped Regions during Runaway Operation. *Energy Convers. Manag.* **2017**, *138*, 375–382. [CrossRef]
7. Yang, W.; Norrlund, P.; Bladh, J.; Yang, J.; Lundin, U. Hydraulic Damping Mechanism of Low Frequency Oscillations in Power Systems: Quantitative Analysis Using a Nonlinear Model of Hydropower Plants. *Appl. Energy* **2018**, *212*, 1138–1152. [CrossRef]
8. Wang, L.X.; Zheng, J.H.; Li, Z.G.; Jing, Z.X.; Wu, Q.H. Order Reduction Method for High-Order Dynamic Analysis of Heterogeneous Integrated Energy Systems. *Appl. Energy* **2022**, *308*, 118265. [CrossRef]
9. Zhao, Z.; Yang, J.; Chung, C.Y.; Yang, W.; He, X.; Chen, M. Performance Enhancement of Pumped Storage Units for System Frequency Support Based on a Novel Small Signal Model. *Energy* **2021**, *234*, 121207. [CrossRef]
10. Lu, X.; Li, C.; Liu, D.; Zhu, Z.; Tan, X. Influence of Water Diversion System Topologies and Operation Scenarios on the Damping Characteristics of Hydropower Units under Ultra-Low Frequency Oscillations. *Energy* **2022**, *239*, 122679. [CrossRef]
11. Guo, W.; Yang, J. Dynamic Performance Analysis of Hydro-Turbine Governing System Considering Combined Effect of Downstream Surge Tank and Sloping Ceiling Tailrace Tunnel. *Renew. Energy* **2018**, *129*, 638–651. [CrossRef]
12. Liu, D.; Wang, X.; Peng, Y.; Zhang, H.; Xiao, Z.; Han, X.; Malik, O.P. Stability Analysis of Hydropower Units under Full Operating Conditions Considering Turbine Nonlinearity. *Renew. Energy* **2020**, *154*, 723–742. [CrossRef]

13. Xu, B.; Luo, X.; Egusquiza, M.; Ye, W.; Liu, J.; Egusquiza, E.; Chen, D.; Guo, P. Nonlinear Modal Interaction Analysis and Vibration Characteristics of a Francis Hydro-Turbine Generator Unit. *Renew. Energy* **2021**, *168*, 854–864. [CrossRef]
14. Liu, Y.; Guo, W. Multi-Frequency Dynamic Performance of Hydropower Plant under Coupling Effect of Power Grid and Turbine Regulating System with Surge Tank. *Renew. Energy* **2021**, *171*, 557–581. [CrossRef]
15. Hu, J.; Yang, J.; Zeng, W.; Zhao, Z.; Yang, J. Hydraulic Interaction of Two Parallel Pump-Turbines in Constant-Speed Oscillation: Measurement, Simulation, and Sensitivity Analysis. *Renew. Energy* **2021**, *176*, 269–279. [CrossRef]
16. Cao, J.; Tian, H.; Ahn, S.-H.; Duo, W.; Bi, H.; Zhao, L.; Zhao, G.; Gao, H.; Wang, M.; Ma, G.; et al. Fatigue Analysis in Rotor of a Prototype Bulb Turbine Based on Fluid-Structure Interaction. *Eng. Fail. Anal.* **2022**, *132*, 105940. [CrossRef]
17. Liao, Y.; Yang, W.; Zhao, Z.; Li, X.; Ci, X.; Alizadeh Bidgoli, M.; Yang, J. Influence Mechanism of Backlash Nonlinearity on Dynamic Regulation Stability of Hydropower Units. *Sustain. Energy Technol. Assess.* **2022**, *51*, 101917. [CrossRef]
18. Brito Junior, G.C.; Machado, R.D.; Chaves Neto, A. Using Simplified Models to Assist Fault Detection and Diagnosis in Large Hydrogenerators. *Int. J. Rotating Mach.* **2017**, *2017*, e9258456. [CrossRef]
19. Liu, X.; Liu, C.Q. Analysis on Oscillations between Two Generators in a Hydro Power Plant and Development of Math Model for a Compound Excitation System. In Proceedings of the Proceedings. International Conference on Power System Technology, Kunming, China, 13–17 October 2002; Volume 2, pp. 1254–1258.
20. Dong, Z.; Tan, J.; Muljadi, E.; Nelms, R.M.; St-Hilaire, A.; Pevarnik, M.; Jacobson, M.D. Developing of Quaternary Pumped Storage Hydropower for Dynamic Studies. *IEEE Trans. Sustain. Energy* **2020**, *11*, 2870–2878. [CrossRef]
21. Giosio, D.R.; Henderson, A.; Walker, J.; Brandner, P. Rapid Reserve Generation from a Francis Turbine for System Frequency Control. *Energies* **2017**, *10*, 496. [CrossRef]
22. Zhou, J.; Zhao, Z.; Zhang, C.; Li, C.; Xu, Y. A Real-Time Accurate Model and Its Predictive Fuzzy PID Controller for Pumped Storage Unit via Error Compensation. *Energies* **2018**, *11*, 35. [CrossRef]
23. Liao, S.; Zhao, H.; Li, G.; Liu, B. Short-Term Load Dispatching Method for a Diversion Hydropower Plant with Multiple Turbines in One Tunnel Using a Two-Stage Model. *Energies* **2019**, *12*, 1476. [CrossRef]
24. Duan, C.; Minglu, Z.; Changbing, Z.; Mengjiao, Y.; Cheng, M.; Chunhe, S. Research on Hydraulic–Electric Interference and Optimisation of Multi-Turbine Hydropower System Based on the Dual Control Mode. *IET Renew. Power Gener.* **2019**, *13*, 1096–1104. [CrossRef]
25. Kong, L.; Cao, J.; Li, X.; Zhou, X.; Hu, H.; Wang, T.; Gui, S.; Lai, W.; Zhu, Z.; Wang, Z.; et al. Numerical Analysis on the Hydraulic Thrust and Dynamic Response Characteristics of a Turbine Pump. *Energies* **2022**, *15*, 1580. [CrossRef]
26. Zhao, J.; Wang, L.; Liu, D.; Wang, J.; Zhao, Y.; Liu, T.; Wang, H. Dynamic Model of Kaplan Turbine Regulating System Suitable for Power System Analysis. *Math. Probl. Eng.* **2015**, *2015*, e294523. [CrossRef]
27. Nanaware, R.A.; Sawant, S.R.; Jadhav, B.T. Modeling of Hydraulic Turbine and Governor for Dynamic Studies of HPP. In Proceedings of the IJCA Proceedings on International Conference on Recent Trends in Information Technology and Computer Science 2012 ICRTITCS 2012, Florence, Italy, 10–14 September 2012pp. 6–11.
28. Guo, P.; Zhang, H.; Gou, D. Dynamic Characteristics of a Hydro-Turbine Governing System Considering Draft Tube Pressure Pulsation. *IET Renew. Power Gener.* **2020**, *14*, 1210–1218. [CrossRef]
29. Choo, Y.C.; Muttaqi, K.M.; Negnevitsky, M. Modelling of Hydraulic Governor-Turbine for Control Stabilisation. *ANZIAM J.* **2007**, *49*, C681–C698. [CrossRef]
30. Yuan, X.; Chen, Z.; Yuan, Y.; Huang, Y. Design of Fuzzy Sliding Mode Controller for Hydraulic Turbine Regulating System via Input State Feedback Linearization Method. *Energy* **2015**, *93*, 173–187. [CrossRef]
31. Mover, W.G.P.; Supply, E. Hydraulic Turbine and Turbine Control Models for System Dynamic Studies. *IEEE Trans. Power Syst.* **1992**, *7*, 167–179. [CrossRef]
32. De Jaeger, E.; Janssens, N.; Malfliet, B.; Van De Meulebroeke, F. Hydro Turbine Model for System Dynamic Studies. *IEEE Trans. Power Syst.* **1994**, *9*, 1709–1715. [CrossRef]
33. Trivedi, C.; Gandhi, B.; Michel, C.J. Effect of Transients on Francis Turbine Runner Life: A Review. *J. Hydraul. Res.* **2013**, *51*, 121–132. [CrossRef]
34. NTNU and LTU Francis-99 Workshop. Available online: https://www.ntnu.edu/nvks/f99-test-case1 (accessed on 13 June 2022).
35. Dynamic Models for Turbine-Governors in Power System Studies. Available online: https://resourcecenter.ieee-pes.org/publications/technical-reports/PESTR1.html (accessed on 5 August 2022).
36. Bao, H.; Yang, J.; Fu, L. Study on Nonlinear Dynamical Model and Control Strategy of Transient Process in Hydropower Station with Francis Turbine. In Proceedings of the 2009 Asia-Pacific Power and Energy Engineering Conference, Wuhan, China, 27–31 March 2009. [CrossRef]
37. Suslov, K.; Shushpanov, I.; Buryanina, N.; Ilyushin, P. Flexible Power Distribution Networks: New Opportunities and Applications. In Proceedings of the 9th International Conference on Smart Cities and Green ICT Systems (SMARTGREENS 2020), Online, 2–4 May 2020; pp. 57–64.
38. Karamov, D.N.; Ilyushin, P.V.; Suslov, K.V. Electrification of Rural Remote Areas Using Renewable Energy Sources: Literature Review. *Energies* **2022**, *15*, 5881. [CrossRef]
39. Rylov, A.; Ilyushin, P.; Kulikov, A.; Suslov, K. Testing Photovoltaic Power Plants for Participation in General Primary Frequency Control under Various Topology and Operating Conditions. *Energies* **2021**, *14*, 5179. [CrossRef]

40. Ilyushin, P. Emergency and Post-Emergency Control in the Formation of Micro-Grids. In *E3S Web Conferences*; EDP Sciences: Les Ulis, France, 2017; Volume 25, p. 02002. [CrossRef]
41. Ilyushin, P.V.; Shepovalova, O.V.; Filippov, S.P.; Nekrasov, A.A. Calculating the Sequence of Stationary Modes in Power Distribution Networks of Russia for Wide-Scale Integration of Renewable Energy Based Installations. *Energy Rep.* **2021**, *7*, 308–327. [CrossRef]
42. Lavrik, A.; Zhukovskiy, Y.; Tcvetkov, P. Optimizing the Size of Autonomous Hybrid Microgrids with Regard to Load Shifting. *Energies* **2021**, *14*, 5059. [CrossRef]
43. Zhukovskiy, Y.; Tsvetkov, P.; Buldysko, A.; Malkova, Y.; Stoianova, A.; Koshenkova, A. Scenario Modeling of Sustainable Development of Energy Supply in the Arctic. *Resources* **2021**, *10*, 124. [CrossRef]

energies

MDPI

*Article*

# Analysis of the Influence of Parameter Condition on Whole Load Power Factor and Efficiency of Line Start Permanent Magnet Assisted Synchronous Reluctance Motor

**Jin Wang, Yan Li *, Shengnan Wu, Zhanyang Yu and Lihui Chen**

National Engineering Research Center for REPM Electrical Machines, Shenyang University of Technology, Shenyang 110178, China; wjcmf11@126.com (J.W.); imwushengnan@163.com (S.W.); ddzhanyang@sina.com (Z.Y.); clh143156@163.com (L.C.)
* Correspondence: eeliyan@126.com

**Abstract:** Line start permanent magnet assisted synchronous reluctance motor (LSPMaSynRM) is an important high-efficiency and high-quality motor. Its parameter matching and operating characteristics are complex, with an increase in salient ratio resulting in a valley in the power factor curve. In this study, the formation principle of power factor curve valley was first deduced by the mathematical model of LSPMaSynRM. Then, the parameter matching principle of power factor curve valley was analyzed in detail. On this basis, the characteristics of load rate corresponding to the critical state of the power factor curve valley were obtained, and its influence on whole load efficiency was analyzed. The design principles for optimal efficiency in wide high-efficiency region and specific load point were obtained. Finally, a 5.5 kW LSPMaSynRM was designed and manufactured to verify the validity of the principle.

**Keywords:** line start permanent magnet assisted synchronous reluctance motor; power factor curve valley; efficiency; whole load region

**Citation:** Wang, J.; Li, Y.; Wu, S.; Yu, Z.; Chen, L. Analysis of the Influence of Parameter Condition on Whole Load Power Factor and Efficiency of Line Start Permanent Magnet Assisted Synchronous Reluctance Motor. *Energies* **2022**, *15*, 3866. https://doi.org/10.3390/en15113866

Academic Editor: Djaffar Ould-Abdeslam

Received: 10 April 2022
Accepted: 19 May 2022
Published: 24 May 2022



## 1. Introduction

High-efficiency, light-weight, and high-quality motor systems are the basic component of high-end equipment in the field of engineering. Permanent magnet synchronous motor (PMSM) has the advantages of high efficiency and high power density [1,2]. It is often used in high-quality motor but requires high cost. Without permanent magnet material, synchronous reluctance motor (SynRMs) relies on reluctance torque to drive the motor. Its cost is low, but its power factor and torque density are also low. Meanwhile, its current is large, and it is difficult for it to be efficient and light.

Permanent magnet assisted synchronous reluctance motor (PMaSynRM) is a special permanent magnet motor that combines the respective characteristics of PMSM and SRM [3]. Reasonable selection and design of rotor blades can improve the salient ratio and power factor of PMaSynRM to reduce the running current. At the same time, the performance of the motor can be optimized under certain constraints by optimizing the matching of permanent magnet torque [4,5].

Many studies have been carried out in this field, including the aspects of power factor and efficiency. In [6], the influence of the shapes of flux barriers and the number of "rotor virtual slots" was investigated based on the multiphysics model, which can achieve low vibration for PMSynRMs. In [7], in order to obtain maximum torque and minimum torque ripple in the design, optimal values of motor parameters were obtained by improving the rotor geometry of the three-phase PMaSynRM. In [8], the influence of permanent magnet flux linkage on power factor was analyzed, and it was proposed that the power factor could be raised to more than 0.8 when the permanent magnet flux linkage was more than 3 times the q-axis. In [9], a PMaSynRM prototype with four poles was designed by placing ferrite magnets inside the rotor of a SynRM, and experimental measurements were performed under various loading conditions. In [10], the optimizing rotor structure was found to improve the power factor, with the power factor of the prototype increasing from 0.879 to 0.918. In [11], the power factor was found to increase from 0.35 to 0.63 by adding AlNiCo on the basis of synchronous reluctance motor. In [12,13], the multilayer magnetic barrier structure was considered, and it was shown that the choice of the first permanent magnet thickness had a great influence on the power factor. In [14], the shape parameters were used to redefine the load rate so as to realize optimization of the load rate to meet the requirements of high efficiency. In [15], a simple structure was proposed with the topology composed of an internally inserted V-shape permanent magnet (IVPM) machine and a synchronous reluctance machine (SynRM). The main novelty was that the PMs in the rotor were diverted so that the reluctance component of the torque and the magnetic component of the torque reached their maximum values at the same load angle, which eventually led to a higher output torque for the same volume. In [16], a PM-assisted-SynRM design was proposed for high torque performance. Although it used the torque components to the fullest, it suffered from high torque ripple and relatively complex rotor geometry. In [17], time-stepping FEM and multiobjective genetic algorithm were used to optimize PMaSynRM, which improved the motor efficiency to more than 92% under rated working conditions and met the IE4 standards. In [18], a two-pole multibarrier ferrite-assisted SynRM for water pumps was designed, with the prototype having high power factor and efficiency. Many scholars have analyzed the operating characteristics of PMaSynRM, such as efficiency and power factor, but most of them have focused on specific rotor structure and geometric parameters. There have been few studies on the efficiency and power factor characteristics of PMaSynRM in the whole load range from the perspective of parameter matching.

In this study, the formation principle of power factor curve valley was deduced by the mathematical model of LSPMaSynRM. The deduction was not confined to any specific rotor structure so that the conclusion could be universal. On this basis, the matching of parameters and the characteristics of the corresponding load rate were analyzed in detail. Then, further analysis of the impact on whole load efficiency was carried out. Finally, a 5.5 kW LSPMaSynRM was designed and manufactured to verify the validity of the principle.

## 2. Analysis of the Principle of Power Factor Curve Valley of PMaSynRM

Power factor is the rate of active power to apparent power, which is essentially the phase relationship between the voltage and current in a specific operating state of the motor determined by parameters under the determination of torque. According to the mathematical model of PMaSynRM, the torque equation can be obtained as follows:

$$T_e = \frac{mpE_0U}{\omega X_\mathrm{d}}\sin\theta + \frac{mpU^2}{2\omega}\left(\frac{1}{X_\mathrm{q}} - \frac{1}{X_\mathrm{d}}\right)\sin 2\theta \tag{1}$$

where $T_\mathrm{e}$ is electromagnetic torque, $m$ is the number of phases, $p$ is poles, $E_0$ is no-load back electromotive force (EMF), $\theta$ is the angle between voltage and no-load back EMF, $U$ is voltage, $\omega$ is angular frequency, $X_\mathrm{d}$ is d-axis reactance, and $X_\mathrm{q}$ is q-axis reactance. The vector diagram was shown in Figure 1.

The torque equation can also be expressed as follows:

$$T_e = p[\psi_\mathrm{f} i_\mathrm{s}\sin\beta + \frac{1}{2}(L_\mathrm{d} - L_\mathrm{q})i_s^2\sin 2\beta] \tag{2}$$

where $\psi_\mathrm{f}$ is permanent magnet flux linkage, $\beta$ is the angle between current and permanent magnet flux linkage, $i_\mathrm{s}$ stator current, $L_\mathrm{d}$ is d-axis inductance, and $L_\mathrm{q}$ is q-axis inductance.

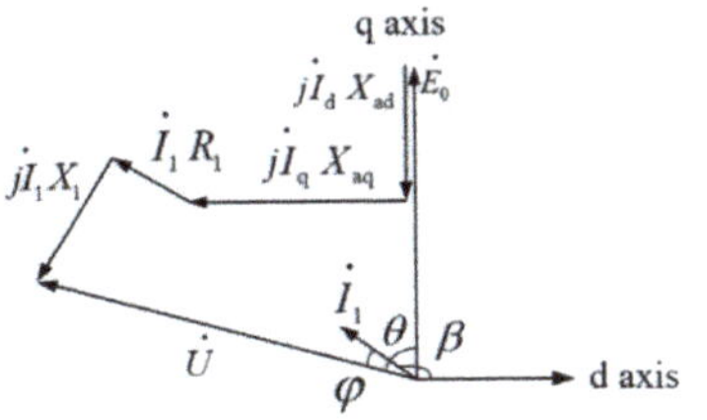


**Figure 1.** The vector diagram.

In sine steady state, the torque equation can be changed as follows:

$$T_e = mp[\frac{E_0}{\omega} I_s \sin\beta + \frac{1}{2}(L_d - L_q) I_s^2 \sin 2\beta] \tag{3}$$

Ignoring resistance, the d-axis and q-axis current are as follows:

$$I_d = \frac{E_0 - U\cos\theta}{X_d} \tag{4}$$

$$I_q = \frac{U\sin\theta}{X_q} \tag{5}$$

The stator current can be expressed as follows:

$$I_s = \sqrt{(\frac{E_0 - U\cos\theta}{X_d})^2 + (\frac{U\sin\theta}{X_q})^2} \tag{6}$$

Putting it into torque Equation (3), the torque equation can be expressed as follows:

$$\begin{aligned} T_e = \tfrac{mpE_0}{\omega}\sqrt{(\tfrac{E_0 - U\cos\theta}{X_d})^2 + (\tfrac{U\sin\theta}{X_q})^2}\sin\beta \\ + \tfrac{mp}{2\omega}(X_d - X_q)[(\tfrac{E_0 - U\cos\theta}{X_d})^2 + (\tfrac{U\sin\theta}{X_q})^2]\sin 2\beta \end{aligned} \tag{7}$$

Torque is a function of $\theta$ and $\beta$. Equation (1) shows that the shape of the torque curve depends on $E_0$, $U$, $X_d$, and $X_q$, and the torque increases as $\theta$ increases. Equation (7) shows that the shape of the torque curve depends on $E_0$, $U$, $X_d$, $X_q$, and $\theta$. For a manufactured motor, $E_0$, $X_d$, and $X_q$ are constants, and $U$ can also be regarded as a fixed value. As the torque increases, there is a one-to-one correspondence between $\theta$ and $\beta$. This corresponding relationship depends on the motor parameters $E_0$, $U$, $X_d$, and $X_q$, which can be attributed to two parameters, namely $\lambda = E_0/U$, which indirectly reflects the amount of permanent magnets, and the salient ratio $\rho = X_q/X_d$. Different parameters have different corresponding relationships between $\theta$ and $\beta$, which ultimately reflect different power factor curve states.

The relationship between $\theta$ and $\beta$ can be obtained by simultaneous Equations (1) and (7). Because the equation is very complicated, it is impossible to obtain the analytical expression of the relationship. One solution of $\theta$ corresponding to two $\beta$ can be obtained by numerical methods. According to the running state of the motor, the true solution and false solution can be judged as shown in Figure 2.

The introduction of the square term of $\theta$ in the process of deriving Equation (7) results in two $\beta$ solutions. According to the voltage and torque equations, the voltage circle and the torque curve are obtained in the current plane of the d–q axis, and the true solution is obtained according to the intersection point, as shown in Figure 3.

In this way, a torque curve with $\theta$ and $\beta$ as independent variables can be obtained. The state of the curve depends on parameters $\lambda$ and $\rho$. Under certain conditions, the two curves will have special states, as shown in Figure 4.

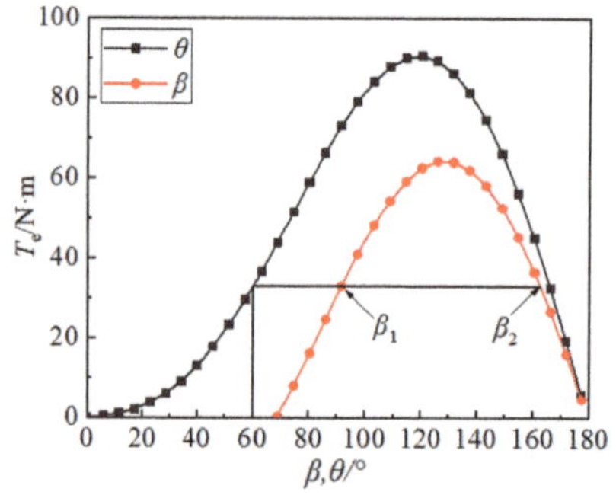


**Figure 2.** $\beta$ corresponding solution at $\theta = 60$.

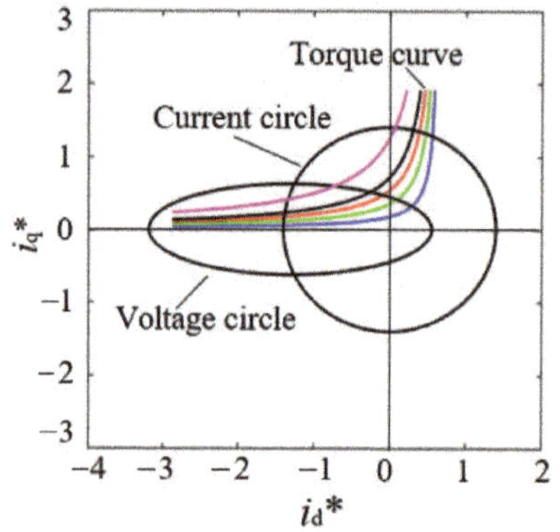


**Figure 3.** The torque curve in plane of the d-q axis with per unit value of $i_d$-$i_q$.

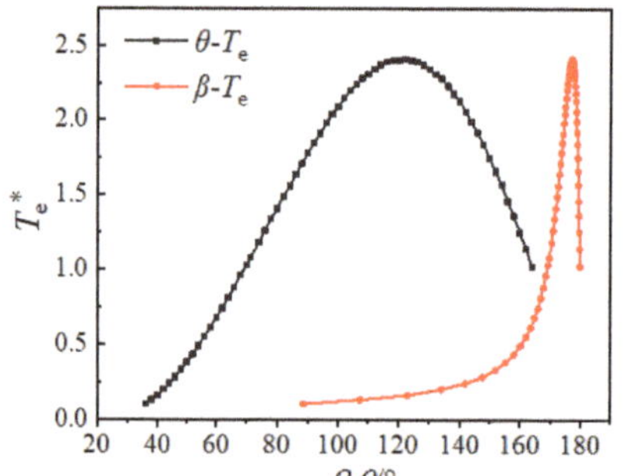


**Figure 4.** The $\theta$-$T_e$ and $\beta$-$T_e$ curves with per unit value.

In order to observe the relationship of the curves more clearly, the $\beta$-$T_e$ curve was shifted to the left by 90°, as shown in Figure 5. It can be seen that there are three intersections between the two curves, and the torque at the rightmost intersection is in the unstable range, so it will not be discussed. At the two intersections on the left, the two torque curves correspond to the same angle, and the power factor is 1. Between the crossing points and on both sides, the two curves of the same torque correspond to different angles, and the power factor is less than 1. This shows that there will be a valley in the power factor curve in the middle of two maximum values.

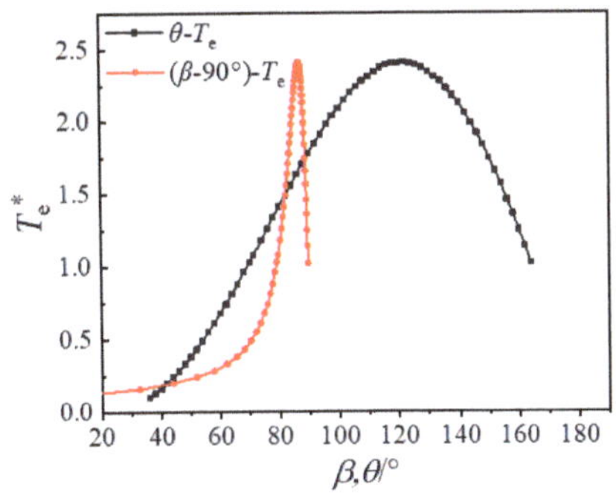


**Figure 5.** The $\theta$-$T_e$ and ($\beta - 90°$)-$T_e$ curves with per unit value.

At the same time, the $\theta$ and $\beta$ relationship curve and the power factor curve can be drawn as the torque increases. As shown in Figure 6, the valley in the power factor curve is apparent.

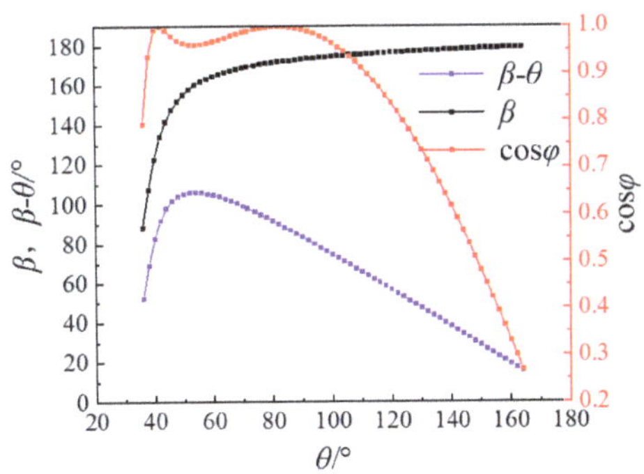


**Figure 6.** The power factor curve.

From the above analysis, it can be seen that the power factor curve valley is caused by two torque curves with two intersection points in the stable operating interval under matching parameters. It can also be understood that $\theta$ and $\beta$ increase at different speeds.

## 3. The Influence of Parameter Matching on Power Factor Curve Valley of PMaSynRM and Its Corresponding Load Rate

### 3.1. The Condition of Power Factor Curve Valley and the Principle of Parameter Matching

The analysis in the previous section shows that the power factor curve valley is caused by the increasing speed of $\theta$ and $\beta$ being different as load increases. Therefore, the condition is that there is a $\beta - \theta > 90°$ state during load increases. Whether there is a state of $\beta - \theta > 90°$ depends on the parameters of the motor. Starting with the matching of $\lambda$ and $\rho$, the principle that produces power factor curve valley are analyzed in this section.

The change curves of $\beta$ with $\theta$ under different salient ratio with $\lambda = 0.2$ are shown in Figure 6. With the increase in salient ratio, the middle part of curve stretches and protrudes to the upper left corner. The slope of the front part increases, but the slope of the back part decreases. This shows that as the salient ratio increases and as the load increases, the growth rate of $\beta$ of the low load zone increases significantly, while the growth rate of $\beta$ of the high load zone decreases. Throughout the whole load range, the value of $\beta - \theta$ increases first and then decreases. There must be a salient ratio state that makes a certain load point of $\beta - \theta = 90°$, which can be called the critical point of the power factor curve valley. Then, there will be a valley on the power factor curve with increasing salient ratio. As can be seen from the rectangular box in Figure 7, the value of $\theta$ at the starting point of the curve is around 80°. As the salient ratio increases, the range of change is relatively small, but the range of change of $\beta$ is relatively large.

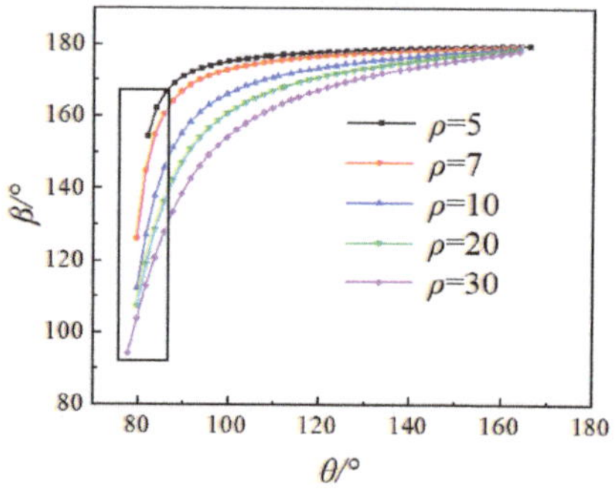


**Figure 7.** The curves of different $\rho$ at $\lambda = 0.2$.

The change curves of $\beta$ with $\theta$ are shown in Figure 8 for the condition of $\lambda = 0.5$. The state of the curves as the salient ratio increases is similar to Figure 7. The difference is that the initial value of $\theta$ in the rectangular box is about 60°.

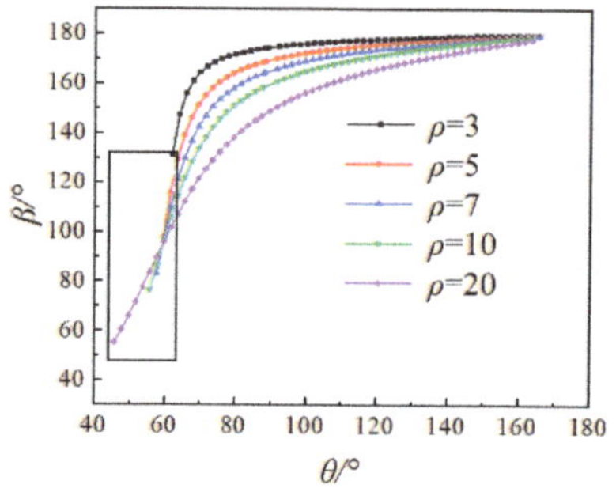


**Figure 8.** The curves of different $\rho$ at $\lambda = 0.5$.

The change curves of $\beta$ with $\theta$ are shown in Figure 9 for the condition of $\lambda = 0.8$. The overall state of the curve is similar to Figures 7 and 8, and the initial value of $\theta$ is further reduced to about 30°.

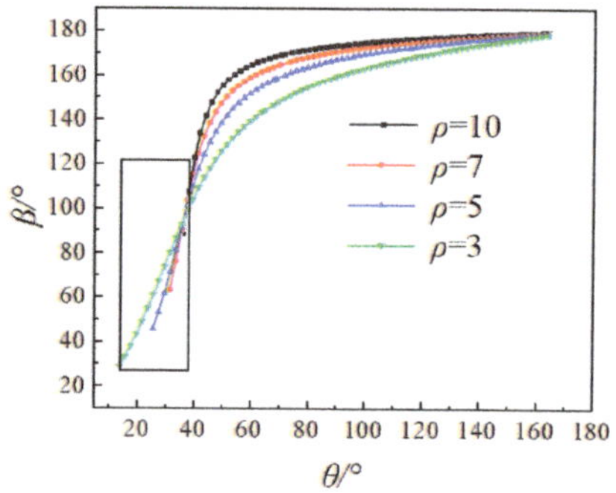


**Figure 9.** The curves of different $\rho$ at $\lambda = 0.8$.

As can be seen from Figures 7–9 as $\lambda$ increases, the initial value of $\theta$ gradually decreases. The smaller the value of $\theta$, the less difficult it is to reach $\beta - \theta > 90°$, which means that it is easier for a valley to be formed on the power factor curve. At the same time, the distribution range of entire curve $\beta$ and $\theta$ increases as $\lambda$ increases.

The above analysis shows that there are different parameter matching principles that result in a valley in the power factor curve. The states of $\beta - \theta = 90°$ can be obtained by calculation, as shown in Figure 10. The required salient ratio increases as $\lambda$ decreases. As the corresponding critical point of the power factor curve valley value of $\theta$ increases, the distance to the initial point is closer and the initial value of $\beta$ and $\theta$ are larger.

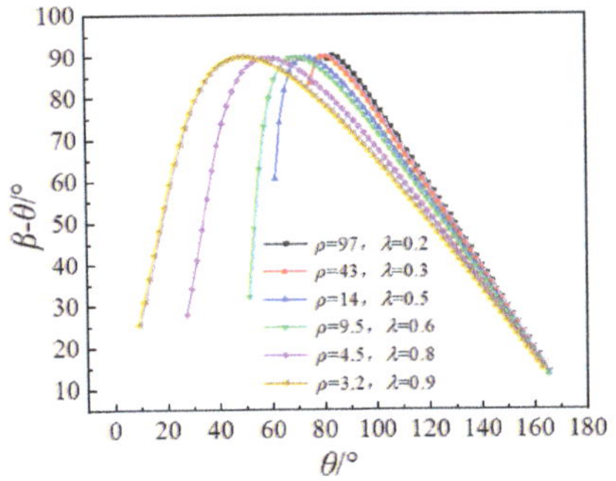


**Figure 10.** The curves under different parameter matching.

By linking the critical point of the power factor curve valley in the $\lambda$-$\rho$ coordinate plane, the power factor curve valley area can be obtained. The parameter matching principles can be obtained as shown in Figure 11.

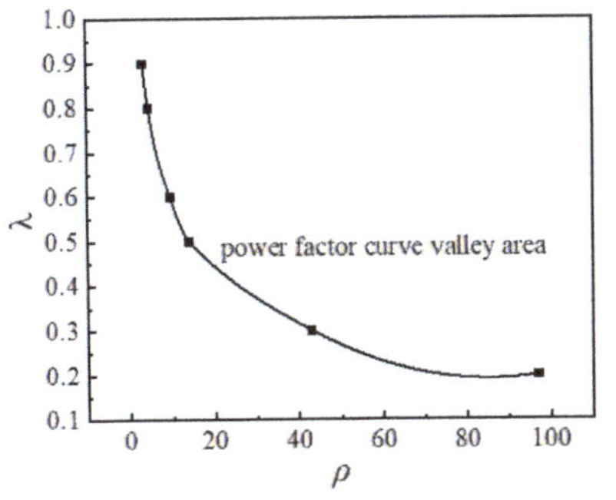


**Figure 11.** Diagram of power factor curve valley area.

The dividing line is not a straight line, and the required $\rho$ increases nonlinearly. The linear relationship is basically between $\lambda$ = 0.6 and 0.9, and the required $\rho$ value increases sharply in the interval less than 0.5. When the value is low, it is difficult to reach the state of power factor curve valley.

### 3.2. The Influence of Parameter Matching on the Load Rate of the Critical Point of Power Factor Curve Valley

In the previous section, the conditions and parameter matching of power factor curve valley were analyzed. From Figure 10, it can be seen that the critical points of different states correspond to different $\theta$ values, indicating that the load rates are different. In order to obtain the principles, the torque curve in different states were calculated, and the $\beta$-$T_e$ curve was moved to the left by 90° to make the relationship clearer, as shown in Figures 12–16.

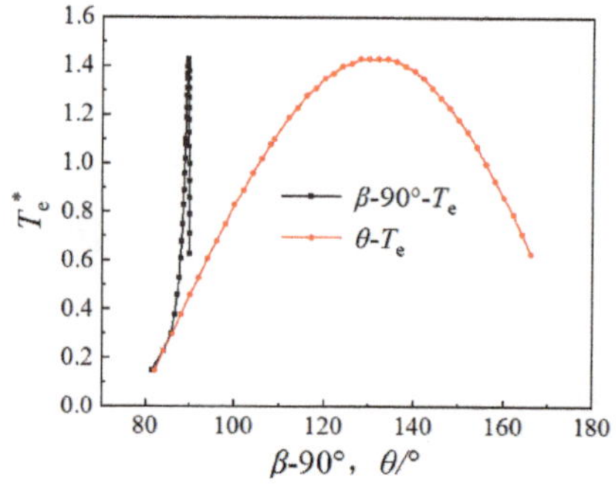


**Figure 12.** The curve of per unit value of torque at $\lambda = 0.2$.

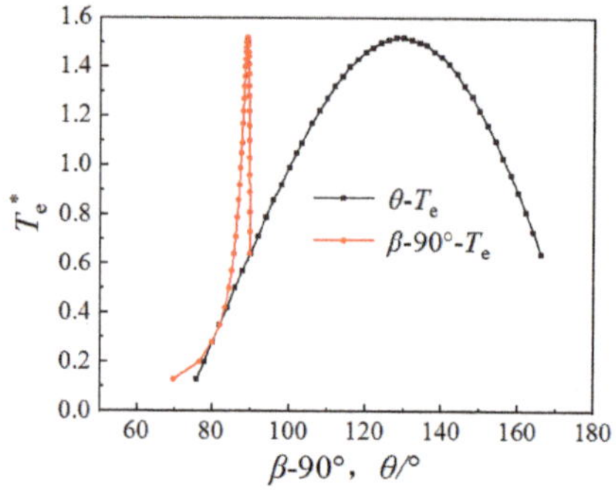


**Figure 13.** The curve of per unit value of torque at $\lambda = 0.3$.

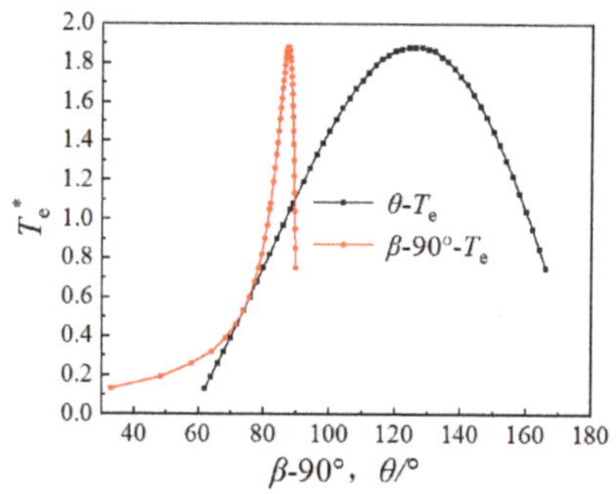


**Figure 14.** The curve of per unit value of torque at $\lambda = 0.5$.

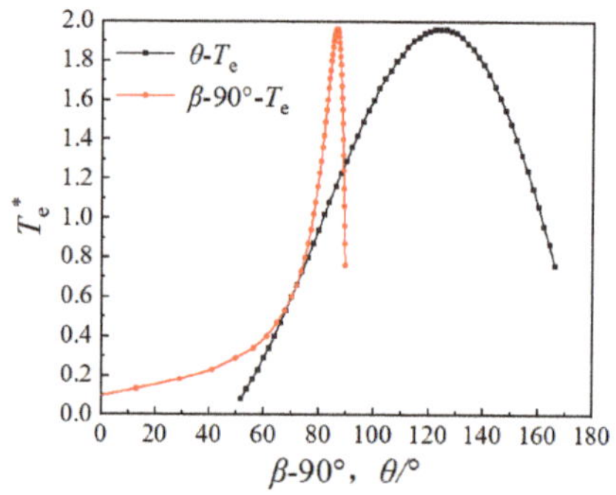


**Figure 15.** The curve of per unit value of torque at $\lambda = 0.6$.

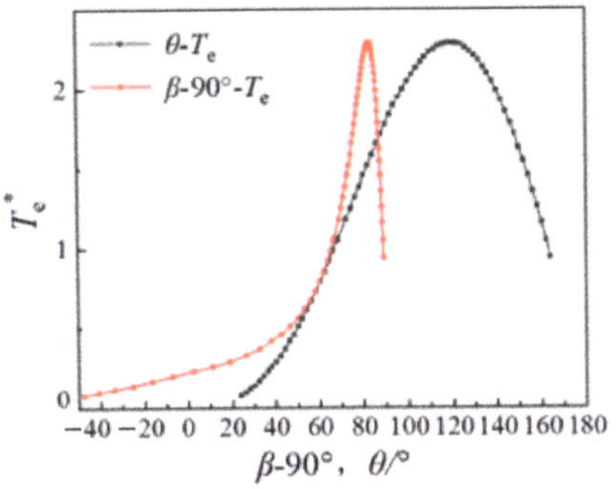


**Figure 16.** The curve of per unit value of torque at $\lambda = 0.8$.

As can be seen from the above figures, the greater the value of $\lambda$, the greater the load rate corresponding to the critical point of the power factor curve valley. When the value of $\lambda$ is small, the $\beta$-$T_e$ curve is distributed in a smaller $\beta$ angle range, while the torque range is relatively large. At the same time, the slope of the rising interval of the $\theta$-$T_e$ curve is relatively small, so the interval between two curves is smaller. When the value of $\lambda$ is larger, the interval between two curves is larger, indicating that the interval of high power factor is larger. The load rate curve corresponding to the critical point of the power factor curve valley under different $\lambda$ is shown in Figure 17.

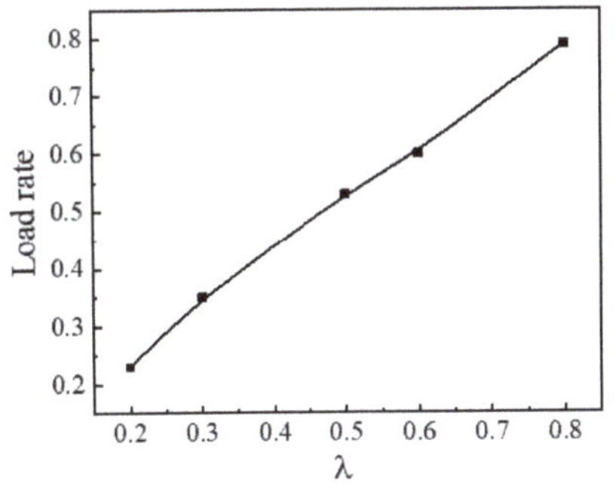


**Figure 17.** The load rate curve with different $\lambda$.

The load rate and value of $\lambda$ basically change linearly, and the minimum current state at any load point can be obtained by selecting the reasonable parameters according to the curve.

### 3.3. Adjustment of Load Rate Corresponding to the Minimum Current Point

The load rate of the power factor curve valley can also be understood as the minimum current at a specific load rate. According to the analysis in the previous section, the minimum current under different load rates can be achieved under specific $\lambda$ and $\rho$ matching. This is a method of adjusting the load rate corresponding to the minimum current point. Its characteristic is to achieve the minimum current with the minimum $\rho$ under each $\lambda$. The high load point requires a larger $\lambda$ and a smaller $\rho$, and the low load point requires a smaller $\lambda$ and a larger $\rho$. This can save the amount of permanent magnets. This is the most reasonable method in theory. However, its disadvantage is that low load requires a large $\rho$, which is difficult to achieve with the existing rotor manufacturing technology in engineering.

From the analysis, it can be seen that two minimum current load points are generated under the power factor curve valley. One point tends toward low load, and the other point tends toward high load. In this way, there can be a second method of adjusting the load rate corresponding to the minimum current point. The curves are shown in Figures 18 and 19 when $\lambda$ is 0.8.

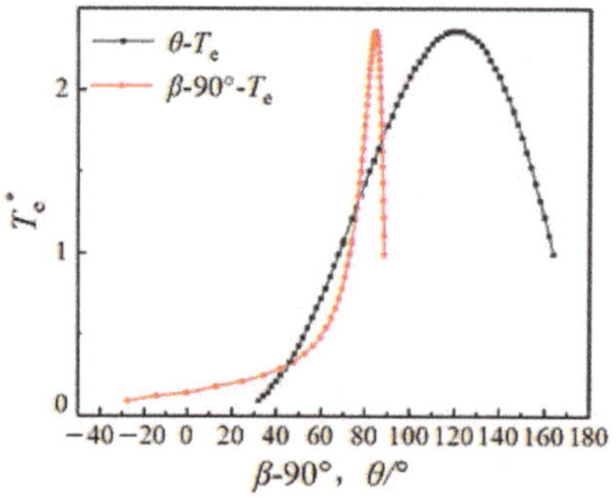


**Figure 18.** The curve of per unit value of torque at $\lambda = 0.8$ $\rho = 7$.

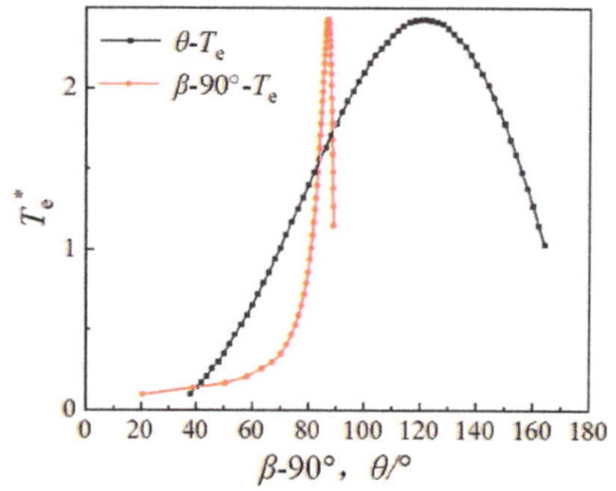


**Figure 19.** The curve of per unit value of torque at $\lambda = 0.8$ $\rho = 20$.

The minimum current point gradually moves to low load as the salient ratio increases, so the minimum current at any load point can also be achieved. Compared with the first method, the second method can achieve the minimum current at the low load point with a smaller salient ratio. The disadvantage of this is that the load point current between two minimum points of current is large.

## 4. Influence of Power Factor Curve Valley of PMaSynRM on Whole Load Efficiency

The analysis of efficiency involves losses and needs to be targeted at specific research objects. A 5.5 kW motor was taken as an example for the present analysis.

### *4.1. The Influence of the Power Factor Curve Valley on the Efficiency of Load Rate Point*

Different parameter matching can realize the power factor curve valley of any load point. In this state, further analysis is needed to determine whether the efficiency of the load point is optimal. $\lambda = 0.5$ and $\lambda = 0.8$ were selected to calculate the efficiency under different salient ratios. The load rate efficiency under the power factor curve valley is shown in Figures 20 and 21.

As shown in Figure 20, the efficiency of the critical point of the power factor curve valley is not the highest. As the salient ratio increases, the efficiency changes from high to low.

As can be seen in Figure 21, the power factor curve valley occurs at a critical point with the highest efficiency, and the efficiency on both sides gradually decreases. The highest efficiency in different states is different. This is because the corresponding load point is the high load point when $\lambda$ is 0.8. Here, copper loss accounts for a larger proportion of the total loss, and the magnitude of the current determines efficiency. Therefore, the power factor curve valley occurs at the critical point with the highest efficiency. The corresponding load point is the low load point when $\lambda$ is 0.5. The copper loss accounts for a smaller proportion of the total loss, and the current cannot completely determine efficiency. The demagnetization field increases as the salient ratio decreases, the iron loss gradually decreases, and the efficiency gradually increases. By combining the loss rate of each load point, a reasonable design of motor parameters can obtain optimal efficiency of any load.

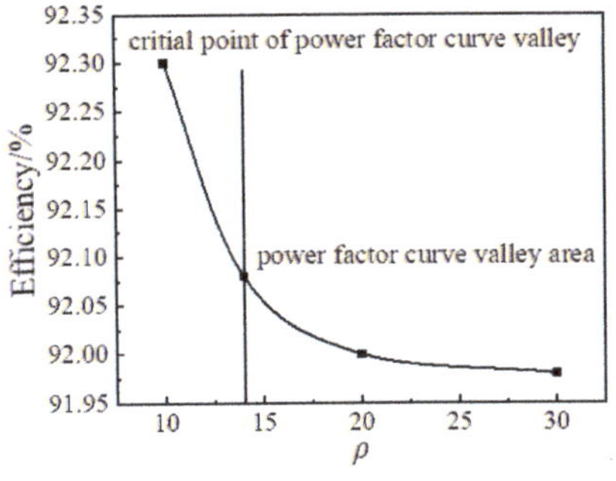


**Figure 20.** Efficiency curve with different $\rho$ at $\lambda = 0.5$.

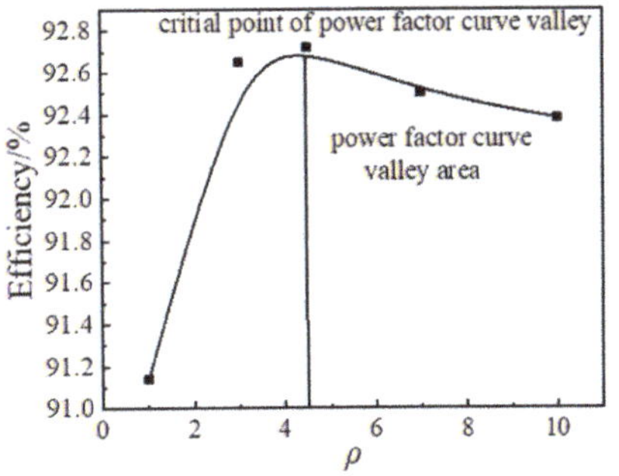


**Figure 21.** Efficiency curve with different $\rho$ at $\lambda = 0.8$.

### *4.2. The Influence of the Corresponding Load Rate Point of Power Factor Curve Valley on Whole Load Efficiency*

Although a reasonable design of motor parameters can obtain optimal efficiency of any load, the effect of the size of the high-efficiency zone in the whole load range needs further research. The conditions of $\lambda = 0.5$ and $\lambda = 0.8$ were again selected for analysis. The power factor and efficiency of the whole load were calculated and are shown in Figures 22–24.

Compared with the state of $\lambda = 0.5$, the power factor of low load is higher and the power factor of high load is lower, as shown in Figure 22. Compared with the state of $\lambda = 0.5$, the low load efficiency is high, and the high load efficiency is low, as shown in Figures 23 and 24. The copper loss accounts for a large proportion of total loss in the high load area. The current state of $\lambda = 0.8$ is small and has high efficiency. The current of $\lambda = 0.5$ in the low load area is small and has high efficiency, but the copper loss is small. Therefore, the low load efficiency difference between the two states is less than that for high load efficiency. As can be seen, the larger the $\lambda$, the larger the range for high efficiency, and the smaller the $\lambda$, the higher the efficiency range for low load.

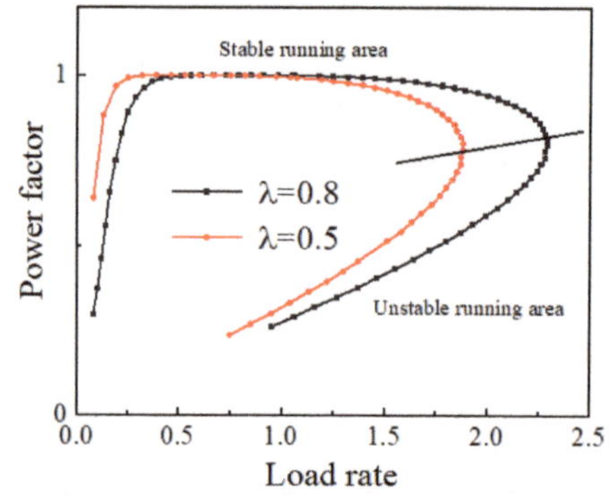


**Figure 22.** Power factor curves.

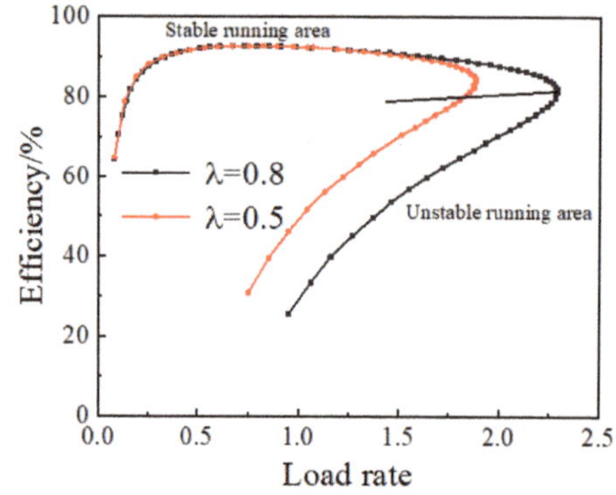


**Figure 23.** Efficiency curves.

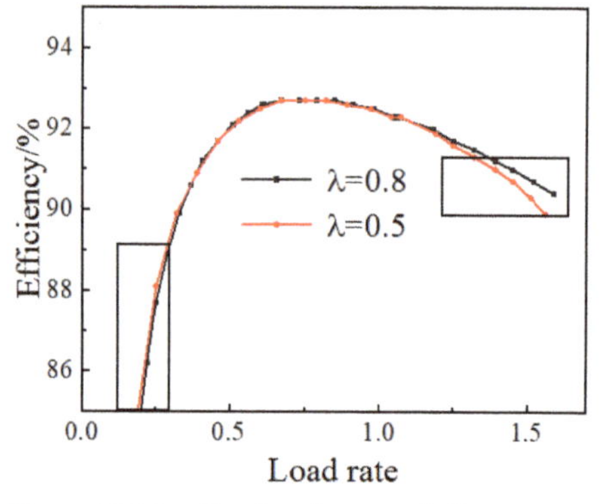


**Figure 24.** Magnification diagram of efficiency.

## 5. Prototype Test

A 5.5 kW PMaSynRM was designed and manufactured. The parameters of the motor are shown in Table 1, and the structure of the rotor is shown in Figure 25. The inductance of the prototype was tested as shown in Figure 26. The d-axis inductance was 41.1 mH, the q-axis inductance was 127.8 mH, and the salient pole rate of the motor was 3.11.

**Table 1.** Main parameters of the prototype.

| Parameters | Value |
|---|---|
| Power | 5.5 kW |
| Rated speed | 1500 r/min |
| Poles | 4 |
| Inner/outer diameter of stator | 210 mm/136 mm |
| Stator slots | 36 |
| Air gap | 0.5 mm |
| Core length | 145 mm |
| Phase resistance | 0.559 Ω |
| d-axis inductance | 41.1 mH |
| q-axis inductance | 127.8 mH |
| Flux linkage of the permanent magnet | 0.723 Wb |
| Leakage stator inductance | 5.54 mH |
| Rotor leakage inductances in q-axis referring to the stator side | 3.06 mH |
| Rotor leakage inductances in d-axis referring to the stator side | 3.06 mH |
| Rotor resistance in d-axis | 0.740 Ω |
| Rotor resistance in q-axis | 0.740 Ω |
| Moment of inertia | 0.05 kg·m$^2$ |
| Maximum moment of inertia with a static load moment torque to the rated value | 3.5 kg·m$^2$ |
| Magnet grade | N38SH |
| Electric steel grade | 35W270 |

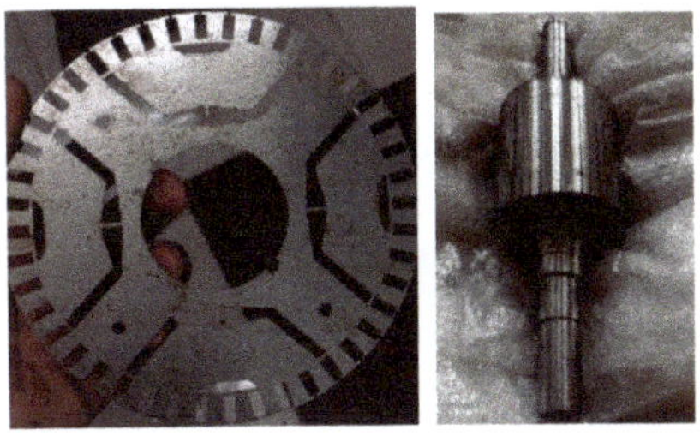

**Figure 25.** The prototype of the rotor.

**Figure 26.** The inductance test of the prototype.

The rated phase voltage of the prototype was 220 V, and the back-EMF was 160.5 V by experimental test. $\lambda$ and $\rho$ were not on the curve of the critical point of the power factor curve valley discussed above, and the back-EMF and $\rho$ could not be changed after the prototype was produced. In order to verify the correctness of the above conclusions, the input voltage was only changed for testing. According to the previous curve, when input voltage is adjusted to 174.3 V, the motor is in a critical state of the power factor curve valley. The input voltage was adjusted to 174.3 V, and the power factor was tested as shown in Figure 27. The PMaSynRM was used as the tested motor linked to a torque sensor, and the load motor was a DC generator. The test power factor curve is shown in Figure 28, and the efficiency curve is shown in Figure 29.

**Figure 27.** The load test of the prototype.

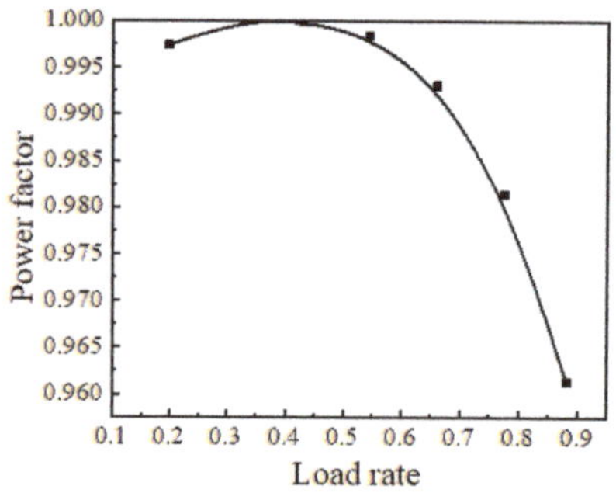


**Figure 28.** The power factor curve ($U$ = 174.3 V).

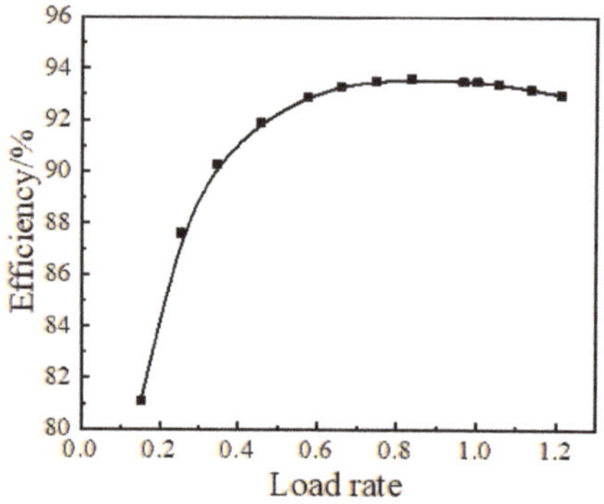


**Figure 29.** The efficiency curve.

The test results showed that, when the power factor reached 1, the load rate was between 0.3 and 0.4, which is the critical state of the power factor curve valley.

## 6. Conclusions

From the LSPMaSynRM mathematical model, the principle of the power factor curve valley was derived, and its parameter matching principle was analyzed. Then, the characteristics of the corresponding load rate and the influence on whole load efficiency were further analyzed. The conclusion reached in this study is not confined to any specific rotor structure. As long as the parameter matching conforms to the analysis carried out here, the corresponding power factor characteristics can be obtained, thus providing a certain theoretical basis for the detection and judgment of the motor running state. The conclusions are as follows:

(1). The principle of the power factor curve valley of LSPMaSynRM is that the matching of specific motor parameters causes the torque curve to deform, and a valley is generated by two intersection points when two torque curves are in the stable operating range. The parameter matching relationship is shown in Figure 11, $\lambda$ is basically a linear relationship between 0.6 and 0.9, and the required $\rho$ value increases sharply in an interval of less than 0.5.

(2). The load rate at the critical point of the power factor curve valley is basically linear to $\lambda$ and increases with increasing $\lambda$.

(3). A reasonable design of motor parameters can obtain the minimum current for any load. In the critical state of power factor curve valley of different parameter matching, the larger the $\lambda$, the larger the high-efficiency range and the higher the efficiency of the high load, whereas the smaller the $\lambda$, the higher the efficiency of the low load, Shengnan Wu, Zhanyang Yu and Lihui Chen.

**Author Contributions:** Conceptualization, J.W. and Z.Y.; methodology, J.W. and Y.L.; software, S.W. and L.C.; validation, J.W., Y.L. and Z.Y.; formal analysis, J.W. and Y.L. All authors have read and agreed to the published version of the manuscript.

**Funding:** This work was supported by the Smart Grid Joint Fund of National Natural Science Foundation of China, No.U2166213.

**Institutional Review Board Statement:** Not applicable.

**Informed Consent Statement:** Not applicable.

**Data Availability Statement:** Not applicable.

**Conflicts of Interest:** The authors declare no conflict of interest.

## References

1. Candelo-Zuluaga, C.; Riba, J.R.; Thangamuthu, D.V.; Garcia, A. Detection of partial demagnetization faults in five-Phase permanent magnet assisted synchronous reluctance machines. *Energies* **2020**, *13*, 3496. [CrossRef]
2. Bianchi, N.; Bolognani, S.; Carraro, E.; Castiello, M.; Fornasiero, E. Electric vehicle traction based on synchronous reluctance motors. *IEEE Trans. Ind. Appl.* **2016**, *52*, 4762–4769. [CrossRef]
3. Du, Z.S.; Lipo, T.A. Design of an improved dual-stator ferrite magnet vernier machine to replace an industrial rare-earth IPM machine. *IEEE Trans. Energy Convers.* **2019**, *34*, 2062–2069. [CrossRef]
4. Liu, H.; Kim, I.; Oh, Y.J.; Lee, J.; Go, S. Design of permanent magnet-assisted synchronous reluctance motor for maximized back-EMF and torque ripple reduction. *IEEE Trans. Magn.* **2017**, *53*, 8202604. [CrossRef]
5. Jin, W.; Yan, L.; Jianguo, J.; Yongteng, J.; Jialin, Z. Analysis of the influence of back-EMF and saliency rate on steady-state characteristics of a high efficiency permanent magnet synchronous reluctance motor. *Trans. China Electrotech. Soc.* **2020**, *35*, 4688–4698.
6. Lu, Y.; Li, J.; Xu, H.; Yang, K.; Sun, J. Comparative study on vibration behaviors of permanent magnet assisted synchronous reluctance machines with different rotor topologies. *IEEE Trans. Ind. Appl.* **2021**, *57*, 1420–1428. [CrossRef]
7. Bozkurt, A.; Baba, A.F.; Oner, Y. Design of outer-rotor permanent-magnet-assisted synchronous reluctance motor for electric vehicles. *Energies* **2021**, *14*, 3739. [CrossRef]
8. Bianchi, N.; Fornasiero, E.; Soong, W. Selection of PM flux linkage for maximum low-speed torque rating in a PM-assisted synchronous reluctance machine. *IEEE Trans. Ind. Appl.* **2015**, *51*, 3600–3608. [CrossRef]
9. Liu, C.T.; Luo, T.Y.; Shih, P.C.; Yen, S.C.; Hwang, C.C. Design and construction assessments of a permanent-magnet-assisted synchronous reluctance motor. *IEEE Trans. Magn.* **2017**, *53*, 2002104. [CrossRef]
10. Chai, W.; Yang, H.M.; Xing, F.; Kwon, B.I. Analysis and design of a PM-assisted wound rotor synchronous machine with reluctance torque enhancement. *IEEE Trans. Ind. Appl.* **2020**, *68*, 2887–2897. [CrossRef]
11. Maroufian, S.S.; Pillay, P. Design and analysis of a novel PM-assisted synchronous reluctance machine topology with AlNiCo magnets. *IEEE Trans. Ind. Appl.* **2019**, *55*, 4733–4742. [CrossRef]
12. Mohanarajah, T.; Nagrial, M.; Rizk, J.; Hellany, A. Permanent Magnet Optimization in PM Assisted Synchronous Reluctance Machines. In Proceedings of the 2018 IEEE 27th International Symposium on Industrial Electronics, Cairns, Australia, 13–15 June 2018; pp. 1347–1351.

13. Huynh, T.A.; Hsieh, M.F.; Shih, K.J.; Kuo, H.F. An Investigation into the effect of PM arrangements on PMa-SynRM performance. *IEEE Trans. Ind. Appl.* **2018**, *54*, 5856–5868. [CrossRef]
14. Jung, D.H.; Kwak, Y.; Lee, J. Study on the optimal design of PMa-SynRM loading rate for achievement of ultrapremium efficiency. In Proceedings of the 2016 IEEE Conference on Electromagnetic Field Computation (CEFC), Miami, FL, USA, 13–16 November 2016; pp. 1–4.
15. Ali, M.; Seyyed, M.M. Design of a novel PM-assisted synchronous reluctance motor topology using V-shape permanent magnets for improvement of torque characteristic. *IEEE Trans. Energy Convers.* **2022**, *37*, 424–432.
16. Zhao, W.; Chen, D.; Lipo, T.A.; Kwon, B.I. Performance improvement of ferrite-assisted synchronous reluctance machines using asymmetrical rotor configurations. *IEEE Trans. Magn.* **2015**, *51*, 8108504. [CrossRef]
17. Lin, Q.; Niu, S.; Fu, W. Design and optimization of a reluctance-torque-assisted synchronous motor with high efficiency and low torque ripple. In Proceedings of the 2019 22nd International Conference on Electrical Machines and Systems, Harbin, China, 11–14 August 2019; pp. 1–4.
18. Baka, S.; Sashidhar, S.; Fernandes, B.G. Design of an energy efficient line-start two-pole ferrite assisted synchronous reluctance motor for water pumps. *IEEE Trans. Energy Convers.* **2020**, *36*, 961–970. [CrossRef]

*energies*

MDPI

*Article*

# Parameter Identification of Asynchronous Load Nodes

**Andrey Kryukov [1,2], Konstantin Suslov [2,3,*], Pavel Ilyushin [3,4] and Azat Akhmetshin [5]**

1 Department of Transport Electric Power, Irkutsk State Transport University, 664074 Irkutsk, Russia
2 Department of Power Supply and Electrical Engineering, Irkutsk National Research Technical University, 664074 Irkutsk, Russia
3 Department of Hydropower and Renewable Energy, National Research University "Moscow Power Engineering Institute", 111250 Moscow, Russia
4 Department of Research on the Relationship between Energy and the Economy, Energy Research Institute of the Russian Academy of Sciences, 117186 Moscow, Russia
5 Department of Power Engineering, Kazan State Power Engineering University, 420066 Kazan, Russia
* Correspondence: dr.souslov@yandex.ru

**Abstract:** Asynchronous loads (AL), because of their low negative-sequence resistance, produce the effect of reduced unbalance at their connection points. Therefore, proper modeling of unbalanced load flows in power supply systems requires properly accounting for AL. Adequate models of the induction motor can be realized in the phase frame of reference. The effective use of such models is possible only if accurate data on the parameters of induction motor equivalent circuits for positive and negative sequences are available. Our analysis shows that the techniques used to determine these parameters on the basis of reference data can yield markedly disparate results. It is possible to overcome this difficulty by applying parameter identification methods that use the phase frame of reference. The paper proposes a technique for parameter identification of models of individual induction motors and asynchronous load nodes. The results of computer-aided simulation allow us to conclude that by using parameter identification, we can obtain an equivalent model of an asynchronous load node, and such a model provides high accuracy for both balanced and unbalanced load flow analysis. By varying load flow parameters, we demonstrate that the model proves valid over a wide range of their values. We have proposed a technique for the identification of asynchronous load nodes with such asynchronous loads, including electrical drives equipped with static frequency converters. With the aid of the AL identification models proposed in this paper, it is possible to solve the following practical tasks of management of electric power systems: increasing the accuracy of modeling their operating conditions; making informed decisions when taking measures to reduce unbalance in power grids while accounting for the balancing adjustment effect of AL.

**Keywords:** power supply systems; unbalanced load flows; unbalanced load; parameter identification

**Citation:** Kryukov, A.; Suslov, K.; Ilyushin, P.; Akhmetshin, A. Parameter Identification of Asynchronous Load Nodes. *Energies* **2023**, *16*, 1893. https://doi.org/10.3390/en16041893

Academic Editors: Yuling He, David Gerada, Conggan Ma and Haisen Zhao

Received: 9 January 2023
Revised: 9 February 2023
Accepted: 11 February 2023
Published: 14 February 2023



## 1. Introduction

The electric power system (EPS) is a set of complex devices that generate, transmit, distribute, and consume electric power. Improving the reliability of operation and efficiency of power system use is impossible without solving a set of problems relating to load dispatching in ordinary and emergency states. Due to the introduction of transient monitoring tools, it became possible to determine the parameters of power system elements in real-time.

Solving the problems of load dispatching in EPSs is based on the use of mathematical models. They are used for steady-state load flow analysis, optimization, state estimation, transient analysis, etc. The basis of the mathematical model of an EPS is an equivalent circuit, which is formed from the circuits of individual components, namely: power transmission lines, power transformers, generating equipment, load nodes, etc. The parameters of the equivalent circuit of each component are usually determined by reference data or nameplate data and are considered immutable, although they depend on wear and tear of components, weather conditions, and other factors. The errors in determining these parameters on the basis of reference data are quite significant. For example, the error of the active resistance of a line can be in the range of +16-–20%, and the representation of corona discharge losses by a constant value of active conductivity can lead to an error of 1.5–3 times the actual value when determining the losses. The most significant errors can occur in the process of building the models of load nodes. This is due to the considerable uncertainty in the mix of consumers served and their operating conditions. The problem of adequate modeling of load nodes can be solved on the basis of identification methods.

Issues of parameter identification of induction machines and nodes with predominantly unbalanced loads have been addressed in a sizeable number of studies. For example, article [1] proposed a technique for identifying the parameters of a three-phase induction motor in the case when the initial values of the estimates change in a wide range. Study [2] presented the results of an experimental study of the efficacy of the technique of adaptive identification of electrical parameters of the induction machine under steady-state conditions on the basis of the power balance. Article [3] dealt with the issues of the identification of load nodes and their stability control. In [4–6], the results of the modeling and identification of an induction machine were presented. Article [7] discussed the identification of the parameters of an induction motor in its operational mode. Study [8] described an algorithm proposed by its authors for the identification of induction machine parameters by a recursive least-squares method. Article [9] studied mathematical methods of identification of lumped parameters of electrical machines. It discussed the basic principles and mathematical foundations of lumped parameter identification methods for various types of electrical machines, including induction machines. Article [10] proposed an approach to parameter identification of induction machines. The excitation input signal was determined by optimization methods. Instrumental variable estimation was introduced to improve the quality of identification by least-squares estimation. A method for magnetizing curve identification of induction machines was described in [11]. The study proposed an experimental method for determining the magnetization curve specifically designed for vector-controlled drives. The method employed an indirect vector controller and a PWM inverter, which were used during normal operation of the drive. The method was verified by extensive experimentation. A novel parameter identification method for the induction motor (IM) was proposed in [12]. The study pointed out that the effect of vector control depends largely on the accuracy of setting its parameters, which change with temperature variation. Based on the relationship between motor winding resistance and temperature, the authors presented a method for calculating dynamic resistance by on-line detection of the winding temperature. The experiment results attested to the high identification accuracy of the approach. Article [13] considers real-time parameter identification algorithms for effective control of electrical machines. Study [14] proposed a technique of flux estimation of induction machines with the linear parameter-varying system identification method. The identification algorithm was tested on data obtained from a nonlinear simulation model with continuous time. Paper [15] described a method for parameter identification of nine-phase induction machines with concentrated windings. Studies [16,17] considered a method of induction machine parameter identification suitable for self-commissioning. An algorithm for parameter identification of electrical machines using numerical simulations was proposed in articles [18,19]. The problem of reducing the electrical energy consumption of fans through parameter identification of the drive was solved in [20]. A method for identification of induction machine parameters, including core loss resistance, using a recursive least mean square algorithm was proposed in [21].

An analysis of the research contributions reviewed above allows us to conclude that the relevance of problems with parameter identification of IMs and asynchronous load nodes has been well established. However, most of the reviewed works dealt with the determination of parameters of individual IMs, whereas load nodes were considered only in [3]. The study relied on a single-line representation of the power system, which significantly hindered the modeling of unbalanced load flows. Adequate models of IMs and asynchronous load nodes in the phase frame of reference, which is the most natural form of representation of multiphase circuits, were proposed in [22,23]. However, effective use of such models is possible only if accurate data on the parameters of IM equivalent circuits for positive and negative sequences are available. The parameter identification techniques discussed below can be used to solve this problem.

## 2. Modeling of Unbalanced Loads in the Phase Frame of Reference

On the basis of phase coordinates, adequate models [22] of asymmetric modes of complex electric power systems at the fundamental frequency and frequencies of higher harmonics can be implemented. So, for example, an experimental verification of the EPS modeling technique in phase coordinates, performed on the basis of comparison with synchronized measurement data for a model containing 619 nodes and 2996 branches, showed that the differences between the calculated and measured values of the asymmetry coefficients do not exceed 0.6%, and for the phase values stresses 2.3%.

Compared to the static elements of an EPS, such as power lines and transformers, an asynchronous load, including a large number of motors, is a more complex object.

The asymmetry of the resistance matrix corresponding to the motor poses challenges to simulation based on the lattice circuit with RLC elements. The difficulties are due to the presence of two magnetic fields rotating clockwise and counterclockwise. When the supply voltages are unbalanced, the induction motor has sine wave processes running at three frequencies: 50 Hz, the frequency of the slip $s$, and about 100 Hz.

The behavior of induction motors under balanced three-phase voltage, when the motor can be represented by a single-line equivalent circuit, has been thoroughly researched. Induction motors can have different parameters of equivalent circuits during starting up and operation with low slip values. Furthermore, there are several variants of equivalent circuits. From the standpoint of load flow analysis in the phase frame of reference, when it is necessary to consider motor parameters at low slip values as well as at slip values close to 2 (electromagnetic brake mode), it is advisable to make the following assumptions.

First, it is convenient to use the equivalent circuit of an induction motor with the external magnetizing circuit placed on the primary terminals, according to Figure 1a. It is assumed that at start-up and the slip value of $2 - s$ (for negative sequence voltage), the equivalent circuit will have different parameters of the rotor circuit, Figure 1b. Figure 1 shows the magnetization branch components $R_\mu$, $X_\mu$, stator resistances $R_1$, $X_1$, and equivalent rotor resistances referred to the stator $\frac{R_2}{s}$, $X_2$, as well as the corresponding starting parameters $\frac{R_{2P}}{2-s}$, $X_{2P}$.

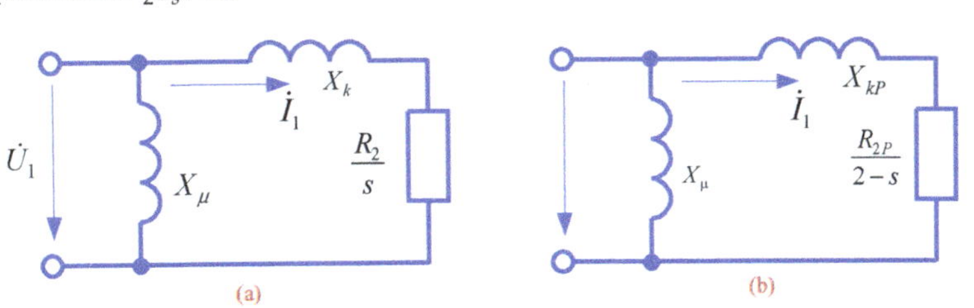


**Figure 1.** Positive (**a**) and negative (**b**) sequence equivalent circuits.

Second, it was assumed that in the start-up and electromagnetic brake modes (for the negative voltage sequence), the square of the reactive resistance is much greater than the square of the active resistance.

Third, with respect to the magnetization branch, a dual approach was adopted. When the no-load parameters are known ($\cos\varphi_x$ and active power $P_x$), its parameters are determined on the basis of the relations presented below, and when the parameters are unknown, the magnetization branch is ignored.

Fourth, the parameters of the circuit components in Figure 1 are determined from the rated values of efficiency η, power factor cos $\varphi_H$, current $I_{GH}$, and the current flowing through the part of the circuit that determines the load flow.

Fifth, the values of the positive and negative sequence voltages and the given mechanical power of the motor are used to determine the positive and negative sequence currents. In this case, the motor is modeled by current sources connected in a star (Figure 2). The values of the source currents are adjusted at each step of the iterative process. The motor neutral is considered to be insulated, and no zero-sequence currents occur in its circuits.

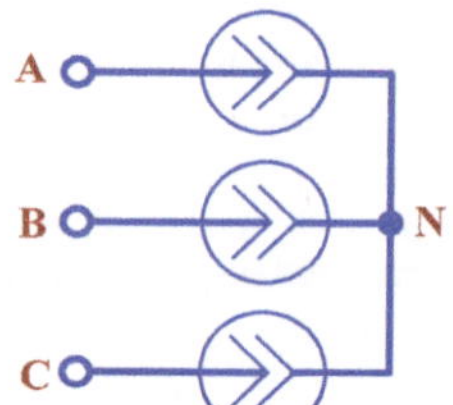


**Figure 2.** Equivalent circuit in the phase frame of reference. (A, B, C—Phase A, Phase B, Phase C, respectively, N—neutral point).

According to Figure 1a, the parameters of the circuit at rated power settings for the positive sequence are determined from the values of efficiency η, rated current $I_{GH}$, and power factor cos $\varphi_H$.

If the active power $P_x$ and cos $\varphi_x$ of the no-load operation of the motor are known, the parameters of the magnetizing branch and the current flowing through it can be determined from them:

$$Z_\mu = \frac{3\ U_1{}^2 \cos\varphi_x}{P_x},\ R_\mu = Z_\mu \cos\varphi_x,\ X_\mu = \sqrt{Z_\mu{}^2 - R_\mu{}^2} \quad (1)$$

The mechanical shaft power of the motor at rated settings is determined by the active power dissipated in the resistance $\frac{R_2(1-s_H)}{s_H}$, where $s_H$ is the rated slip. The efficiency factor is made up of the following loss components:

- mechanical losses in the rotor $\Delta P_M$;
- steel losses in the stator $\frac{U_1{}^2}{Z_\mu{}^2}R_\mu$;
- additional losses in the stator $\Delta P_d$;
- copper losses due to stator and rotor resistances $R_1$ and $R_2$.

The efficiency at rated power settings is defined as the ratio of shaft power to gross power:

$$\eta = \frac{P_H}{P_H + \Delta P_M + \Delta P_d + 3I_{CH}{}^2(R_1+R_2) + \frac{3U_1{}^2}{Z_\mu{}^2}R_\mu},$$
$$R_0 = R_1 + R_2, \quad (2)$$
$$R_0 = \frac{P_H}{3I_{GH}{}^2}\left(\frac{1}{\eta} - 1 - \frac{\Delta P_M + \Delta P_d}{P_H} - \frac{3U_1{}^2 R_\mu}{P_H Z_\mu{}^2}\right).$$

The magnetizing branch current when the vectors are counted from the voltage $\dot{U}_1$ is

$$\dot{I}_\mu = \frac{U_1}{R_\mu + j\ X_\mu} = I_\mu{}' + j\ I_\mu{}'' = \frac{U_1\ R_\mu}{R_\mu{}^2 + X_\mu{}^2} - j\frac{U_1\ X_\mu}{R_\mu{}^2 + X_\mu{}^2}. \quad (3)$$

At rated load, the motor current is

$$\dot{I}_H = I_H{}' + j\ I_H{}'' = \frac{P_H}{3\,U_1\,\eta} - j\frac{P_H}{3\,U_1\,\eta}\sqrt{\frac{1}{\cos^2\varphi_H} - 1}, \quad (4)$$

where $\frac{P_H}{\eta}$ is the active power consumed at rated power settings.

The rated current and power factor are determined by the expressions:

$$I_{GH} = \sqrt{(I_H' - I_\mu')^2 + (I_H'' - I_\mu'')^2}; \quad (5)$$

$$\cos\varphi_{GH} = \frac{I_H' - I_\mu'}{I_{GH}}. \quad (6)$$

However, if the no-load operation parameters of the motor are unknown, it is possible to assume, by way of approximation, that $I_{GH} = \frac{P_H}{3\ U_1\ \eta\ \cos\varphi_H}$ neglecting the magnetizing current and considering that $\cos\ \varphi_{GH} = \cos\ \varphi_H$.

According to the equivalent circuit of Figure 1a, we can write:

$$\dot{I}_G = \frac{\dot{U}_1}{\left(R_1 + \frac{R_2}{s}\right) + j\ X_k},\ X_k = X_1 + X_2,$$

$$I_{GH} = \frac{U_1}{\sqrt{\left(R_1 + \frac{R_2}{s_H}\right)^2 + {X_k}^2}}, \quad (7)$$

$$\cos\varphi_{GH} = \frac{R_1 + R_2/s_H}{\sqrt{\left(R_1 + \frac{R_2}{s_H}\right)^2 + {X_k}^2}} = \frac{R}{\sqrt{R^2 + {X_k}^2}}, \quad (8)$$

$$R = R_1 + R_2/s_H. \quad (9)$$

The system of Equations (2), (7) and (8) is solved by simple substitution. From Equation (8) we determine

$${X_k}^2 = R^2\left(\frac{1}{\cos^2\varphi_{GH}} - 1\right),$$

and it follows from Equation (7) that $Z^2 = \left(\frac{U_1}{I_{GH}}\right)^2 = R^2 + {X_k}^2$, so that $R = Z\cos\varphi_{GH}$.

From relations (2) and (9) we can determine components $R_1$ and $R_2$:

$$R_2 = \frac{s_H(R - R_0)}{1 - s_H};\ R_1 = R_0 - R_2.$$

When determining $R_0$ from relation (3), it can be assumed that the added losses are 0.5% of the input power and that the mechanical losses are 1.0% of the rated power.

Denoting $X_1 + X_{2p} = X_{kp}$ and assuming that $(R_1 + R_{2p})^2 << {X_{kp}}^2$, we obtain

$$X_{kp} = \frac{U_1}{K_P I_{GH}},$$

where $K_P$ is the locked-rotor current ratio. From the equation of the electromagnetic locked-rotor torque when the magnetization branch is ignored, we get the relation

$$M_P = \frac{3\ U_1^2 R_{2p} p}{X_{kp}^2\ \omega},\ \text{or}\ R_{2p} = \frac{\omega}{p}\frac{MX_p^2}{3\ U_1^2},$$

where $p$ is the number of motor pole pairs.

The locked-rotor torque was determined from the locked-rotor torque ratio $k_{MP} = \frac{M_P}{M_H}$ and the rated torque $M_H = \frac{2P_H p}{(1+\eta)2\pi f}$, from which it follows that

$$R_{2p} = \frac{2\ k_{MP} P_H X_{kp}^2}{3\ U_1^2(1+\eta)}.$$

The multiplier $\frac{2}{1+\eta}$ allows one to convert the shaft power to the electromagnetic power of the motor with a small error.

Effective use of the described model is possible only if accurate data on the parameters $X_k$, $R_2$, $X_{kP}$, $R_{2P}$, $X_\mu$ of IM equivalent circuit for positive and negative sequences are available. This problem can be solved by applying the parameter identification technique [22], described below. Identification results can be used in the unbalanced load flow analysis of complex power supply systems. In this case, the asynchronous load node can be represented by an equivalent IM according to the technique detailed in [24].

## 3. Induction Motor Identification Technique

The problem of adequate modeling of load nodes can be solved on the basis of identification methods. In the context of power system control, it is advisable for identification purposes to use information about parameters of operating conditions obtained from information and measurement systems built using PMUs (Figure 3).

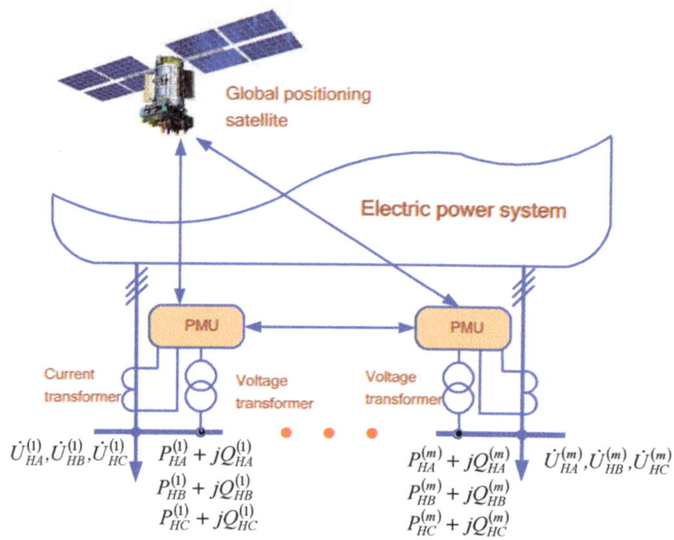


**Figure 3.** Synchronized measurements.

The problem of topology and parameter identification of load nodes can be formalized as follows [22,25]. To this end, we can introduce a class of models $\Im = (\Im_1 \quad \Im_2 \quad \ldots \quad \Im_m)$ describing the processes occurring at load nodes. Each of the models is represented as

$$\Im_i = \Im_i(\mathbf{X}, \mathbf{Z}, \mathbf{P}, \mathbf{\Sigma}, \mathbf{L})$$

where $x_k \in \mathbf{X}$, $k = 1 \ldots n_X$—state variables; $z_k \in \mathbf{Z}, k = 1 \ldots n_Z$—input variables; $p_k \in \mathbf{P}$, $k = 1 \ldots n_P$—model parameters subject to identification; $\sigma_k \in \mathbf{\Sigma}$, $k = 1 \ldots n_\Sigma$—internal relations defining the model structure; $l_k \in \mathbf{L}$, $k = 1 \ldots n_X$—functional relationships acting as mathematical relations operators allowing to find the parameters describing the object state $x_k \in \mathbf{X}$, $k = 1 \ldots n_X$ by inputs $z_k \in \mathbf{Z}$, $k = 1 \ldots n_Z$, with the required degree of certainty.

Then we can write

$$\mathbf{X} = \mathbf{L}(\mathbf{Y}, \mathbf{P}, \mathbf{\Sigma}). \tag{10}$$

This relationship is called the rule governing the functioning of the model. To form the relationship (10), it is necessary to choose from a class of models $\Im = (\Im_1 \quad \Im_2 \quad \ldots \quad \Im_m)$ a model $\Im_k \in \Im$ with the rule

$$\mathbf{L}^* : \left(\Im^{(O)}\right) \to \Im_k.$$

The parentheses in the last relation denote that $\mathbf{L}^*$ is a partially defined relation; that is, not all characteristics of the original $\Im^{(O)}$ are captured by the model, but only those that are deemed significant in solving the stated problem of modeling power system conditions.

The functional transformation $\mathbf{L}^*$ can be chosen subject to the following condition

$$\|\mathbf{X} - \mathbf{L}^*(\mathbf{Y}, \mathbf{P}, \boldsymbol{\Sigma})\| \to \min_{\substack{p_k \in \mathbf{P} \\ \sigma_k \in \boldsymbol{\Sigma}}}$$

in some parts of the chosen class of functions.

In addition, the choice of $\mathbf{L}^*$ can be made subject to the condition that there be a minimum of some criterion of discrepancy between the model and the original:

$$\aleph_{\mathbf{L}^*} \to \min_{\mathbf{L}^* \in \mathbf{L}}.$$

As a rule, the choice of the functional transformation $\mathbf{L}^*$, carried out at the stage of structural identification, is subjective and does not lend itself easily to rigorous formalization. Figure 4 is a diagram showing possible types of load node models.

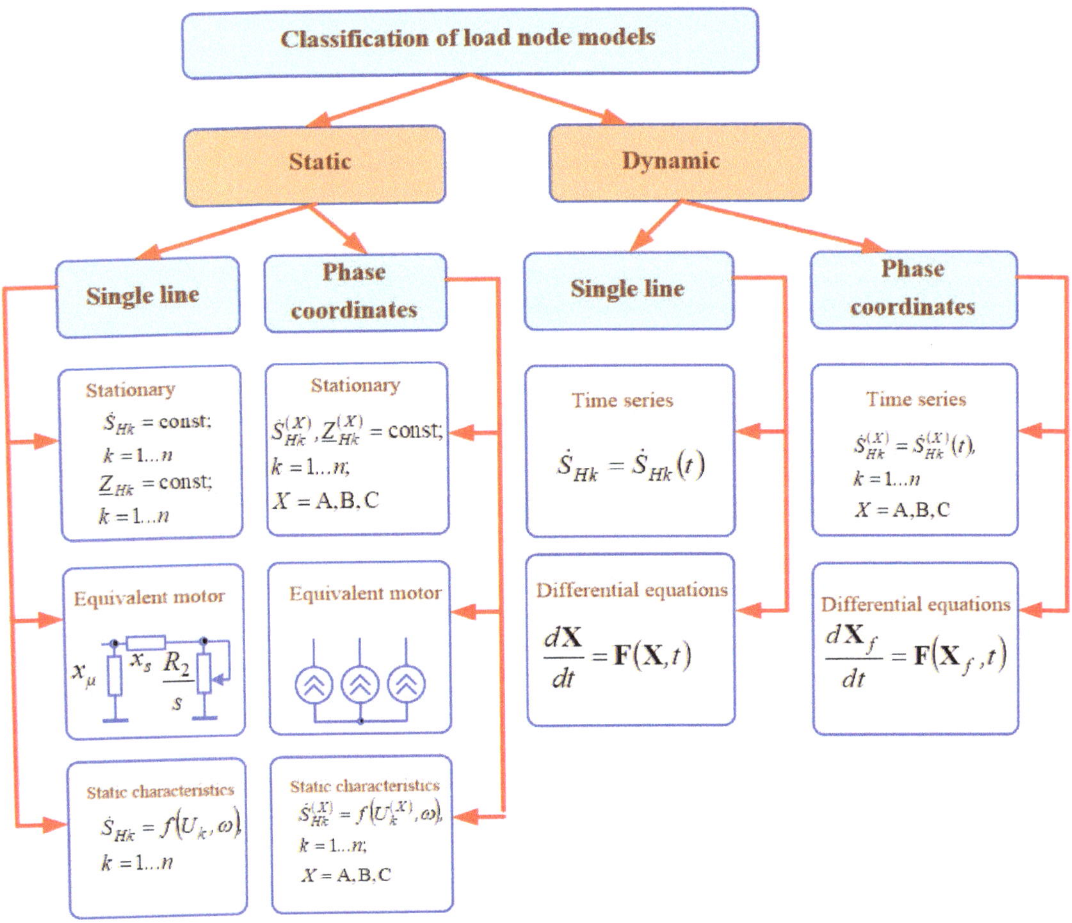


**Figure 4.** On the problem of structural identification of load nodes.

Parameters $X_k$, $R_2$, $X_{kP}$, $R_{2P}$ can be determined on the basis of measurements of complexes of currents consumed by the motor and voltages at its terminals, as well as its rotation speed. To solve this problem, it is necessary to know the resistance of the magnetizing branch $X_\mu$. This parameter can be found on the basis of additional measurements, e.g., under no-load conditions, or determined by the indirect technique described below.

If the value of $X_\mu$ is known, the parameters of the equivalent circuit of the positive sequence can be found based on measurements of the moduli and phases of the IM currents and voltages, as well as the speed of rotation (slip $s$) based on the following relation:

$$\underline{Z}_{D1} = \frac{jX_\mu \underline{Z}_k}{jX_\mu + \underline{Z}_k}, \tag{11}$$

where $\underline{Z}_k = \frac{R_2}{s} + jX_k$; $\underline{Z}_{D1} = \frac{\dot{U}_1}{\dot{I}_1}$; $\dot{U}_1$, $\dot{I}_1$ are complexes of positive sequence voltage and current, determined on the basis of measurements of phase currents $\dot{I}_A$, $\dot{I}_B$, $\dot{I}_C$ and voltages $\dot{U}_A$, $\dot{U}_B$, $\dot{U}_C$ according to the known relations of the method of symmetrical components. Measurements can be made under both balanced and unbalanced load flows.

The main practical focus of the research presented in the article is to create methods for adequately taking into account load nodes when modeling stationary modes of electric power systems. Therefore, for further consideration, the model of the load node in phase coordinates in the form of an equivalent asynchronous electric motor was adopted. To solve the identification problem, it is necessary to measure the parameters of load nodes, which can be determined on the basis of PMU-WAMS devices (Figures 3 and 5).

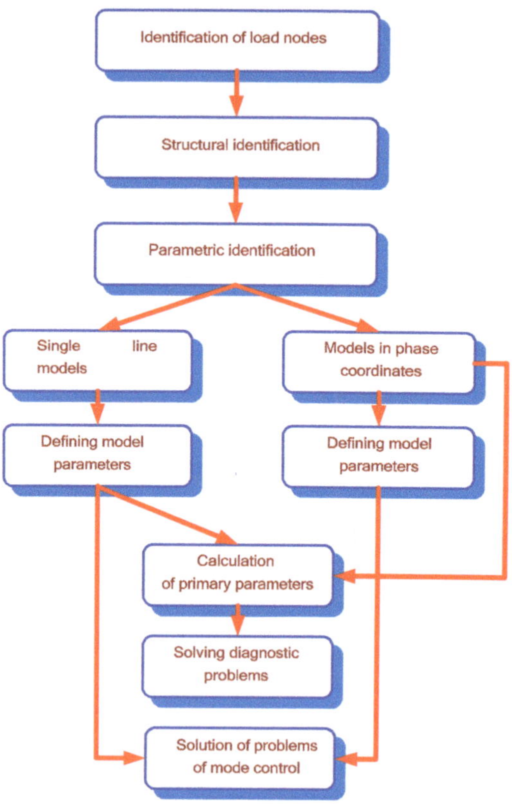


**Figure 5.** A set of problems for identifying load nodes in an electric power system.

Based on (11), we can write the following expression:

$$\underline{Z}_k = \frac{jX_\mu \underline{Z}_{D1}}{jX_\mu - \underline{Z}_{D1}}.$$

If the slip $s$ is known, the parameter $R_2$ can be determined from the above equation.

The parameters of the equivalent circuit of the negative sequence can be found as per equations similar to those given above:

$$\underline{Z}_{kp} = \frac{R_{2p}}{2-s} + jX_{kp} = \frac{jX_\mu \underline{Z}_{D2}}{jX_\mu - \underline{Z}_{D2}};\ \underline{Z}_{D2} = \frac{\dot{U}_2}{\dot{I}_2},$$

where $\dot{U}_2$, $\dot{I}_2$— complexes of negative sequence currents and voltages are determined by measurements of phase currents and voltages. To obtain acceptable accuracy, the parameters of start-up conditions should be found in the power flow with a voltage unbalance ($k_{2U}$ of about 10%).

The resistance $X_\mu$ can be determined by the data provided in reference books. All that is required is information about the rated voltage and the rated motor power. An acceptable accuracy of calculation of $X_\mu$ can be obtained on the basis of a nonlinear approximation of the following kind:

$$X_{\mu *} = X_{\mu\ 0}[1 + \Delta X_\mu (1 - e^{-\ \alpha\ P})]. \quad (12)$$

Parameters $X_{\mu 0}$, $\Delta X_\mu$, and $\alpha$, for IM powers exceeding 5 kW are given in Table 1.

**Table 1.** Parameters of the approximation of the relationship $X_{\mu *} = X_{\mu *}(P)$.

| Parameter | 750 rpm | 1000 rpm | 1500 rpm | 3000 rpm |
|---|---|---|---|---|
| $X_{\mu 0}$, per unit | 1.4 | 1.7 | 2.0 | 2.3 |
| $\Delta X_\mu$, per unit | 1.0 | 1.2 | 1.4 | 1.4 |
| $\alpha$, $kW^{-1}$ | 0.04 | 0.05 | 0.045 | 0.04 |

The obtained value $X_{\mu *}$ should be multiplied by the basic resistance, determined by the rated parameters of the IM.

## 4. Identification Results

Input information in the form of moduli and angles of current and voltage, as well as those of slip, was formed on the basis of computer-aided simulation using the software package Fazonord [22]. For this purpose, an equivalent circuit of an IM with a rated power of 90 kW was created. In the obtained currents and voltages of the calculated load flow, the errors corresponding to the accuracy classes of measuring instruments (0.1, 0.2, 0.5, and 1) were introduced. The resistance $X_\mu$ was calculated on the basis of expression (12). The results of the identification are shown in Figure 6. The parameter $R_2$ was determined with an error close to zero.

The results obtained show that in order to achieve acceptable identification accuracy, it is necessary to use measuring instruments with an accuracy class that provides a maximum error of no more than 0.2%.

The proposed technique can be used to solve the problem of parameter identification for a group of IMs connected to a node in an electrical network. To confirm this possibility, we performed identification of the AL node, the circuit of which is shown in Figure 7. The IM parameters are summarized in Table 2. The equivalent circuit is shown in Figure 8.

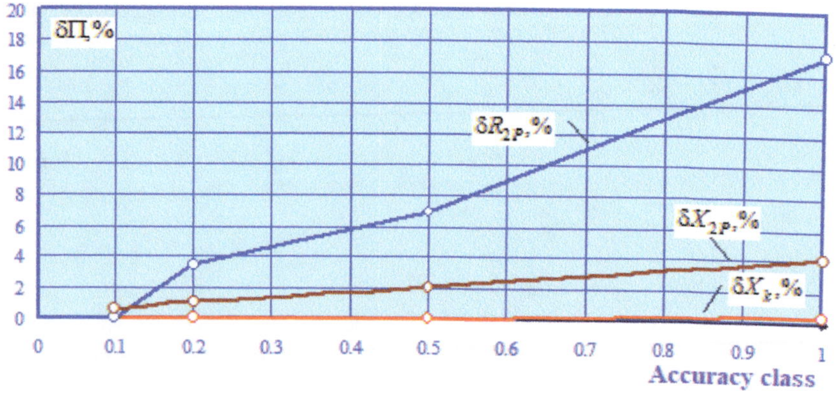


**Figure 6.** Identification errors.

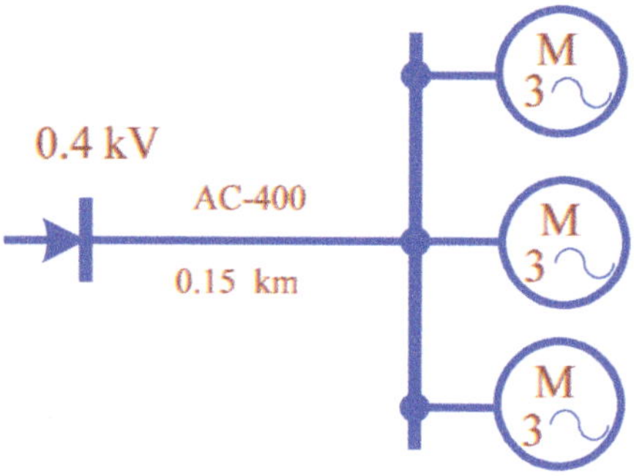


**Figure 7.** Original circuit.

**Table 2.** Parameters of the nodal IM.

| IM No. | $P_H$, kW | Efficiency, % | cosφ, p.u. | $R_2$, p.u. | $X_k$, p.u. | $R_{2P}$, p.u. | $X_{kP}$, p.u. | $X_{\mu *}$, p.u. |
|---|---|---|---|---|---|---|---|---|
| 1 | 45 | 92 | 0.9 | 0.017 | 0.222 | 0.034 | 0.16 | 4.6 |
| 2 | 110 | 92.5 | 0.9 | 0.019 | 0.282 | 0.048 | 0.21 | 4.9 |
| 3 | 160 | 93.5 | 0.91 | 0.017 | 0.257 | 0.045 | 0.19 | 4.6 |

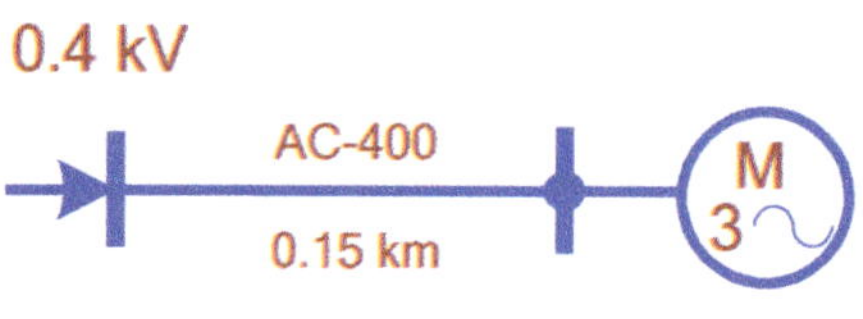


**Figure 8.** Equivalent circuit.

In the process of identification, the slip value was set on the basis of the data for the equivalent IM given in [25]. The load flow analysis errors that arise when using the equivalent model of an asynchronous load node, obtained on the basis of parameter identification, are shown in Figures 9–12.

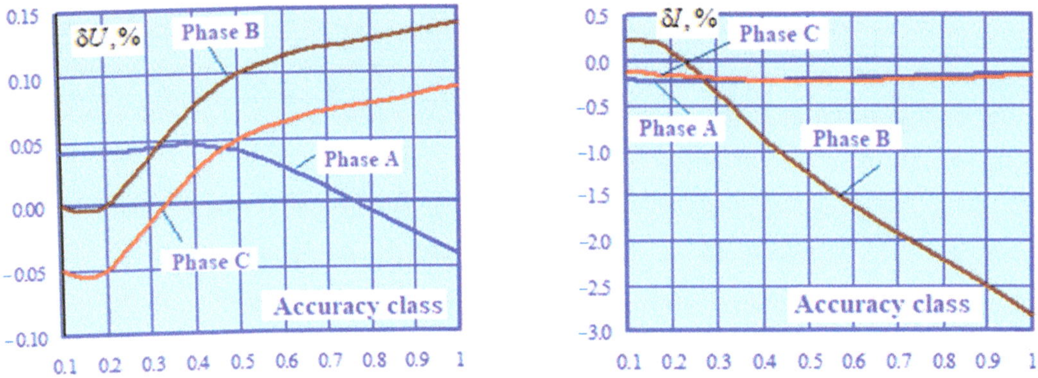


(a) (b)

**Figure 9.** Errors in determining the moduli of voltages (**a**) and currents (**b**).

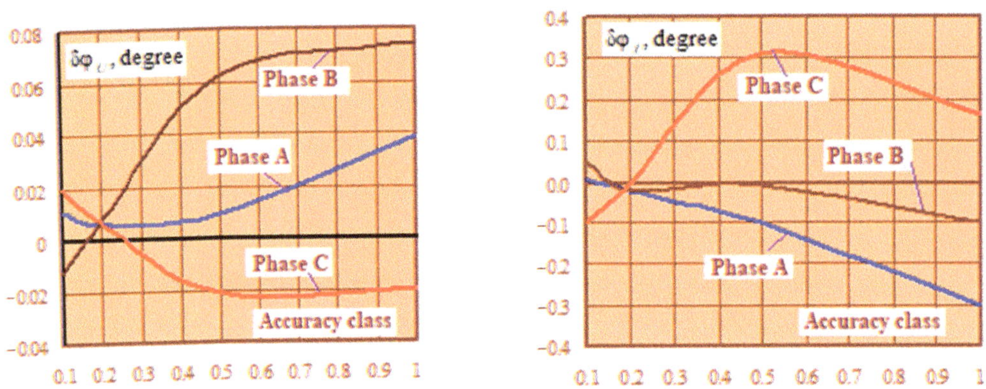


(a) (b)

**Figure 10.** Errors in determining phases of voltages (**a**) and currents (**b**).

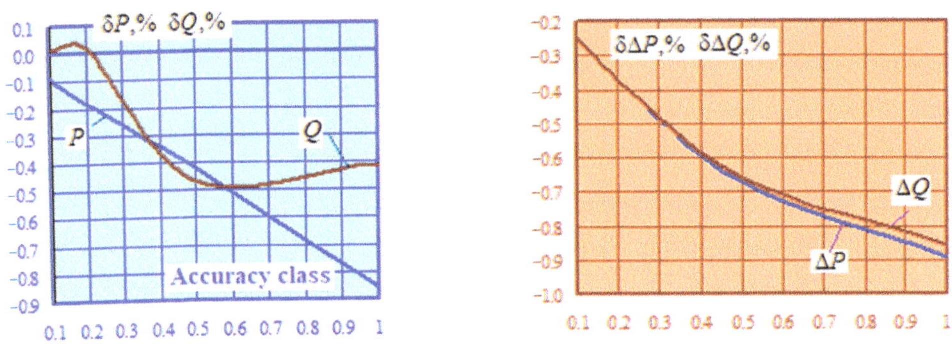


(a) (b)

**Figure 11.** Errors in determining the active and reactive power consumed by the AL node (**a**), and power losses in the transmission line (**b**).

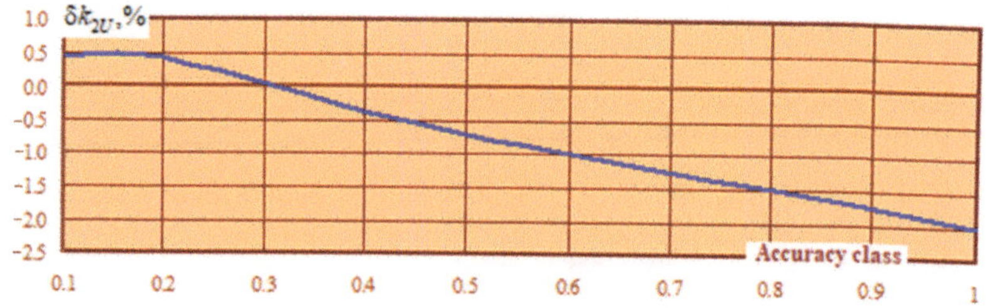


**Figure 12.** Errors in determining the unbalance ratio in the negative sequence.

Table 3 and Figure 13 show the static characteristics of $P = P(U_{PH})$ and $Q = Q(U_{PH})$, where $P$, $Q$ stands for active and reactive powers consumed by the AL node; $U_{PH}$ is the phase voltage. These dependencies were plotted for the original and equivalent AL node models. Our analysis of the results thus obtained allows us to conclude that the AL node model, formed on the basis of parameter identification, provides for valid simulation of the asynchronous load node within a wide range of changes in network operating conditions.

**Table 3.** Static load characteristics.

| $U_{PH}$, kV | Original Model | | Equivalent Model | | Discrepancy | |
|---|---|---|---|---|---|---|
| | $P$ kW | $Q$ kVar | $P$ kW | $Q$ kVar | $\delta P$ % | $\delta Q$ % |
| 0.18 | 325 | 179 | 324 | 180 | 0.19 | −0.72 |
| 0.19 | 325 | 166 | 324 | 167 | 0.19 | −0.31 |
| 0.21 | 325 | 160 | 324 | 160 | 0.19 | −0.01 |
| 0.22 | 325 | 158 | 324 | 157 | 0.19 | 0.23 |
| 0.23 | 325 | 157 | 324 | 156 | 0.19 | 0.43 |
| 0.24 | 325 | 159 | 324 | 158 | 0.19 | 0.61 |
| 0.25 | 325 | 162 | 324 | 161 | 0.19 | 0.77 |
| 0.26 | 325 | 166 | 324 | 165 | 0.19 | 0.90 |

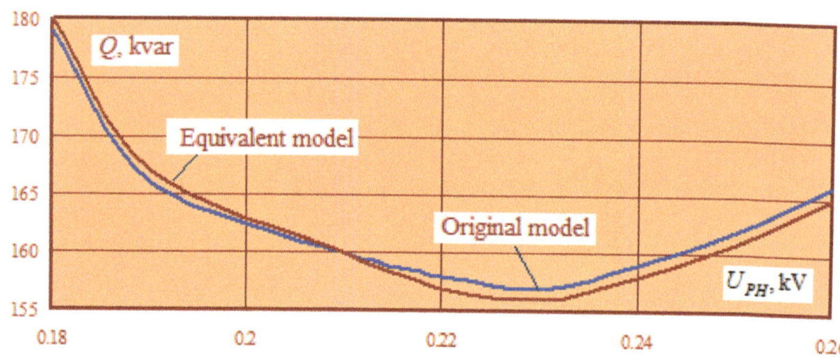


**Figure 13.** Static reactive power characteristic.

The results obtained also allow us to conclude that, with the aid of parameter identification, we can arrive at an equivalent model of an asynchronous load node that provides high accuracy for both balanced and unbalanced load flow analysis. It should be emphasized that the model was validated in a wide range of changing power flow parameters.

The proposed technique is also valid for AL circuits of a more general type, the models of which are shown in Figure 14. In these circuits, induction motors are connected to the buses of the node through cable lines. In addition, the node was powered by a busbar trunking system.

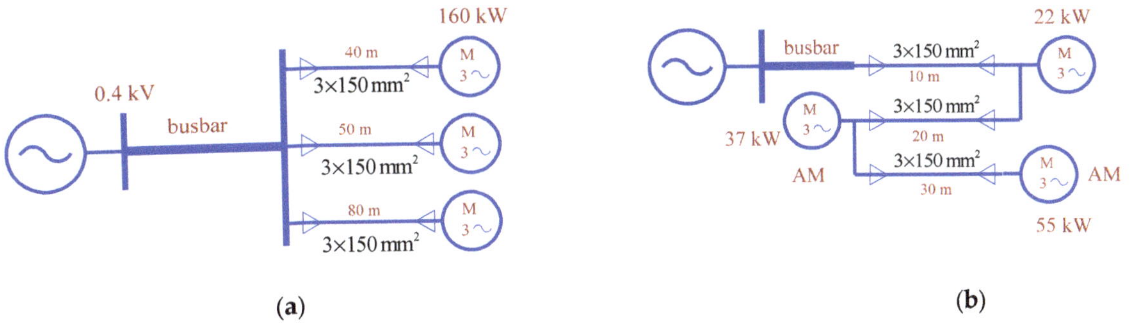


**Figure 14.** Models of AL node circuits (**a**)—first type of connection, (**b**)—second type of connection).

The simulation results are presented in Table 4, which shows that equivalent models with a structure similar to the one presented in Figure 8 provided acceptable accuracy for unbalanced load flow analysis.

**Table 4.** Identification errors.

| Parameters | Error, % | |
|---|---|---|
| | **Circuit of Figure 1a** | **Circuit of Figure 1b** |
| $\delta\, U_A, \%$ | –0.04 | −1.51 |
| $\delta\, U_B, \%$ | –0.05 | −0.05 |
| $\delta\, U_C, \%$ | 0.00 | 0.00 |
| $\delta\, I_A, \%$ | 2.56 | 0.33 |
| $\delta\, I_B, \%$ | 8.71 | 6.96 |
| $\delta\, I_C, \%$ | 1.88 | −2.08 |
| $\delta P, \%$ | 3.26 | 0.25 |
| $\delta Q, \%$ | 2.82 | −0.32 |
| $\delta \Delta P,\ L\%$ | 2.40 | 3.21 |
| $\delta \Delta Q,\ L\%$ | 3.94 | 4.09 |
| $\delta\, k_{2U},\ \%$ | –0.18 | −0.17 |

Figures 15 and 16 show the results of parameter identification for the case when, in addition to the IM, a static load was connected to the nodal point of the network, with such a load specified by the amount of power $\dot{S}_C = P + jQ$ drawn by it. Percentage of stationary load

$$\lambda = \frac{S_C}{S_{AH}} \cdot 100$$

ranged from 0–80%. Here $S_{AH}$ is the total power of the IM group.

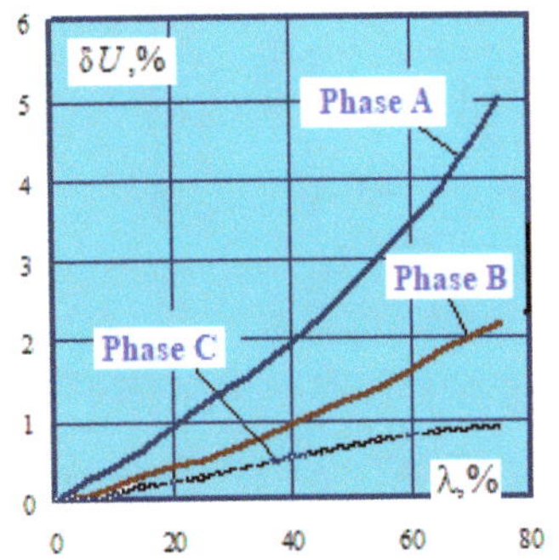


**Figure 15.** Relationship $\delta\, U = \delta\, U(\lambda)$.

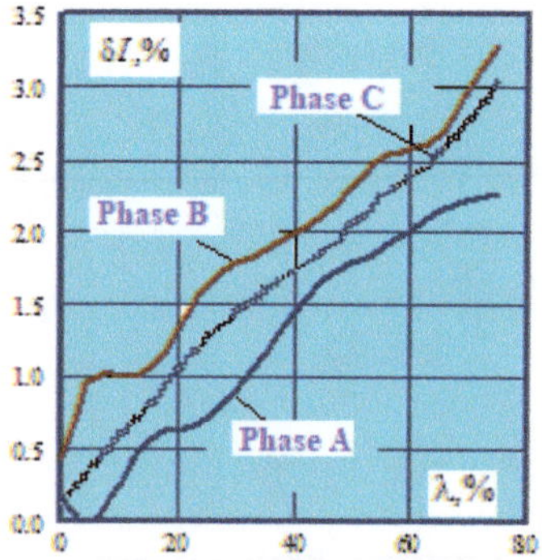


**Figure 16.** Relationship $\delta\, I = \delta\, I(\lambda)$.

The obtained relationships $\delta\, U = \delta\, U(\lambda)$ and $\delta\, I = \delta\, I(\lambda)$ attested to the fact that with an increase of the parameter $\lambda$, the errors of load flow analysis using the equivalent IM model increased but remained quite acceptable for practical applications over a sufficiently wide range of variation of $\lambda$.

The following conclusions can be drawn on the basis of the results obtained:

1. The technique of parameter identification of an asynchronous load node allows one to obtain adequate models of IMs that provide high accuracy for balanced load flow analysis. In the numeric example presented in the paper, the calculation error of phase voltage moduli for different motor connection schemes did not exceed 1.5%;

2. If a static load is present at the node, the error of the equivalent model increases; in the numeric example above, when the value of the parameter $\lambda =$ was equal to 75%, the error of the voltage moduli increased to 5%, and the error of the currents increased to 3.3%.

**5. Parameter Identification of Asynchronous Load Nodes with Variable Frequency Drives**

At modern production facilities, variable-frequency induction-motor drives equipped with static frequency converters (SFC) are widely used. Therefore, the problem of identifying AL nodes that contain, along with conventional IMs, frequency-controlled asynchronous electric drives that can create harmonic distortions in networks becomes relevant [26]. The technique described above can be used to solve this problem.

The verification of its efficacy and accuracy in the presence of electric drives at the load node that are equipped with an SFC was carried out for the circuit shown in Figure 17 as follows. Based on the SimPowerSystems MATLAB package, a load node model was generated.

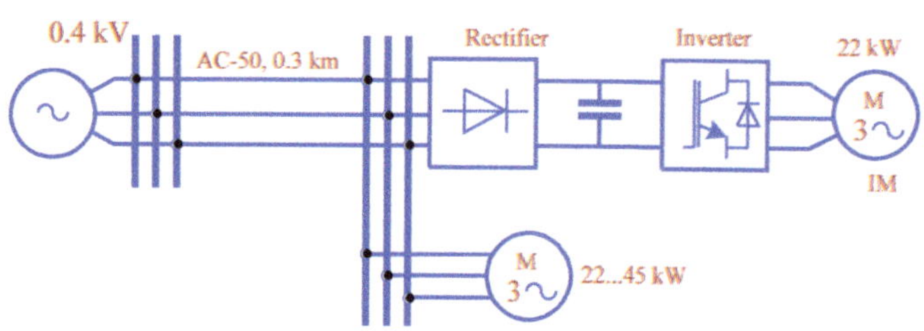


**Figure 17.** Network circuit.

The power of the IM controlled by the SFC was assumed to be 22 kW. The power of the fixed-speed IM ranged from 22 kW to 45 kW. The power ratio was set by the coefficient

$$\alpha = \frac{P_{spch}}{P_{aed}},$$

where $P_{spch}$—power of the motor equipped with an SFC; $P_{aed}$—power of the fixed-speed IM. The power supply system was fed from a source with unbalanced voltage ($k_{2U} = 3\%$), which corresponded to real-life conditions that hold true for many facilities connected to district windings of traction substations on mainline AC railroads. The results of the simulation are presented in Table 5.

**Table 5.** Voltages and currents at load node buses.

| α | Voltages | | | Currents | | |
|---|---|---|---|---|---|---|
| | Phase A | Phase B | Phase C | Phase A | Phase B | Phase C |
| | Modulus, B<br>Angle, Deg. | Modulus, B<br>Angle, Deg. | Modulus, B<br>Angle, Deg. | Modulus, A<br>Angle, Deg. | Modulus, A<br>Angle, Deg. | Modulus, A<br>Angle, Deg. |
| 1 | 215.4<br>27.8 | 209.2<br>–91.3 | 215.5<br>149.6 | 79.97<br>2.1 | 69.59<br>–102.8 | 90.59<br>134.7 |
| 0.6 | 209.7<br>27.5 | 203.8<br>–91.5 | 209.1<br>149.4 | 105.7<br>7.5 | 96.52<br>–98.9 | 120.7<br>137.6 |
| 0.5 | 204.7<br>27.4 | 198.7<br>–91.7 | 204.2<br>149.3 | 125.3<br>1.7 | 113.9<br>–107.9 | 137.8<br>130.7 |

In accordance with the identification technique described above, we determined the parameters of AL equivalent circuits (Table 6). Next, we created a model of the power supply system in the software package Fazonord, in which the load node was represented by an equivalent induction motor (see Figure 18). With the aid of this model, a load flow analysis was performed using the parameters of the AL node obtained as a result of identification. Comparative results of a simulation run in MATLAB and Fazonord are shown in Table 7. Errors in determining the active and reactive powers drawn from the network, as well as the unbalance ratio $k_{2U}$, are shown in Figures 19 and 20.

**Table 6.** Equivalent circuit parameters.

| α | $R_2$, Ohm | $X_k$, Ohm | $R_{2P}$, Ohm | $X_{kP}$, Ohm | $X_\mu$, Ohm |
|---|---|---|---|---|---|
| 1 | 0.056 | 0.224 | 0.430 | 0.250 | 12.005 |
| 0.6 | 0.04 | 0.023 | 0.315 | 0.216 | 8.991 |
| 0.5 | 0.034 | 0.237 | 0.313 | 0.236 | 7.566 |

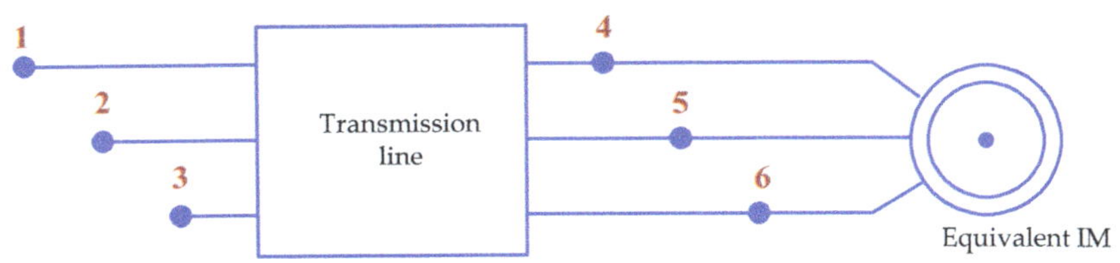


**Figure 18.** Load node model formed based on identification results (1–6—numbers of nodes).

**Table 7.** Parameters characterizing the load flow of the load node.

| α | P, kW | | Q, kVar | | $k_{2U}$,% | |
|---|---|---|---|---|---|---|
| | MATLAB | Fazonord | MATLAB | Fazonord | MATLAB | Fazonord |
| 1 | 48.76 | 48.85 | 16.52 | 15.49 | 1.90 | 1.93 |
| 0.6 | 65.29 | 65.25 | 15.88 | 15.47 | 1.84 | 1.82 |
| 0.5 | 71.58 | 71.54 | 27.11 | 26.57 | 1.90 | 1.89 |

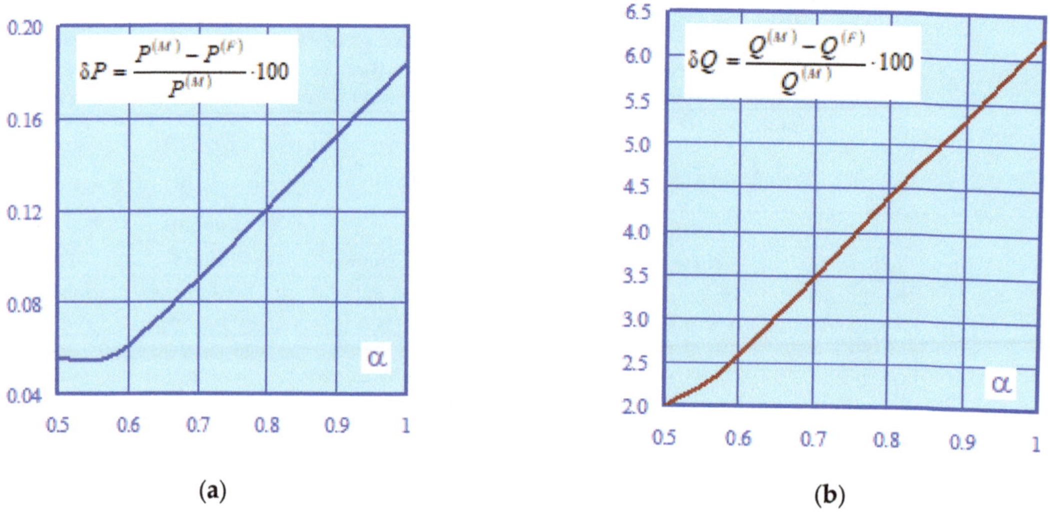


**Figure 19.** Errors of active (**a**) and reactive (**b**) powers as functions of the parameter.

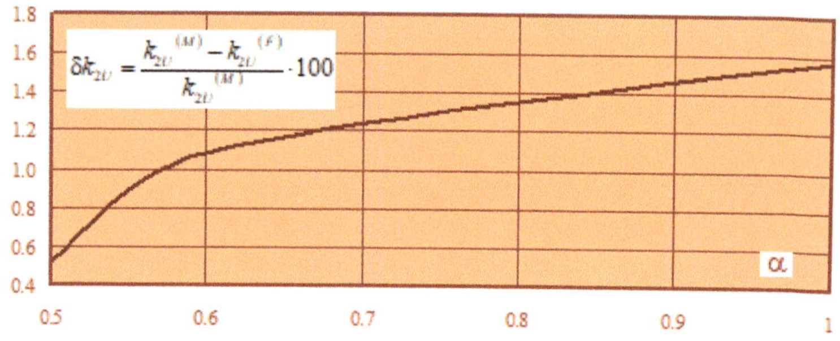


**Figure 20.** Unbalance ratio error as a function of the parameter $k_{2U}$, % α.

In Figures 19 and 20, the index "*M*" refers to the results obtained using the SimPowerSystems package, and the index "*F*" refers to the data calculated using the Fazonord software package.

The results obtained allow us to draw the following conclusions:

1. Our technique of parameter identification in the phase frame of reference is based on the substitution of a load node with an equivalent induction motor. The technique allows one to obtain high accuracy in unbalanced load flow analysis in the presence of conventional induction machines and electric motors with variable-frequency drives at the node. The error in determining the unbalance ratio of the negative sequence did not exceed two percent.

2. As the share of static frequency converters at the load node increased, the errors in determining the unbalance ratios $k_{2U}$ also increased but remained within the limits deemed acceptable for solving practical problems;

3. The error in determining the reactive power, which reaches 6.5%, can be explained by differences in approaches to its determination adopted in the MATLAB and Fazonord software systems. In calculations aided by the Fazonord package, only one of its segments was used, which provided the fundamental harmonic for the unbalanced load flow analysis. When using the MATLAB system, the simulation was carried out so as to take into account the non-linear current-voltage characteristics of SFC components, and the reactive power was determined by factoring in the higher harmonics.

The reactive power error can be reduced by additional non-sine load flow modeling using the technique reported in [22]. After determining the higher harmonic voltages, it is possible to recalculate the reactive power, e.g., using the technique of equivalent sine waves.

## 6. Conclusions

Based on the evidence provided in this study, we can claim we have solved a currently relevant scientific and engineering problem of enhancing the accuracy of modeling unbalanced load flows in electric power systems. Our solution is based on an adequate model of induction motors and a technique of parameter identification for asynchronous loads. The following results were obtained:

1. We have developed a technique for modeling complex load nodes. The technique stands out from other known solutions for its use of the phase frame of reference. Its application scope covers the problems of load dispatching in smart grids;
2. The study has contributed a technique for parameter identification of load nodes. The technique is applicable to the problems of power system load dispatching. A key defining feature of the technique is the structure of the model, which is made up of three sources of current with parameters that are refined in the process of iterative load flow analysis of the power system;
3. We have proposed a technique for the identification of asynchronous load nodes with such asynchronous loads, including electrical drives equipped with static frequency converters;
4. With the aid of the asynchronous load identification models proposed in this paper, it is possible to solve the following practical tasks of electric power system management: increasing the accuracy of load flow modeling; making informed decisions when taking measures to reduce unbalance in power grids; and accounting for the balancing adjustment effect of asynchronous loads.

**Author Contributions:** Conceptualization, A.K., P.I. and A.A.; methodology, A.K. and K.S.; software A.K.; validation, A.K., A.A., P.I. and K.S.; formal analysis, A.K. and K.S.; investigation, A.K., A.A., P.I. and K.S.; resources, K.S.; data curation, A.K.; writing—original draft preparation, A.K. and K.S.; writing—review and editing, A.K., A.A. and K.S.; visualization, A.A. and P.I.; supervision, A.K. and K.S.; project administration, A.A. and K.S.; funding acquisition, K.S. All authors have read and agreed to the published version of the manuscript.

**Funding:** The research was carried out within the framework of the state task "Conducting applied scientific research" on the topic "Development of methods, algorithms and software for modeling the modes of traction power supply systems for DC railways and electromagnetic fields at traction substations for AC railways".

**Data Availability Statement:** Data sharing not applicable. No new data were created or analyzed in this study. Data sharing is not applicable to this article.

**Conflicts of Interest:** The authors declare no conflict of interest. The funders had no role in the design of the study; in the collection, analyses, or interpretation of data; in the writing of the manuscript, or in the decision to publish the results.

## References

1. Afanasyev, A.Y.; Makarov, V.G.; Hannanova, V.N. Identification of three-phase induction motor parameters in the case when initial values of estimates vary in a wide range. *Power Eng. Res. Equip. Technol.* **2015**, *11–12*, 87–96. (In Russian)
2. Rakov, I.V. Experimental study of efficacy of the technique of adaptive identification of electrical parameters of the induction machine with an open-circuit rotor winding under steady-state conditions on the basis of the power balance. *Elektrotekhnicheskie I Inf. Kompleks. I Sist.* **2022**, *18*, 63–76. (In Russian)
3. Nagaitsev, A.L.; Semenov, A.V.; Fedyukov, R.V.; Fishov, A.G.; Chershova, V.O. Online parameter identification of the equivalent circuit and control of load stability. *Nauchnye Probl. Transp. Sib. I DAL'NEGO Vost.* **2015**, *3*, 198–203. (In Russian)
4. Khemliche, M.; Latreche, S.; Khellaf, A. Modelling and identification of the asynchronous machine. In Proceedings of the First International Symposium on Control, Communications and Signal Processing 2004, Hammamet, Tunisia, 21–24 March 2004.
5. Janisch, G.; Kugi, A.; Kemmetmüller, W. Model calibration strategy for energy-efficient operation of induction machines. *IFAC-PapersOnLine* **2022**, *55*, 307–312. [CrossRef]

6. Bulatov, Y.; Kryukov, A.; Suslov, K. Using Group Predictive Voltage and Frequency Regulators of Distributed Generation Plants in Cyber-Physical Power Supply Systems. *Energies* **2022**, *15*, 1253. [CrossRef]
7. Frolov, M.; Dulov, I.; Yunusova, I. Identification of Asynchronous Motor Parameters in Operational Mode. In Proceedings of the 2019 International Ural Conference on Electrical Power Engineering (UralCon), Chelyabinsk, Russia, 1–3 October 2019.
8. Sun, Y.; Kang, Z.; Liu, J. Research on Parameter Identification Method of Asynchronous Motor Considering Load Characteristics. In Proceedings of the 2022 IEEE 5th International Electrical and Energy Conference (CIEEC), Nanjing, China, 27–29 May 2022.
9. Henrotte, F.; Heidt, J.; Hameyer, K. The mathematics of lumped parameter identification in electrical machines. In Proceedings of the 2008 IET 7th International Conference on Computation in Electromagnetics, Brighton, UK, 7–10 April 2008.
10. Li, Y.M.; Bergmann, C.; Feuvrie, B. An approach of parameter identification for asynchronous machine. In Proceedings of the 1996 IEEE IECON. 22nd International Conference on Industrial Electronics, Control, and Instrumentation, Taipei, Taiwan, 9 August 1996; Volume 3.
11. Touhami, O.; Fadel, M.; Levi, E.; Sokola, M.; Vukosavic, S.N. A method for magnetizing curve identification in rotor flux oriented induction machines. *IEEE Trans. Energy Convers.* **2000**, *15*, 2.
12. Wang, L.; Deng, X.; Hu, K.; Zhang, X.; Wang, K. A Novel Parameter Identification Method for Induction Motor. In Proceedings of the International Conference on Measuring Technology and Mechatronics Automation, Changsha, China, 13–14 March 2010; Volume 1.
13. Uzunović, T.; Montoya, F.G.; Osmanović, A.; Arrabal-Campos, F.M.; Alcayde, A.; Eid, A.H.; Šabanović, A. Combining Real-time Parameter Identification and Robust Control Algorithms for Effective Control of Electrical Machines. In Proceedings of the International Conference on Electrical Machines (ICEM), Valencia, Spain, 5–8 September 2022.
14. Pan, J.; Westwick, D.; Nowicki, E. Flux estimation of induction machines with the linear parameter-varying system identification method. In Proceedings of the Canadian Conference on Electrical and Computer Engineering, Niagara Falls, ON, Canada, 2–5 May 2004; Volume 4.
15. Liu, Z.; Zheng, Z.; Li, Y.; Peng, L. Parameter identification of nine-phase induction machines with concentrated windings. In Proceedings of the 17th International Conference on Electrical Machines and Systems (ICEMS), Hangzhou, China, 22–25 October 2014.
16. Košťál, T. Induction machine parameters identification method suitable for self-commissioning. In Proceedings of the XXVI International Scientific Conference Electronics (ET), Sozopol, Bulgaria, 13–15 September 2017.
17. Košťál, T. Offline induction machine parameters identification suitable for self-commissioning. In Proceedings of the International Conference on Applied Electronics (AE), Pilsen, Czech Republic, 5–6 September 2017.
18. Heidler, B.; Brune, K.; Doppelbauer, M. High-frequency model and parameter identification of electrical machines using numerical simulations. In Proceedings of the IEEE International Electric Machines & Drives Conference (IEMDC), Coeur d'Alene, ID, USA, 10–13 May 2015.
19. Sarma, N.; Tuohy, P.; Djurović, S. Condition monitoring of rotating electrical machines. In *Encyclopedia of Electrical and Electronic Power Engineering*; Elsevier: Amsterdam, Netherlands, 2022; pp. 143–154.
20. Košťál, T. Reducing electrical energy consumption of AHU fans through parameter identification of the drive. In Proceedings of the 3rd International Conference on Intelligent Green Building and Smart Grid (IGBSG), Yilan, Taiwan, 22–25 April 2018.
21. Siddavatam, R.P.R.; Loganathan, U. Identification of Induction Machine parameters including Core Loss Resistance using Recursive Least Mean Square Algorithm. In Proceedings of the IECON 2019—45th Annual Conference of the IEEE Industrial Electronics Society, Lisbon, Portugal, 14–17 October 2019.
22. Zakaryukin, V.P.; Kryukov, A.V.; Kong, D.L. *Modeling and Parameter Identification of Load Nodes of Electric Power Systems*; Irkutsk state transport university: Irkutsk, Russia, 2016; 158p. (In Russian)
23. Kryukov, A.; Suslov, K.; Van Thao, L.; Hung, T.D.; Akhmetshin, A. Power Flow Modeling of Multi-Circuit Transmission Lines. *Energies* **2022**, *15*, 8249. [CrossRef]
24. Gurevich Yu., E.; Libova, L.E.; Khachatryan, E.A. *Load Stability of Electrical Systems*; Energoizdat: Moscow, Russia, 1981; 208p. (In Russian)
25. Deich, A.M. *Methods for Identification of Dynamic Objects*; Energiya: Moscow, Russia, 1979; 240p. (In Russian)
26. Kuwałek, P.; Wiczyński, G. Problem of Total Harmonic Distortion Measurement Performed by Smart Energy Meters. *Meas. Sci. Rev.* **2022**, *22*, 1–10. [CrossRef]

MDPI
St. Alban-Anlage 66
4052 Basel
Switzerland
www.mdpi.com

*Energies* Editorial Office
E-mail: energies@mdpi.com
www.mdpi.com/journal/energies

MDPI

www.ingramcontent.com/pod-product-compliance
Lightning Source LLC
Chambersburg PA
CBHW061507210926
51850CB00008B/1847
* 9 7 8 3 0 3 6 5 9 3 9 7 5 *